BIG IDEAS MATH®
Advanced 2

Record and Practice Journal

- Fair Game Review Worksheets
- Activity Recording Journal
- Practice Worksheets
- Glossary
- Activity Manipulatives

Erie, Pennsylvania

About the Record and Practice Journal

Fair Game Review

The Fair Game Review corresponds to the Pupil Edition Chapter Opener. Here you have the opportunity to practice prior skills necessary to move forward.

Activity Recording Journal

The Activity pages correspond to the Activity in the Pupil Edition. Here you have room to show your work and record your answers.

Practice Worksheets

Each section of the Pupil Edition has an additional Practice page with room for you to show your work and record your answers.

Glossary

This student-friendly glossary is designed to be a reference for key vocabulary, properties, and mathematical terms. Several of the entries include a short example to aid your understanding of important concepts

Activity Manipulatives

Manipulatives needed for the activities are included in the back of the Record and Practice Journal.

Printed in the United States

ISBN 13: 978-1-60840-530-5
ISBN 10: 1-60840-530-3

10-VLP-17

Contents

Contents

Contents

Contents

Contents

Contents

Contents

Contents

Name____ Date__________

Fair Game Review

Simplify the expression.

1. $18x - 6x + 2x$

2. $4b - 7 - 15b + 3$

3. $15(6 - g)$

4. $-24 + 2(y - 9)$

5. $9m + 4(12 - m)$

6. $16(a - 2) + 3(10 - a)$

7. You are selling lemonade for $1.50, a bag of kettle corn for $3, and a hot dog for $2.50 at a fair. Write and simplify an expression for the amount of money you receive when p people buy one of each item.

Name ______________________________ Date __________

Chapter 1 Fair Game Review (continued)

Add or subtract.

8. $-1 + (-3)$

9. $0 + (-12)$

10. $-5 + (-3)$

11. $-4 + (-4)$

12. $5 - (-2)$

13. $-5 - 2$

14. $0 - (-6)$

15. $-9 - 3$

16. In a city, the record monthly high temperature for July is 88°F. The record monthly low temperature is 30°F. What is the range of temperatures for July?

Name____ Date__________

Solving Simple Equations
For use with Activity 1.1

Essential Question How can you use inductive reasoning to discover rules in mathematics? How can you test a rule?

ACTIVITY: Sum of the Angles of a Triangle

Work with a partner. Use a protractor to measure the angles of each triangle. Complete the table to organize your results.

a.

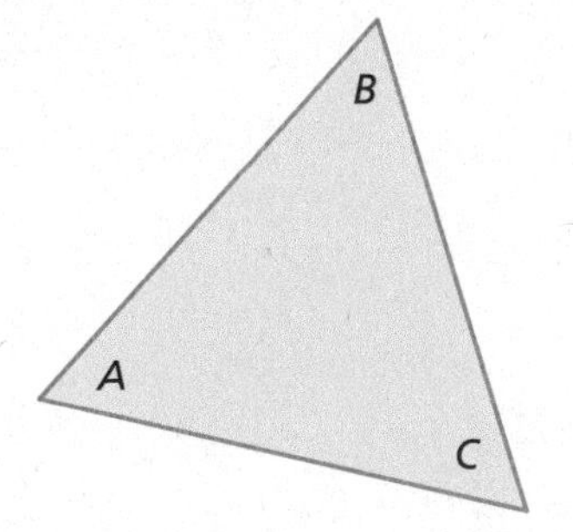

b.

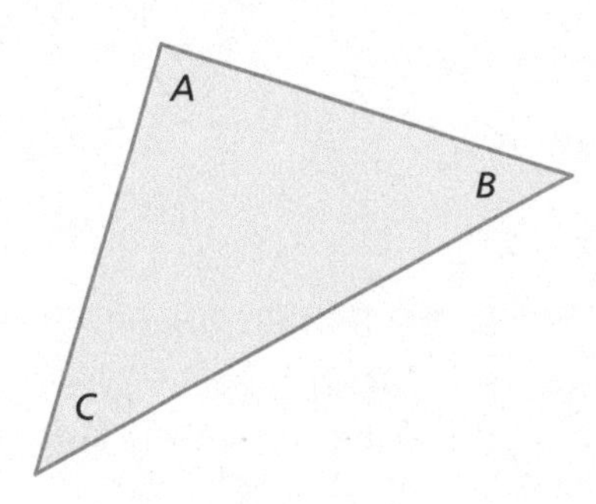

c.

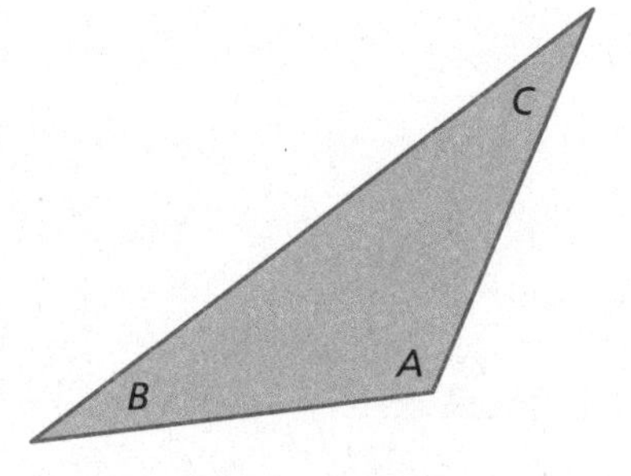

d.

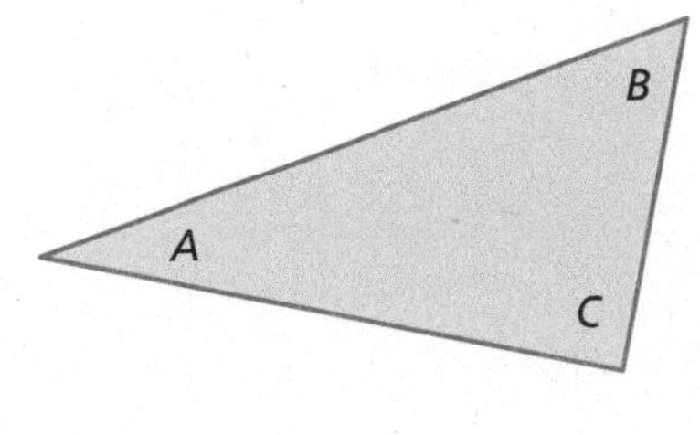

Triangle	Angle A (degrees)	Angle B (degrees)	Angle C (degrees)	$A + B + C$
a.				
b.				
c.				
d.				

Name ____________________ Date ________

2 ACTIVITY: Writing a Rule

Work with a partner. Use inductive reasoning to write and test a rule.

a. **STRUCTURE** Use the completed table in Activity 1 to write a rule about the sum of the angle measures of a triangle.

b. **TEST YOUR RULE** Draw four triangles that are different from those in Activity 1. Measure the angles of each triangle. Organize your results in a table. Find the sum of the angle measures of each triangle.

Name_______________________________ Date__________

1.1 Solving Simple Equations (continued)

3 ACTIVITY: Applying Your Rule

Work with a partner. Use the rule you wrote in Activity 2 to write an equation for each triangle. Then solve the equation to find the value of *x*. Use a protractor to check the reasonableness of your answer.

a.

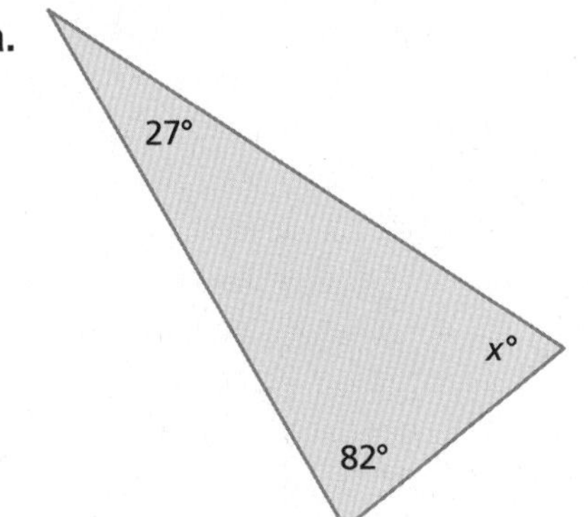

b.

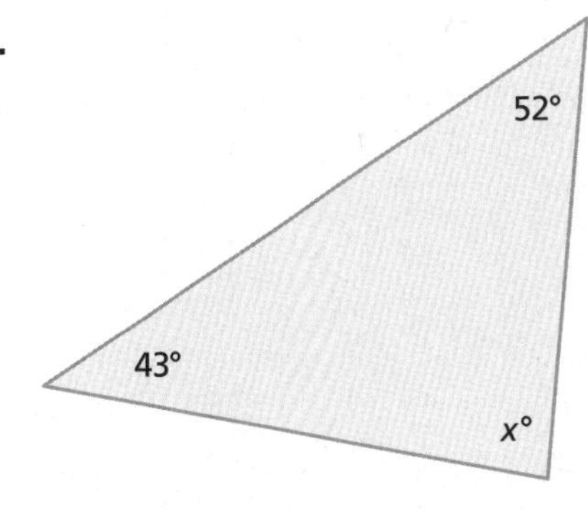

c.

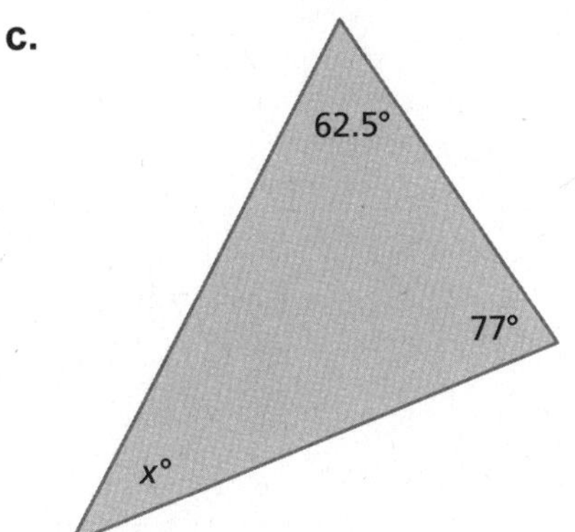

d.

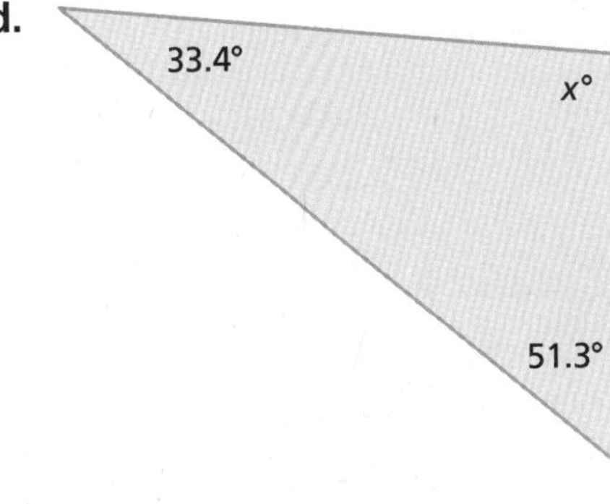

What Is Your Answer?

4. **IN YOUR OWN WORDS** How can you use inductive reasoning to discover rules in mathematics? How can you test a rule? How can you use a rule to solve problems in mathematics?

Name ______________________ Date __________

1.1 Practice

For use after Lesson 1.1

Solve the equation. Check your solution.

1. $x + 5 = 16$

2. $11 = w - 12$

3. $\frac{3}{4} + z = \frac{5}{6}$

4. $3y = 18$

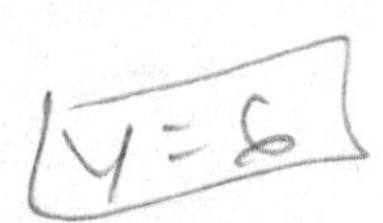

5. $\frac{k}{7} = 10$

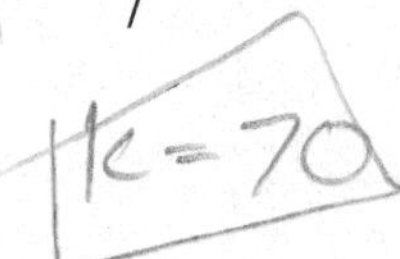

6. $\frac{4}{5}n = \frac{9}{10}$

7. $x - 12 \div 6 = 9$

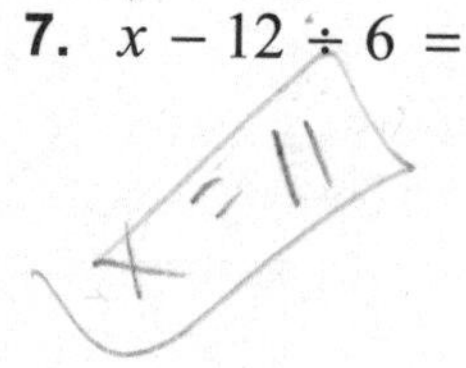

8. $h + |-8| = 15$

9. $1.3(2) + p = 7.9$

10. A coupon subtracts \$5.16 from the price p of a shirt. You pay \$15.48 for the shirt after using the coupon. Write and solve an equation to find the original price of the shirt.

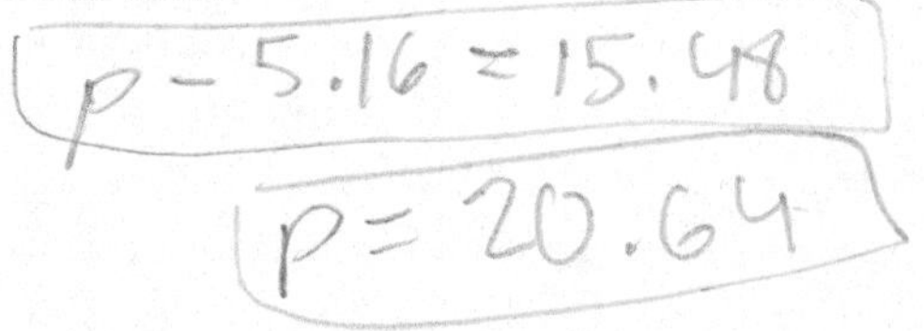

Name 0.556 Date ______

1.2 Solving Multi-Step Equations

For use with Activity 1.2

Essential Question How can you solve a multi-step equation? How can you check the reasonableness of your solution?

1 ACTIVITY: Solving for the Angles of a Triangle

Work with a partner. Write an equation for each triangle. Solve the equation to find the value of the variable. Then find the angle measures of each triangle. Use a protractor to check the reasonableness of your answer.

a.

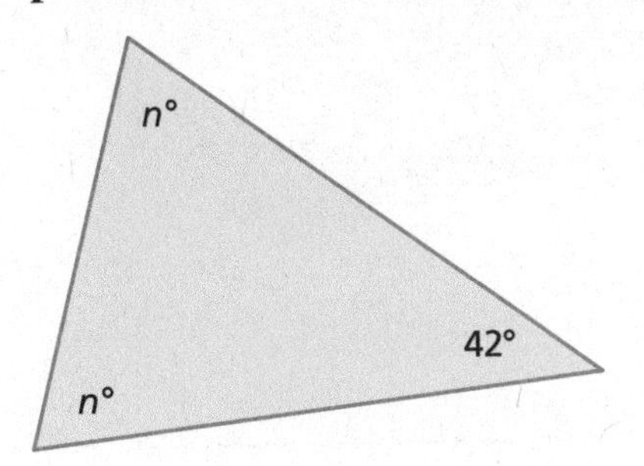

b.

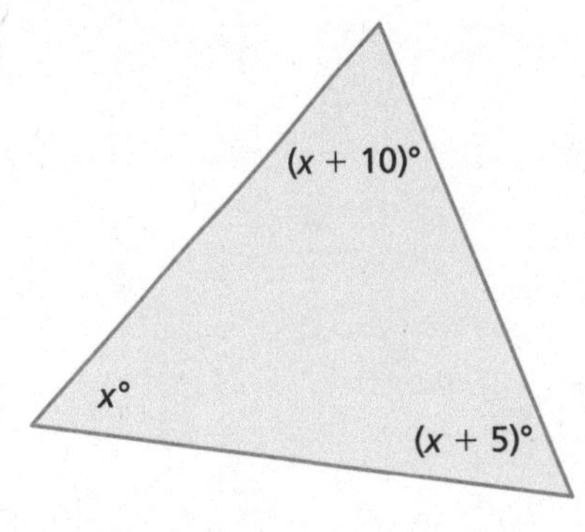

c.

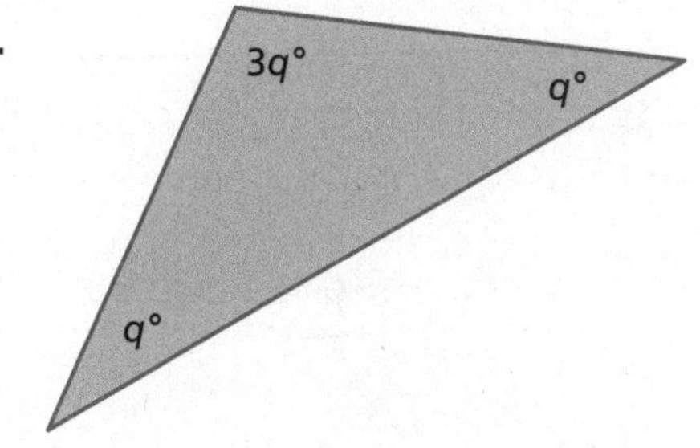

d.

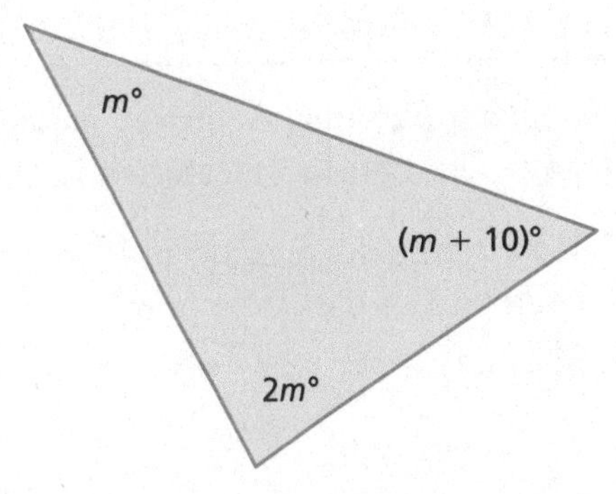

e.

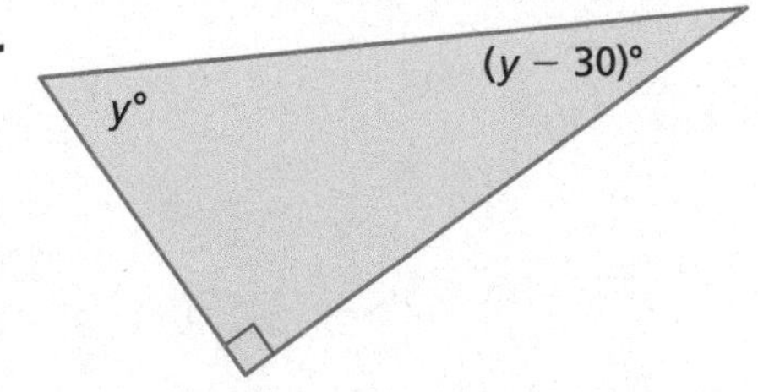

f.

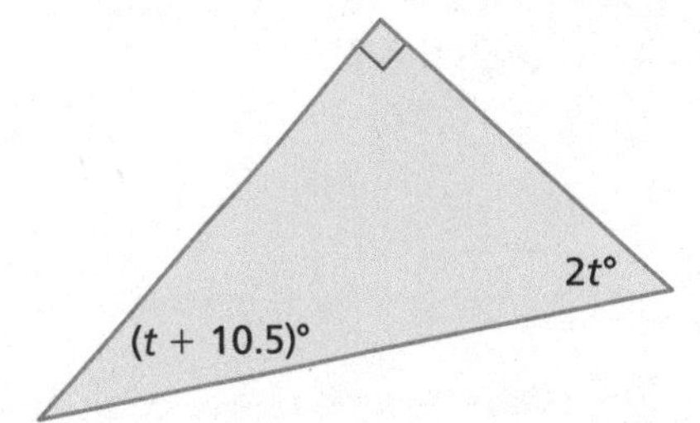

2 ACTIVITY: Problem Solving Strategy

Work with a partner.

The six triangles form a rectangle.

Find the angle measures of each triangle. Use a protractor to check the reasonableness of your answers.

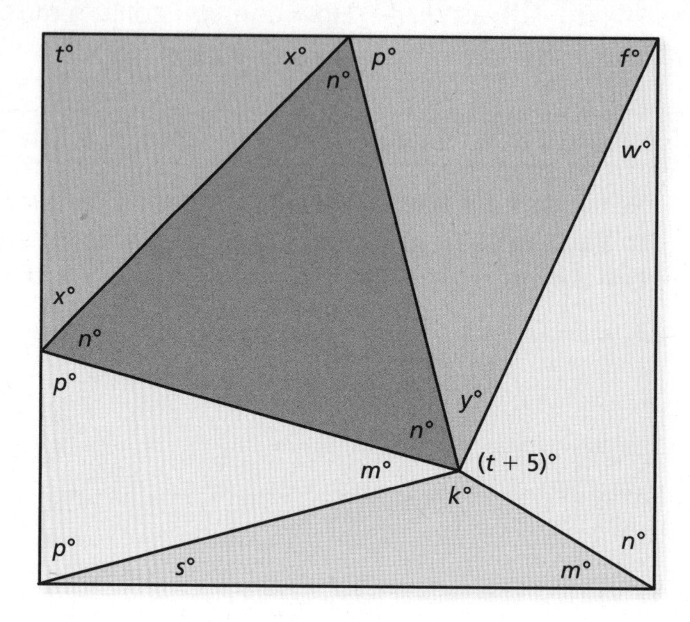

3 ACTIVITY: Puzzle

Work with a partner. A survey asked 200 people to name their favorite weekday. The results are shown in the circle graph.

a. How many degrees are in each part of the circle graph?

b. What percent of the people chose each day?

c. How many people chose each day?

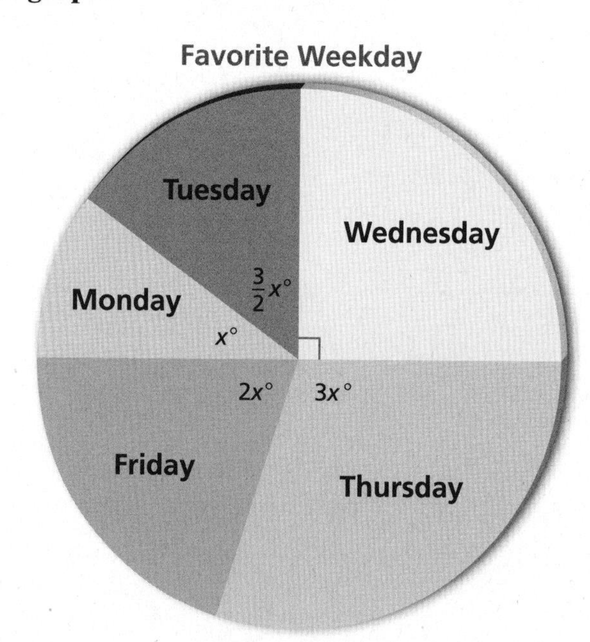

Name____ Date__________

d. Organize your results in a table.

What Is Your Answer?

4. **IN YOUR OWN WORDS** How can you solve a multi-step equation? How can you check the reasonableness of your solution?

Name ______________________________ Date __________

1.2 Practice

For use after Lesson 1.2

Solve the equation. Check your solution.

1. $3x - 11 = 22$

2. $24 - 10b = 9$

3. $2.4z + 1.2z - 6.5 = 0.7$

4. $\frac{3}{4}w - \frac{1}{2}w - 4 = 12$

5. $2(a + 7) - 7 = 9$

6. $20 + 8(q - 11) = -12$

7. Find the width of the rectangular prism when the surface area is 208 square centimeters.

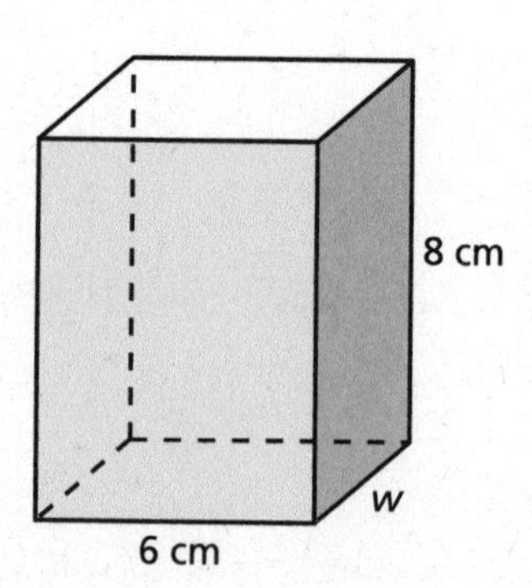

Name____ Date__________

Solving Equations with Variables on Both Sides

For use with Activity 1.3

Essential Question How can you solve an equation that has variables on both sides?

1 ACTIVITY: Perimeter and Area

Work with a partner.

- **Each figure has the unusual property that the value of its perimeter (in feet) is equal to the value of its area (in square feet). Write an equation for each figure.**
- **Solve each equation for x.**
- **Use the value of x to find the perimeter and the area of each figure.**
- **Describe how you can check your solution.**

a.

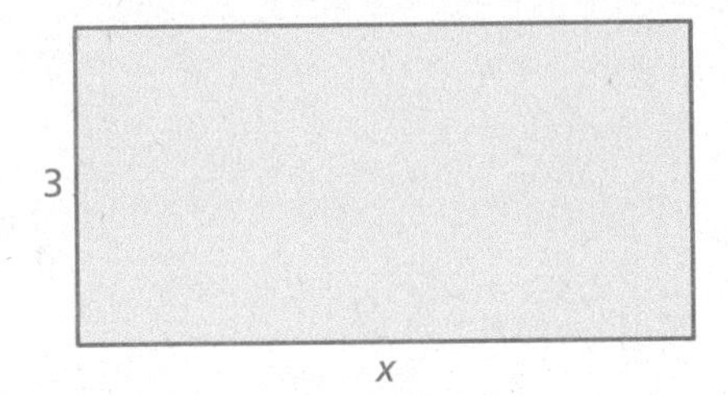

b.

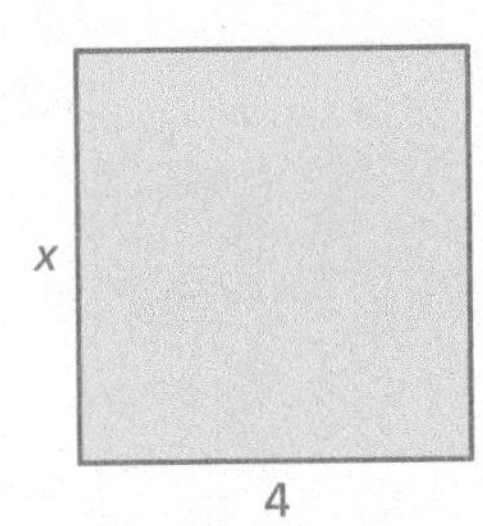

c.

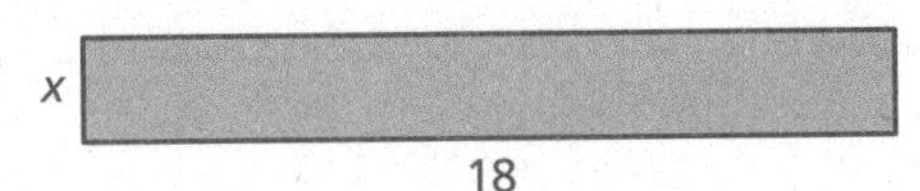

d.

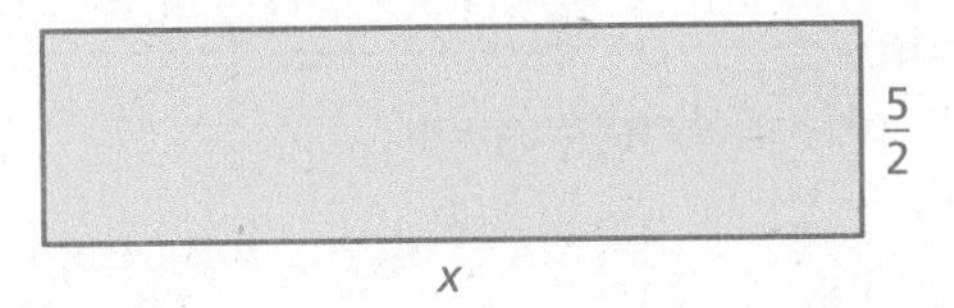

e.

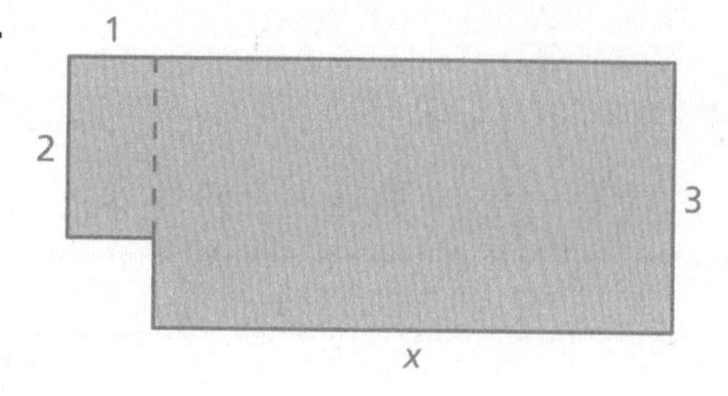

f.

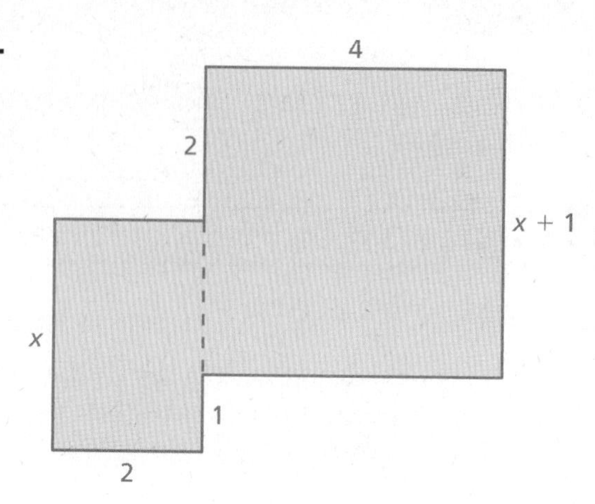

g.

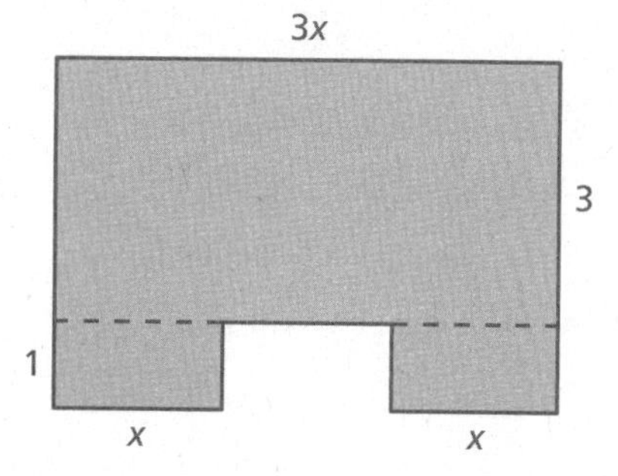

2 ACTIVITY: Surface Area and Volume

Work with a partner.

- **Each solid on the next page has the unusual property that the value of its surface area (in square inches) is equal to the value of its volume (in cubic inches). Write an equation for each solid.**
- **Solve each equation for x.**
- **Use the value of x to find the surface area and the volume of each solid.**
- **Describe how you can check your solution.**

Name____ Date__________

a.

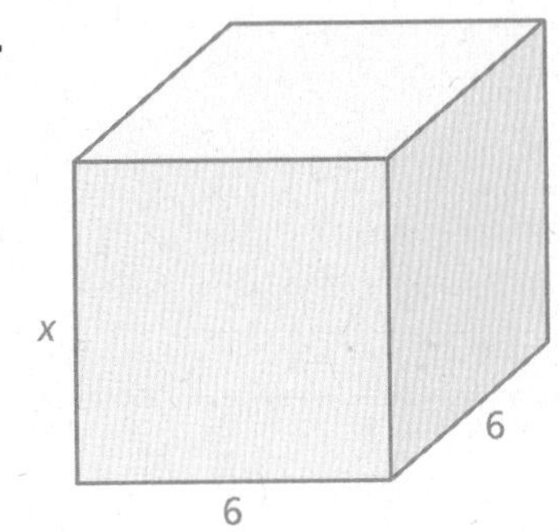

b.

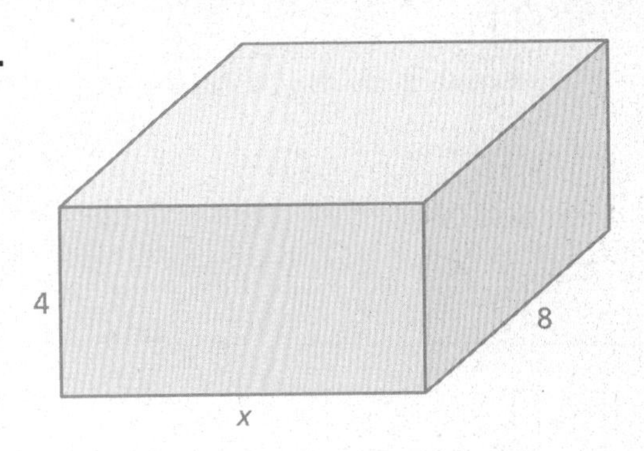

3 ACTIVITY: Puzzle

Work with a partner. The perimeter of the larger triangle is 150% of the perimeter of the smaller triangle. Find the dimensions of each triangle.

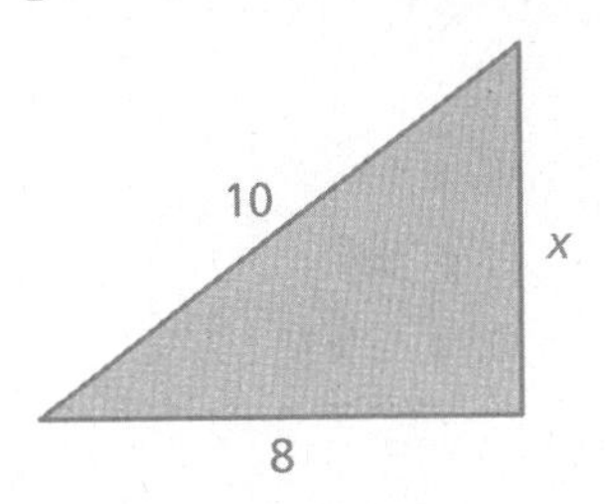

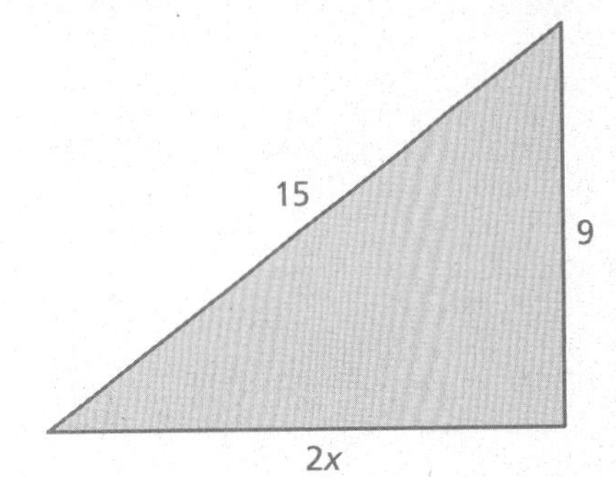

What Is Your Answer?

4. **IN YOUR OWN WORDS** How can you solve an equation that has variables on both sides? How do you move a variable term from one side of the equation to the other?

5. Write an equation that has variables on both sides. Solve the equation.

Name ______________________ Date ________

1.3 Practice

For use after Lesson 1.3

Solve the equation. Check your solution.

1. $x + 16 = 9x$

2. $4y - 70 = 12y + 2$

3. $5(p + 6) = 8p$

4. $3(g - 7) = 2(10 + g)$

5. $1.8 + 7n = 9.5 - 4n$

6. $\frac{3}{7}w - 11 = -\frac{4}{7}w$

7. One movie club charges a \$100 membership fee and \$10 for each movie. Another club charges no membership fee but movies cost \$15 each. Write and solve an equation to find the number of movies you need to buy for the cost of each movie club to be the same.

Name ____________________ Date ________

1.4 Rewriting Equations and Formulas

For use with Activity 1.4

Essential Question How can you use a formula for one measurement to write a formula for a different measurement?

1 ACTIVITY: Using Perimeter and Area Formulas

Work with a partner.

a. • Write a formula for the perimeter P of a rectangle.

• Solve the formula for w.

• Use the new formula to find the width of the rectangle.

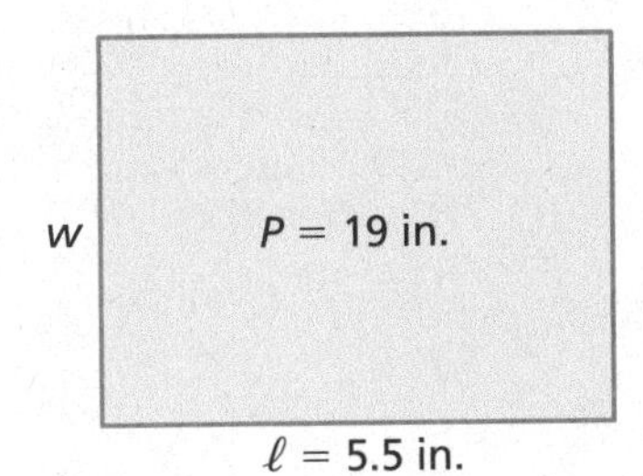

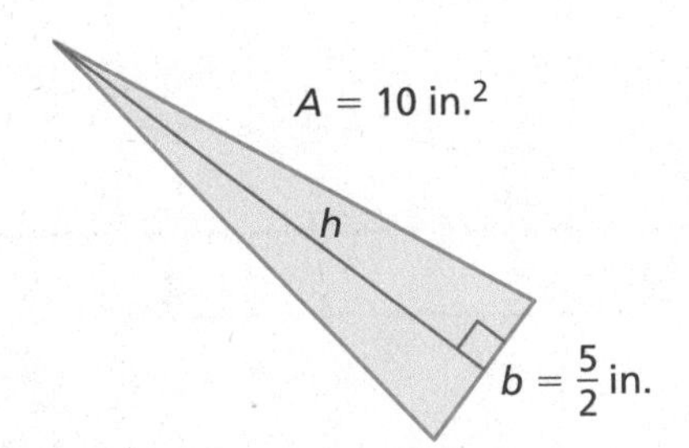

b. • Write a formula for the area A of a triangle.

• Solve the formula for h.

• Use the new formula to find the height of the triangle.

c. • Write a formula for the circumference C of a circle.

• Solve the formula for r.

• Use the new formula to find the radius of the circle.

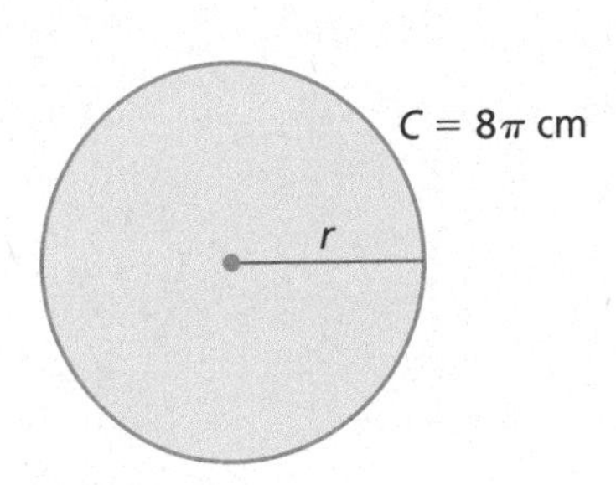

Name ______________________________ Date __________

- **Write a formula for the area A.**
- **Solve the formula for h.**
- **Use the new formula to find the height.**

d.

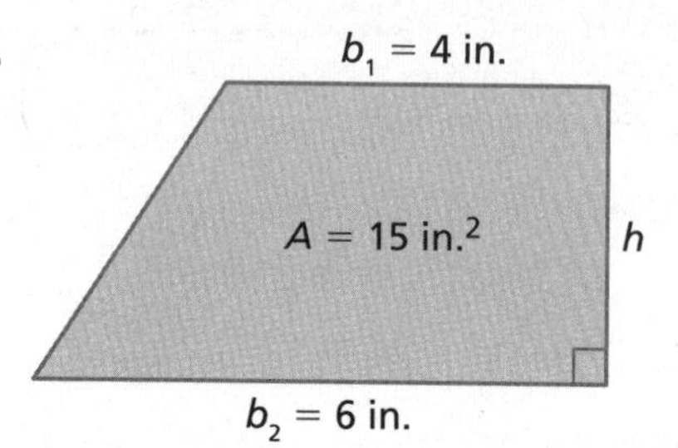

e.

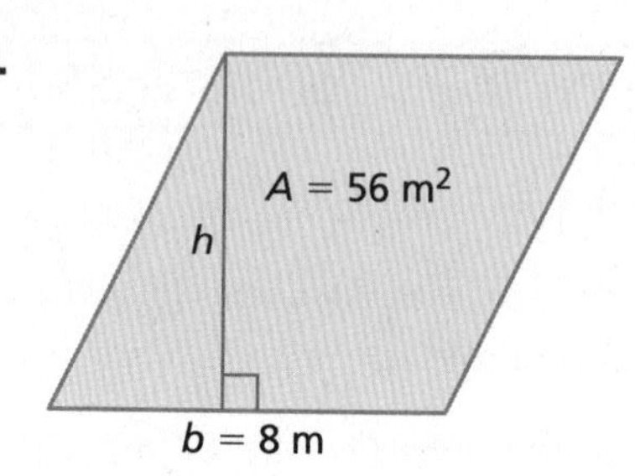

2 ACTIVITY: Using Volume and Surface Area Formulas

Work with a partner.

a.
- Write a formula for the volume V of a prism.
- Solve the formula for h.
- Use the new formula to find the height of the prism.

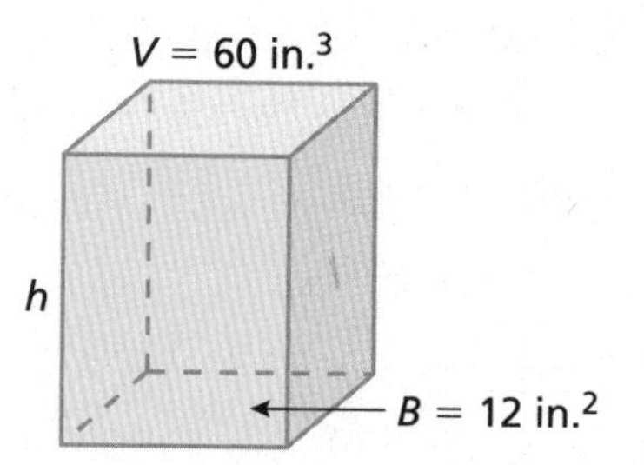

Name______________________________ Date__________

b. • Write a formula for the volume V of a pyramid.

• Solve the formula for B.

• Use the new formula to find the area of the base of the pyramid.

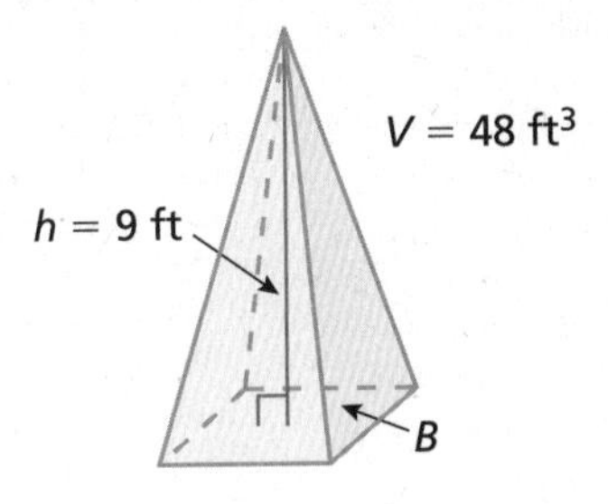

c. • Write a formula for the lateral surface area S of a cylinder.

• Solve the formula for h.

• Use the new formula to find the height of the cylinder.

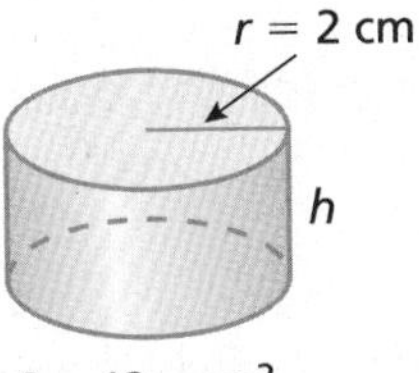

d. • Write a formula for the surface area S of a rectangular prism.

• Solve the formula for ℓ.

• Use the new formula to find the length of the rectangular prism.

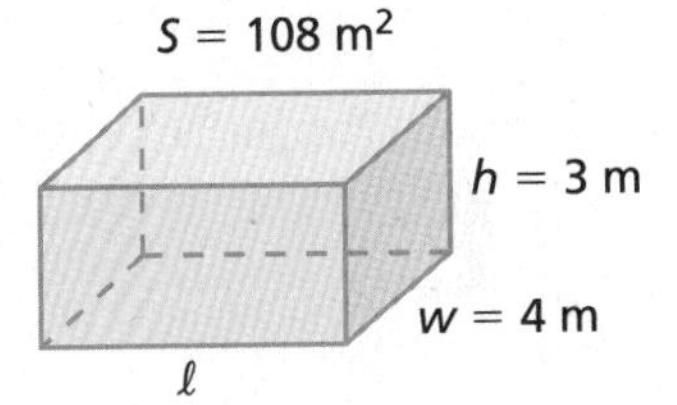

What Is Your Answer?

3. IN YOUR OWN WORDS How can you use a formula for one measurement to write a formula for a different measurement? Give an example that is different from the examples on these three pages.

Name ______________________________ Date __________

1.4 Practice

For use after Lesson 1.4

Solve the equation for *y*.

1. $2x + y = -9$

2. $4x - 10y = 12$

3. $13 = \frac{1}{6}y + 2x$

Solve the formula for the bold variable.

4. $V = \ell w \mathbf{h}$

5. $f = \frac{1}{2}(\mathbf{r} + 6.5)$

6. $S = 2\pi r^2 + 2\pi r\mathbf{h}$

7. The formula for the area of a triangle is $A = \frac{1}{2}bh$.

a. Solve the formula for h.

b. Use the new formula to find the value of h.

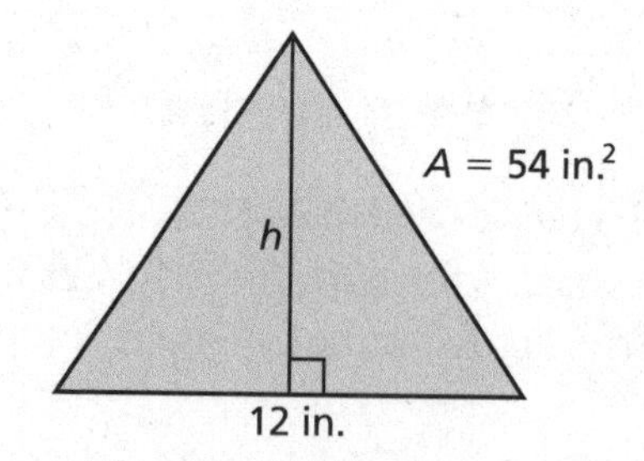

Name____ Date__________

Chapter 2 Fair Game Review

Reflect the point in (a) the x-axis and (b) the y-axis.

1. $(1, 1)$

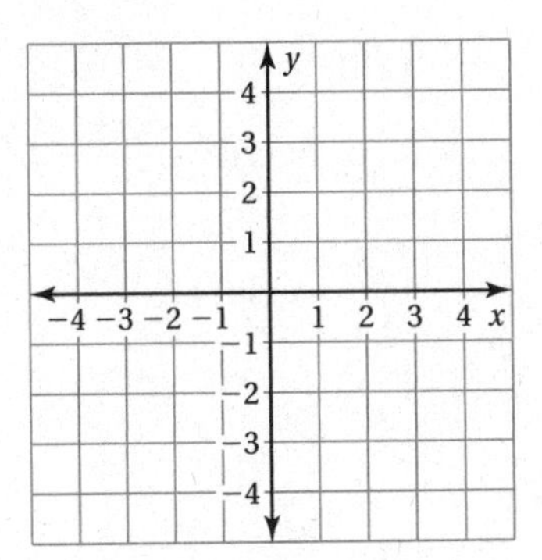

2. $(-2, -4)$

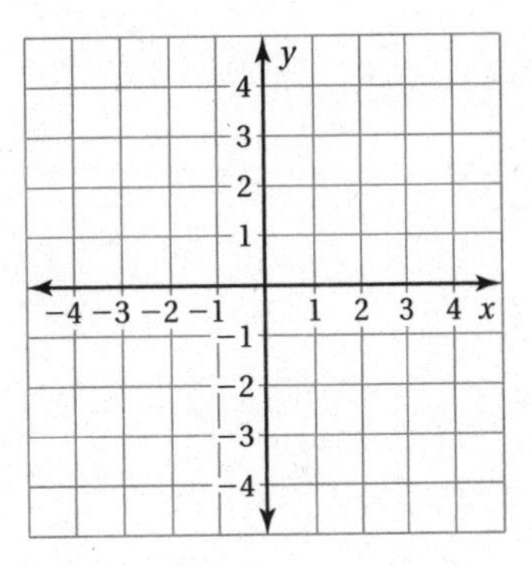

3. $(-3, 3)$

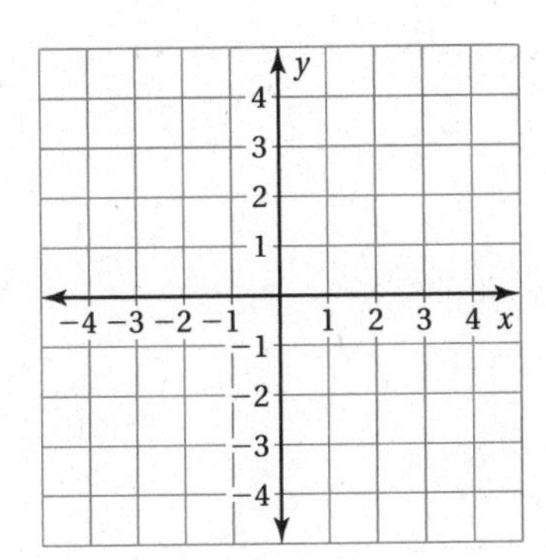

4. $(4, -3)$

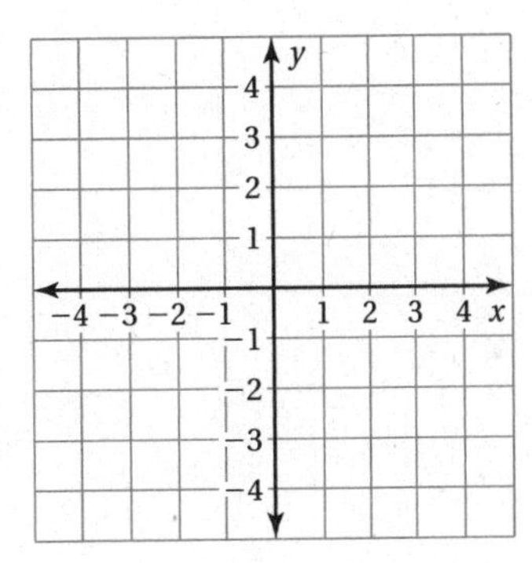

5. $(-1, 2)$

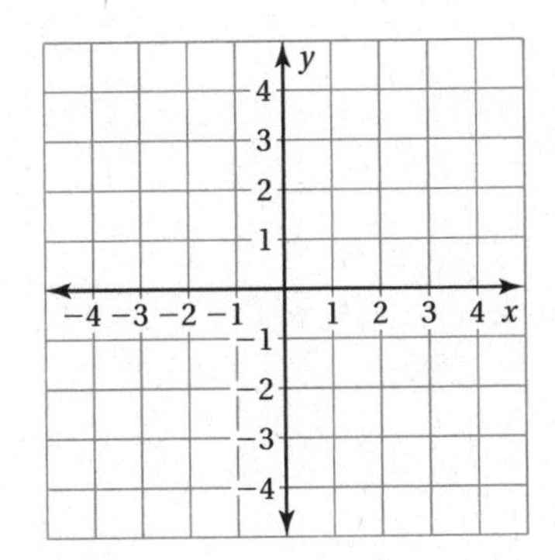

6. $(3, 2)$

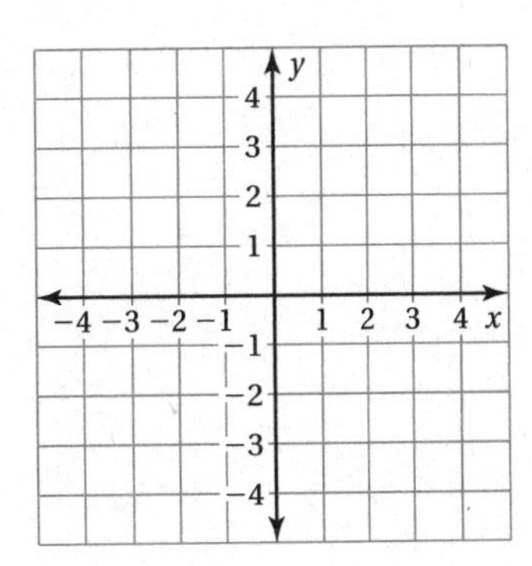

Name ______________________________ Date __________

Fair Game Review (continued)

Draw the polygon with the given vertices in a coordinate plane.

7. $A(2, 2)$, $B(2, 7)$, $C(6, 7)$, $D(6, 2)$

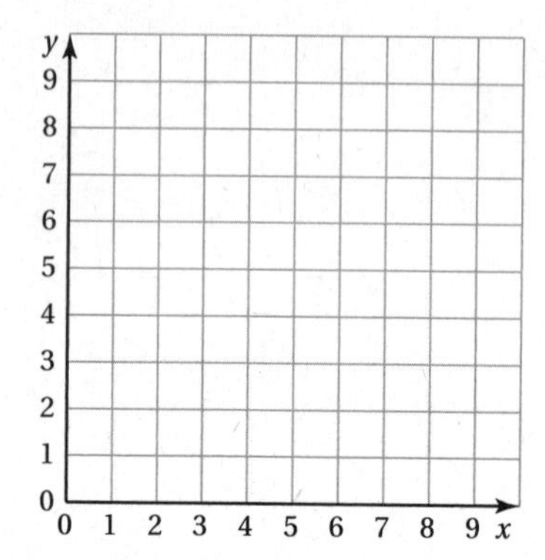

8. $E(3, 8)$, $F(3, 1)$, $G(6, 1)$, $H(6, 8)$

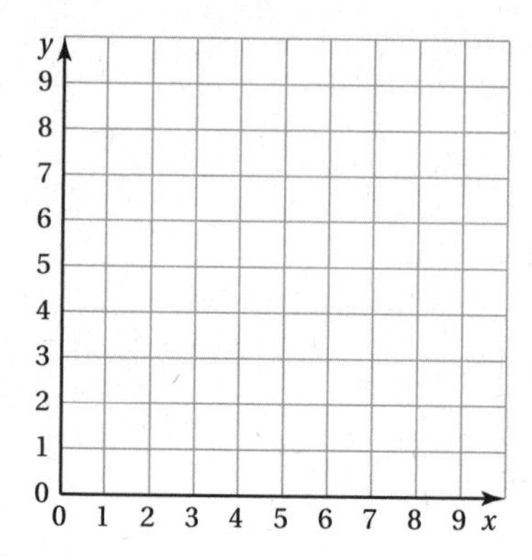

9. $I(7, 6)$, $J(5, 2)$, $K(2, 4)$

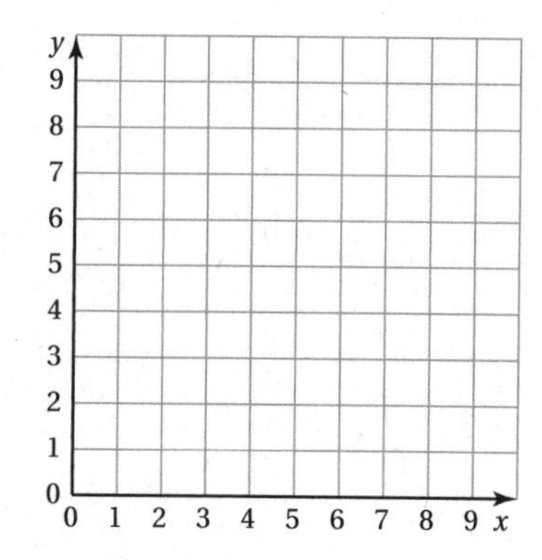

10. $L(1, 5)$, $M(1, 2)$, $N(8, 2)$

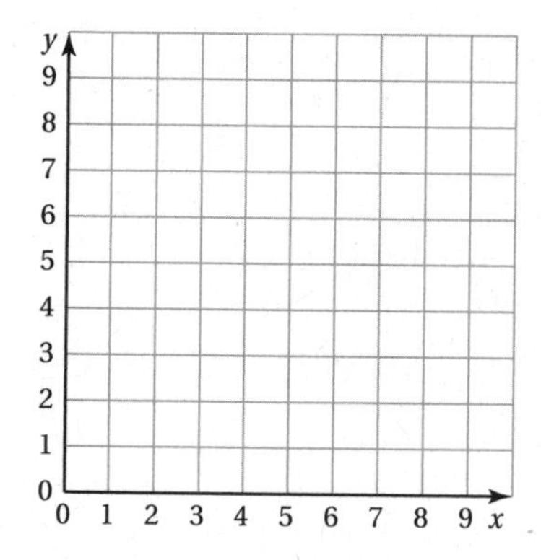

11. $O(3, 7)$, $P(6, 7)$, $Q(9, 3)$, $R(1, 3)$

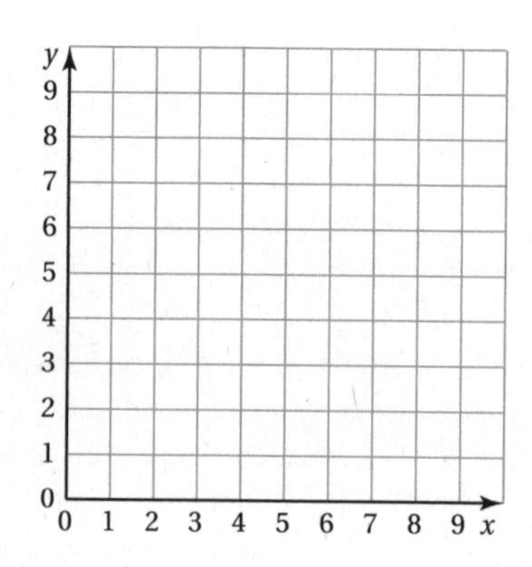

12. $S(9, 9)$, $T(7, 1)$, $U(2, 4)$, $V(4, 7)$

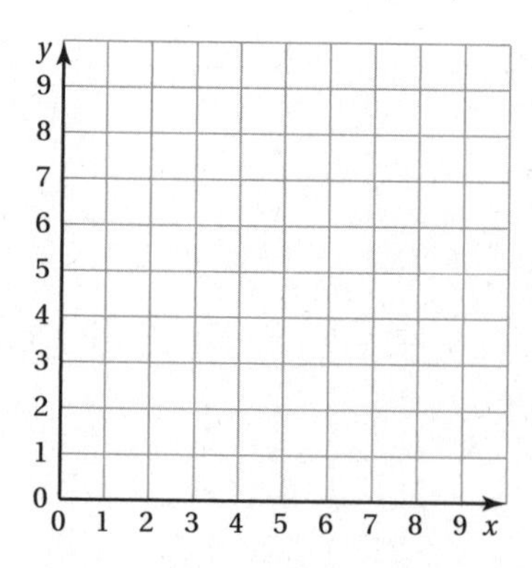

Name________________________________ Date__________

2.1 Congruent Figures
For use with Activity 2.1

Essential Question How can you identify congruent triangles?

Two figures are congruent when they have the same size and the same shape.

1 ACTIVITY: Identifying Congruent Triangles

Work with a partner.

- **Which of the geoboard triangles below are congruent to the geoboard triangle at the right?**

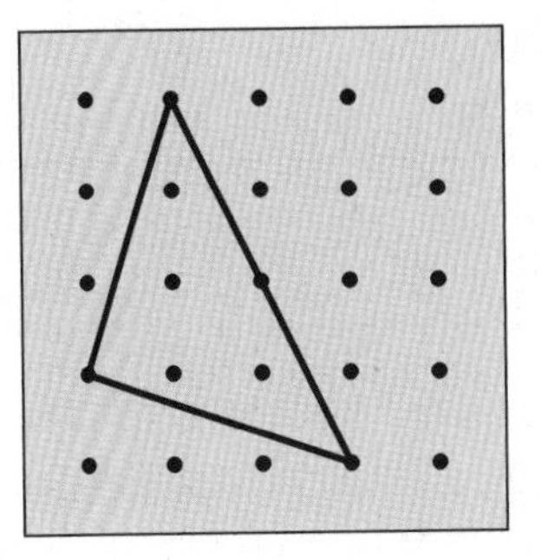

a. 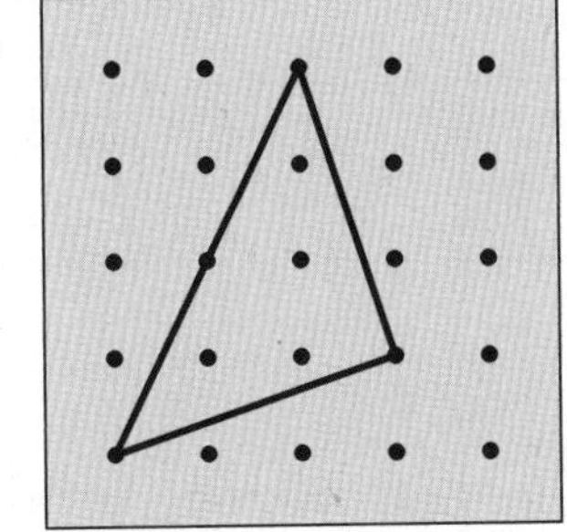

b. 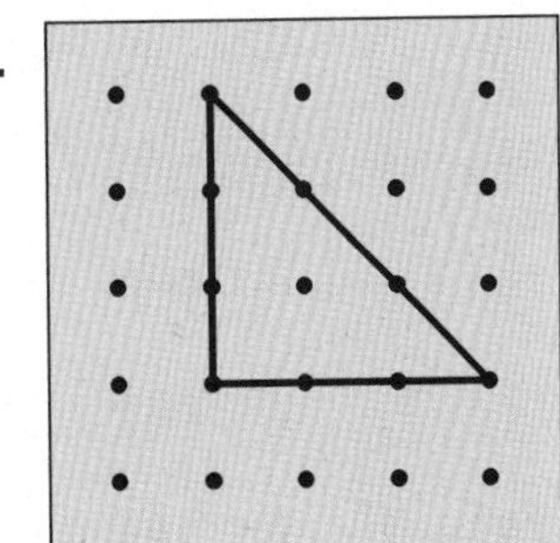

c. 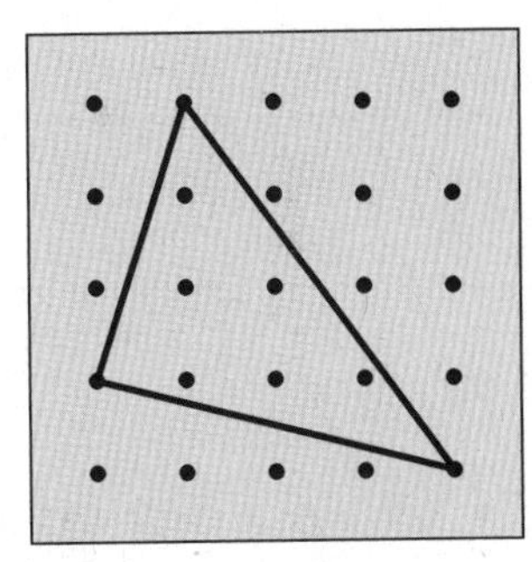

d. 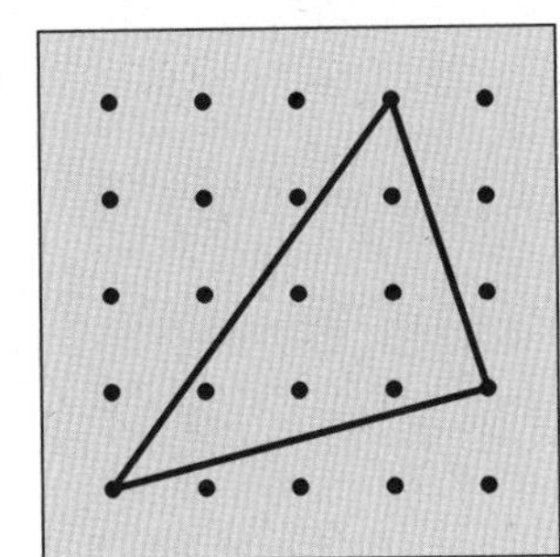

e. 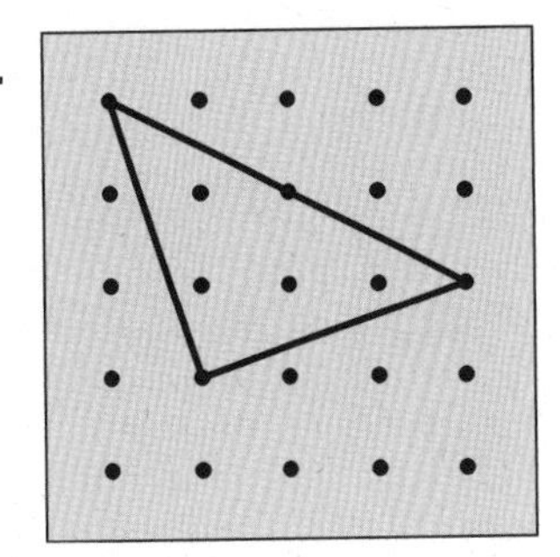

f. 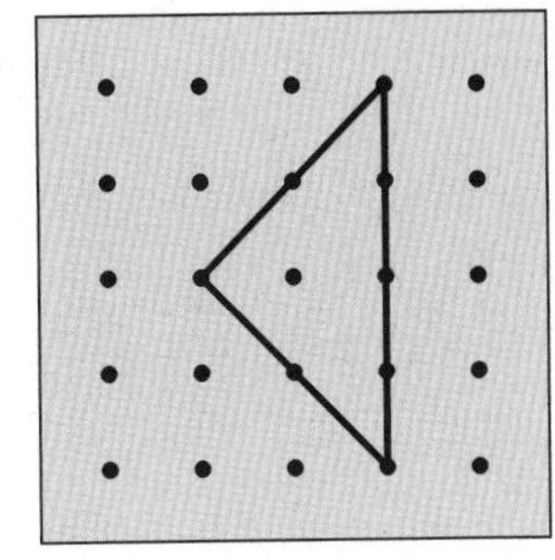

Name ______________________________ Date __________

2.1 Congruent Figures (continued)

- **Form each triangle on a geoboard.**

- **Measure each side with a ruler. Record your results in the table.**

	Side 1	Side 2	Side 3
Given Triangle			
a.			
b.			
c.			
d.			
e.			
f.			

- **Write a conclusion about the side lengths of triangles that are congruent.**

Name____ Date__________

2.2 Translatior

For use with Activ 2.2

Essential Question How n you arrange tiles to make a tessellation?

1 ACTIVITY: Describing Te ellations

Work with a partner. Can you ake the tessellation by translating single tiles that are all of the same sh and design? If so, show how.

a. Sample:

Tile Pattern **Single Tiles**

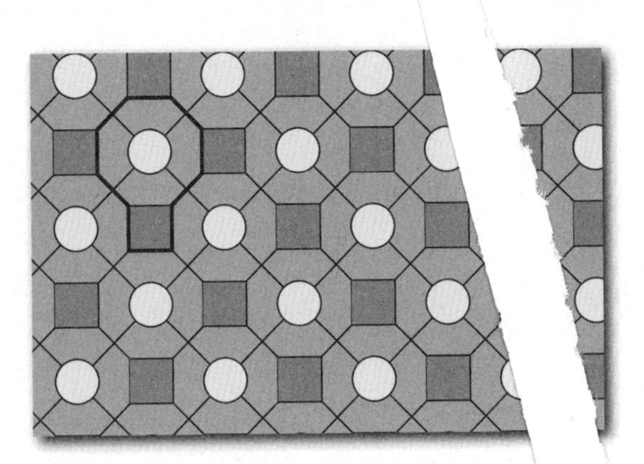

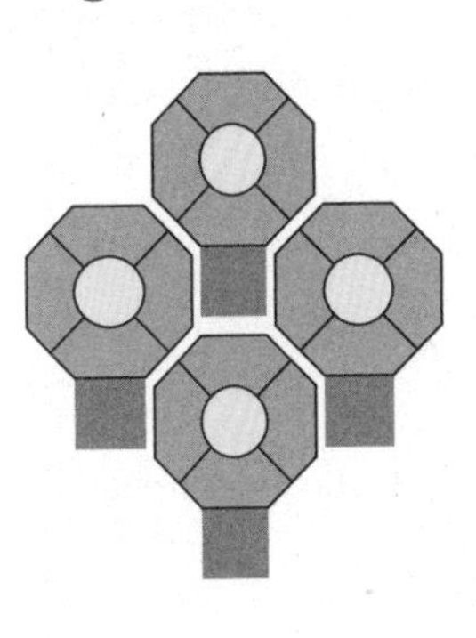

b.

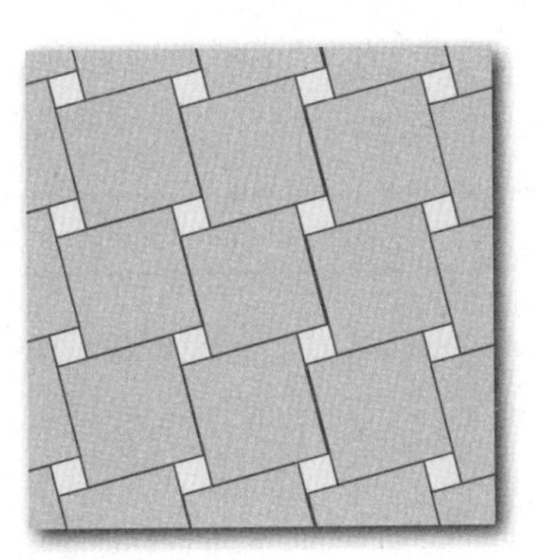

Name ______________________________ Date ________

2 ACTIVITY: Tessellations and Basic Shapes

Work with a partner.

a. Which pattern blocks can you use to make a tessellation? For each one that works, draw the tessellation.

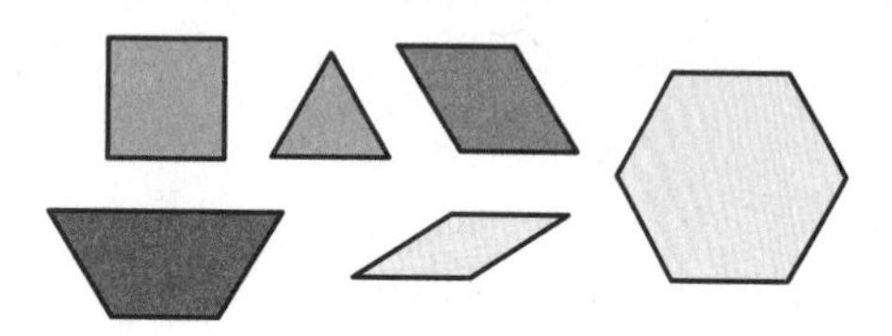

b. Can you make the tessellation by translating? Or do you have to rotate or flip the pattern blocks?

3 ACTIVITY: Designing Tessellations

Work with a partner. Design your own tessellation. Use one of the basic shapes from Activity 2.

Sample:

Step 1: Start with a square.

Step 2: Cut a design out of one side.

Step 3: Tape it to the other side to make your pattern.

Step 4: Translate the pattern to make your tessellation.

Step 5: Color the tessellation.

Name____ Date__________

4 ACTIVITY: Translating in the Coordinate Plane

Work with a partner.

a. Draw a rectangle in a coordinate plane. Find the dimensions of the rectangle.

b. Move each vertex 3 units right and 4 units up. Draw the new figure. List the vertices.

c. Compare the dimensions and the angle measures of the new figure to those of the original rectangle.

d. Are the opposite sides of the new figure parallel? Explain.

e. Can you conclude that the two figures are congruent? Explain.

f. Compare your results with those of other students in your class. Do you think the results are true for any type of figure?

What Is Your Answer?

5. **IN YOUR OWN WORDS** How can you arrange tiles to make a tessellation? Give an example.

6. **PRECISION** Explain why any parallelogram can be translated to make a tessellation.

Name ______________________________ Date __________

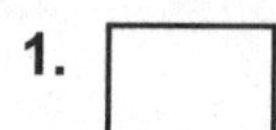

Practice

For use after Lesson 2.2

Tell whether the shaded figure is a translation of the nonshaded figure.

1. 

2. 

3. 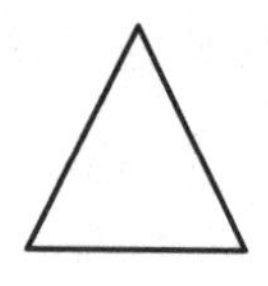 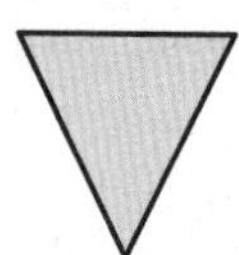

4. Translate the figure 4 units left and 1 unit down. What are the coordinates of the image?

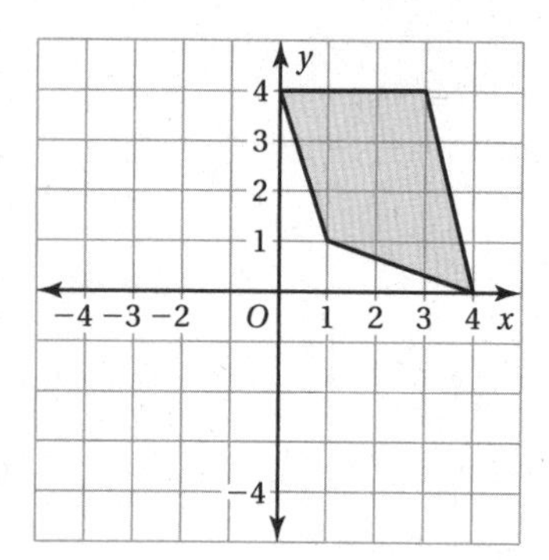

5. Translate the triangle 5 units right and 4 units up. What are the coordinates of the image?

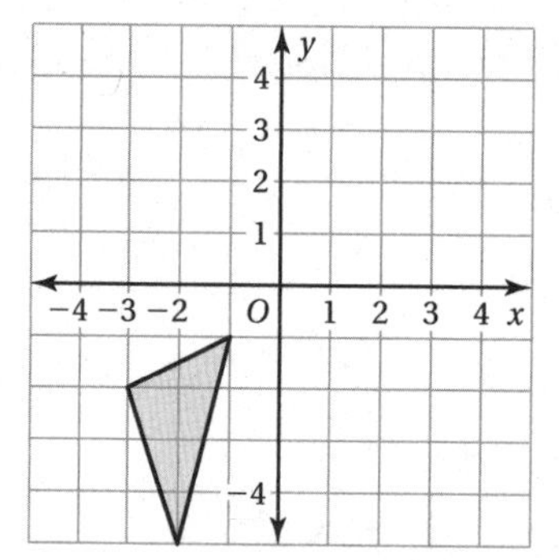

6. Describe the translation from the shaded figure to the nonshaded figure.

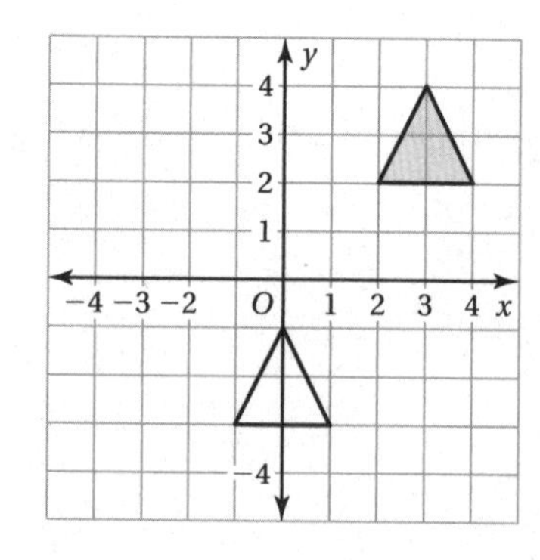

Name______________________________ Date__________

2.3 Reflections

For use with Activity 2.3

Essential Question How can you use reflections to classify a frieze pattern?

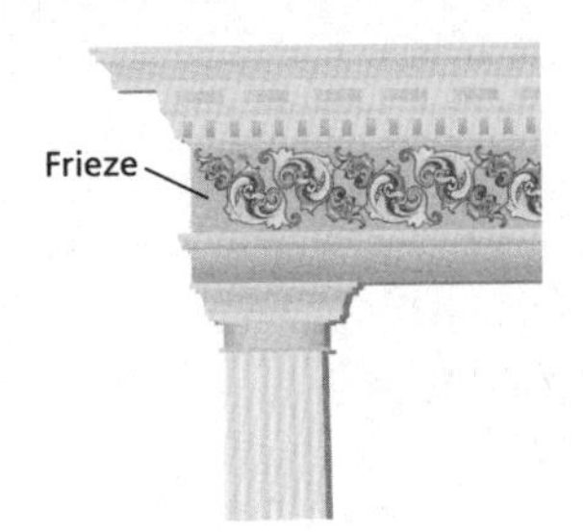

A *frieze* is a horizontal band that runs at the top of a building. A frieze is often decorated with a design that repeats.

- All frieze patterns are translations of themselves.
- Some frieze patterns are reflections of themselves.

1 ACTIVITY: Frieze Patterns and Reflections

Work with a partner. Consider the frieze pattern shown. *

a. Is the frieze pattern a reflection of itself when folded horizontally? Explain.

b. Is the frieze pattern a reflection of itself when folded vertically? Explain.

*Cut-outs are available in the back of the Record and Practice Journal.

Name ______________________________ Date ________

2 ACTIVITY: Frieze Patterns and Reflections

Work with a partner. Is the frieze pattern a reflection of itself when folded *horizontally*, *vertically*, or *neither*?

a. 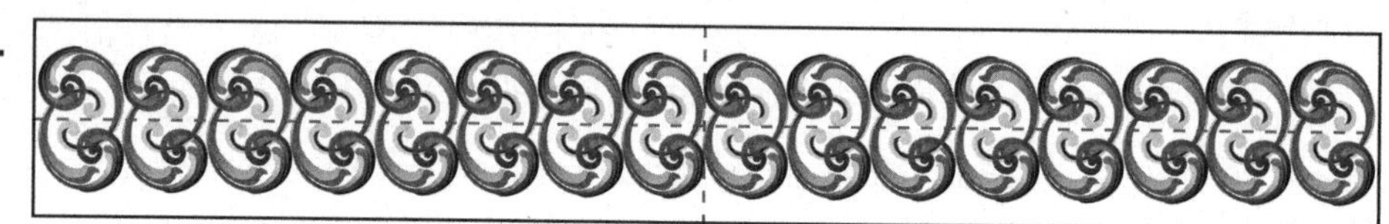

b. 

3 ACTIVITY: Reflecting in the Coordinate Plane

Work with a partner.

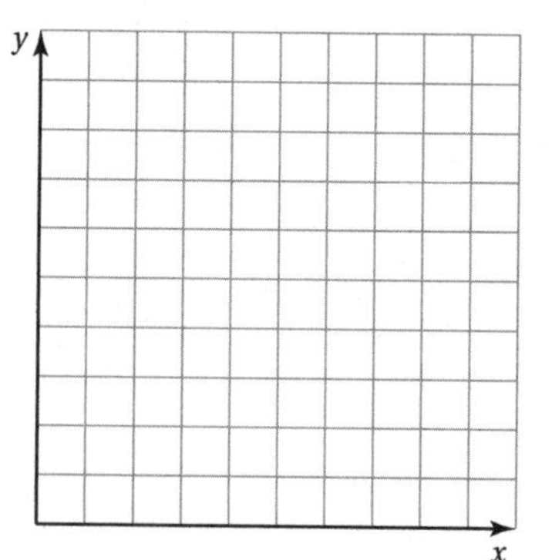

a. Draw a rectangle in Quadrant I of a coordinate plane. Find the dimensions of the rectangle.

b. Copy the axes and the rectangle onto a piece of transparent paper.

Flip the transparent paper once so that the rectangle is in Quadrant IV. Then align the origin and the axes with the coordinate plane.

Draw the new figure in the coordinate plane. List the vertices.

Name____ Date__________

2.4 Rotations

For use with Activity 2.4

Essential Question What are the three basic ways to move an object in a plane?

1 ACTIVITY: Three Basic Ways to Move Things

There are three basic ways to move objects on a flat surface.

________ the object. ________ the object. ________ the object.

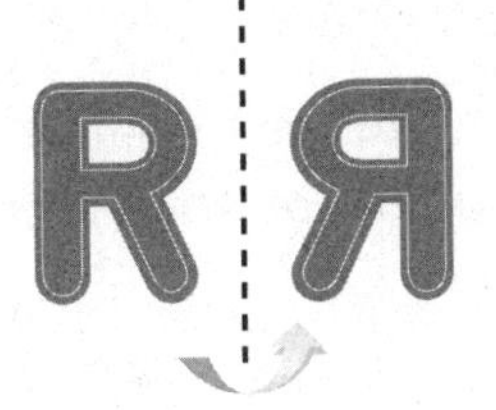

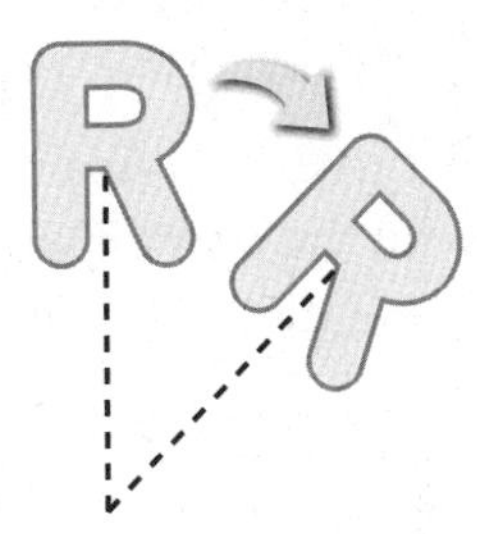

Work with a partner.

a. What type of triangle is the shaded triangle? Is it congruent to the other triangles? Explain.

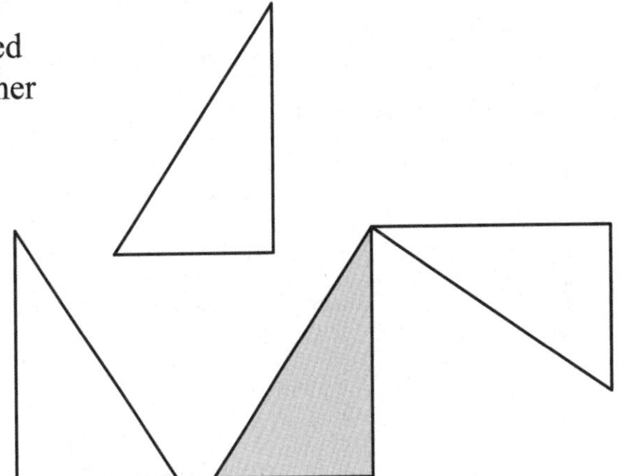

b. Decide how you can move the shaded triangle to obtain each of the other triangles.

c. Is each move a *translation*, a *reflection*, or a *rotation*?

2 ACTIVITY: Rotating in the Coordinate Plane

Work with a partner.

a. Draw a rectangle in Quadrant II of a coordinate plane. Find the dimensions of the rectangle.

b. Copy the axes and the rectangle onto a piece of transparent paper.

Align the origin and the vertices of the rectangle on the transparent paper with the coordinate plane. Turn the transparent paper so that the rectangle is in Quadrant I and the axes align.

Draw the new figure in the coordinate plane. List the vertices.

c. Compare the dimensions and the angle measures of the new figure to those of the original rectangle.

d. Are the opposite sides of the new figure still parallel? Explain.

e. Can you conclude that the two figures are congruent? Explain.

f. Turn the transparent paper so that the original rectangle is in Quadrant IV. Draw the new figure in the coordinate plane. List the vertices. Then repeat parts (c)–(e).

Name____ Date__________

2.5 Similar Figures
For use with Activity 2.5

Essential Question How can you use proportions to help make decisions in art, design, and magazine layouts?

Original Photograph

In a computer art program, when you click and drag on a side of a photograph, you distort it.

But when you click and drag on a corner of the photograph, the dimensions remain proportional to the original.

Distorted

Distorted

Proportional

1 ACTIVITY: Reducing Photographs

Work with a partner. You are trying to reduce the photograph to the indicated size for a nature magazine. Can you reduce the photograph to the indicated size without distorting or cropping? Explain your reasoning.

a.

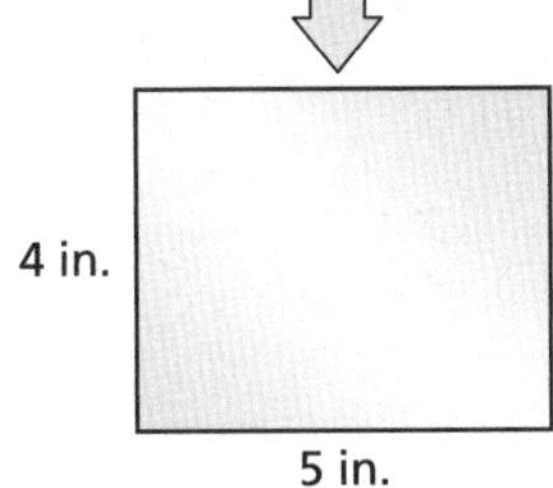

b.

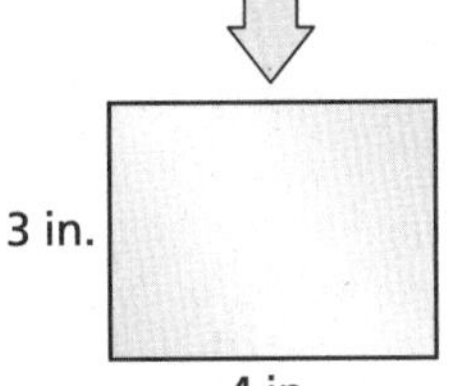

2.5 Similar Figures (continued)

2 ACTIVITY: Creating Designs

Work with a partner.

a. Tell whether the dimensions of the new designs are proportional to the dimensions of the original design. Explain your reasoning.

Original	**Design 1**	**Design 2**
8, 8, 7	7, 7, 6	$6\frac{6}{7}$, $6\frac{6}{7}$, 6

b. Draw two designs whose dimensions are proportional to the given design. Make one bigger and one smaller. Label the sides of the designs with their lengths.

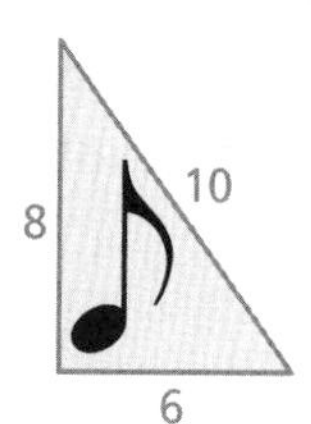

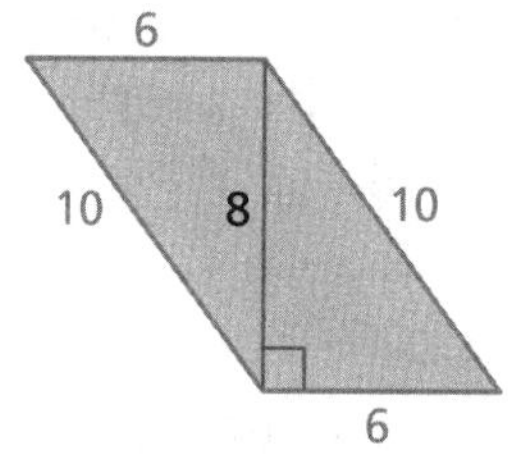

What Is Your Answer?

3. **IN YOUR OWN WORDS** How can you use proportions to help make decisions in art, design, and magazine layouts? Give two examples.

4. **a.** Use a computer art program to draw two rectangles whose dimensions are proportional to each other.

   b. Print the two rectangles on the same piece of paper.

   c. Use a centimeter ruler to measure the length and the width of each rectangle. Record your measurements here.

"I love this statue. It seems similar to a big statue I saw in New York."

   d. Find the following ratios. What can you conclude?

$$\frac{\text{Length of larger}}{\text{Length of smaller}} \qquad \frac{\text{Width of larger}}{\text{Width of smaller}}$$

Name ____ Date __________

2.5 Practice
For use after Lesson 2.5

Tell whether the two figures are similar. Explain your reasoning.

1.

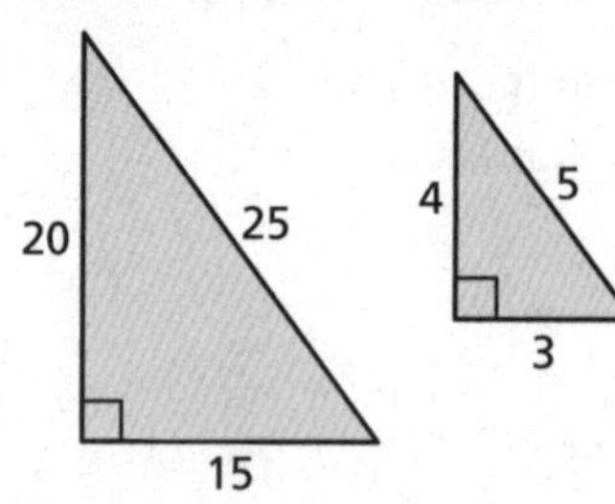

2.

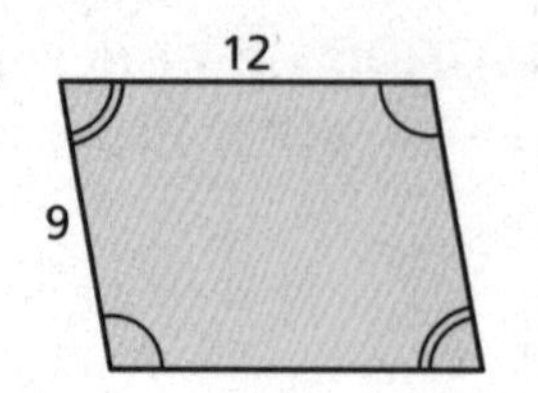

3. In your classroom, a dry erase board is 8 feet long and 4 feet wide. Your teacher makes individual dry erase boards for you to use at your desk that are 11.5 inches long and 9.5 inches wide. Are the boards similar?

4. You have a 4 x 6 photo of you and your friend.

a. You order a 5 x 7 print of the photo. Is the new photo similar to the original?

b. You enlarge the original photo to three times its size on your computer. Is the new photo similar to the original?

Name____ Date__________

2.6 Perimeters and Areas of Similar Figures

For use with Activity 2.6

Essential Question How do changes in dimensions of similar geometric figures affect the perimeters and the areas of the figures?

1 ACTIVITY: Creating Similar Figures

Work with a partner. Use pattern blocks to make a figure whose dimensions are 2, 3, and 4 times greater than those of the original figure.*

a. Square

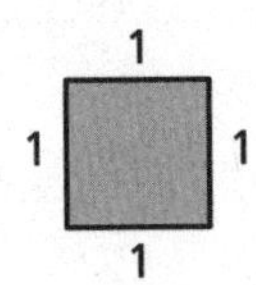

b. Rectangle

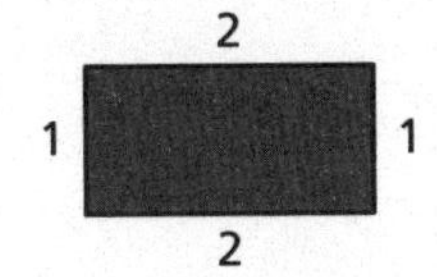

2 ACTIVITY: Finding Patterns for Perimeters

Work with a partner. Complete the table for the perimeter *P* of each figure in Activity 1. Describe the pattern.

Figure	Original Side Lengths	Double Side Lengths	Triple Side Lengths	Quadruple Side Lengths
	$P =$ _____			
	$P =$ _____			

*Cut-outs are available in the back of the Record and Practice Journal.

3 ACTIVITY: Finding Patterns for Areas

Work with a partner. Complete the table for the area *A* of each figure in Activity 1. Describe a pattern.

Figure	Original Side Lengths	Double Side Lengths	Triple Side Lengths	Quadruple Side Lengths
	$A =$ _____			
	$A =$ _____			

4 ACTIVITY: Drawing and Labeling Similar Figures

Work with a partner.

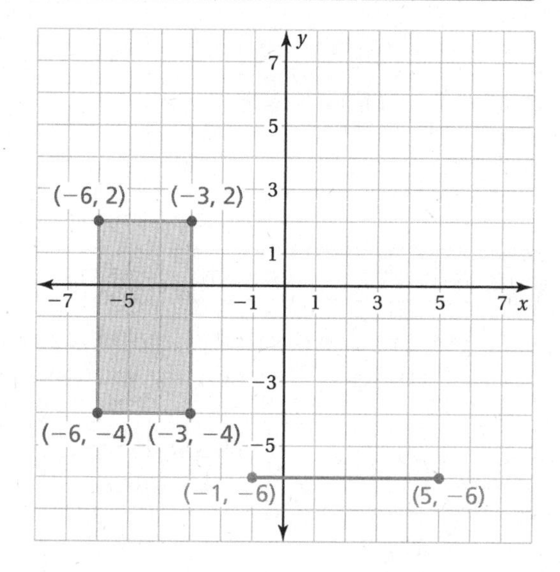

a. Find another rectangle that is similar and has one side from $(-1, -6)$ to $(5, -6)$. Label the vertices.

Check that the two rectangles are similar by showing that the ratios of corresponding sides are equal.

$$\frac{\text{Shaded Length}}{\text{Unshaded Length}} \stackrel{?}{=} \frac{\text{Shaded Width}}{\text{Unshaded Width}}$$

$$\frac{\text{change in } y}{\text{change in } y} \stackrel{?}{=} \frac{\text{change in } x}{\text{change in } x}$$

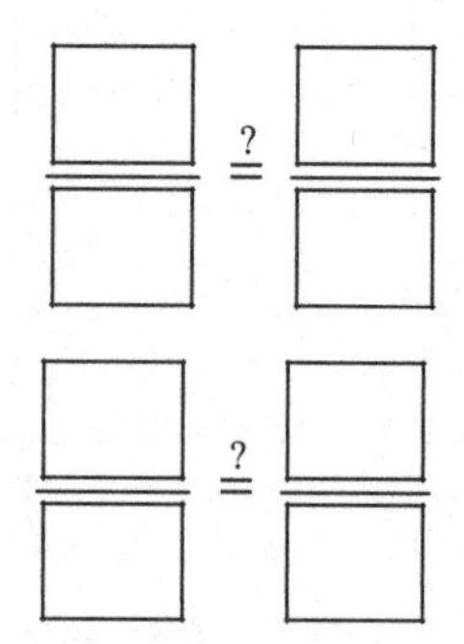

The ratios are ____________. So, the rectangles are ____________.

Name____ Date__________

b. Compare the perimeters and the areas of the figures. Are the results the same as your results from Activities 2 and 3? Explain.

c. There are three other rectangles that are similar to the shaded rectangle and have the given side.

- Draw each one. Label the vertices of each.

- Show that each is similar to the original shaded rectangle.

What Is Your Answer?

5. IN YOUR OWN WORDS How do changes in dimensions of similar geometric figures affect the perimeters and the areas of the figures?

6. What information do you need to know to find the dimensions of a figure that is similar to another figure? Give examples to support your explanation.

Name ______________________________ Date __________

2.6 Practice

For use after Lesson 2.6

The two figures are similar. Find the ratios (shaded to nonshaded) of the perimeters and of the areas.

1.

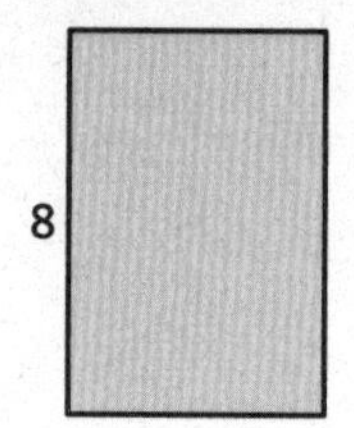

2.

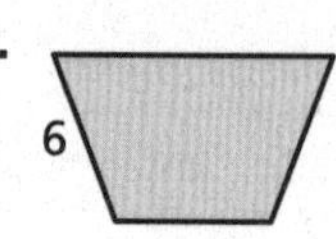

The polygons are similar. Find *x*.

3.

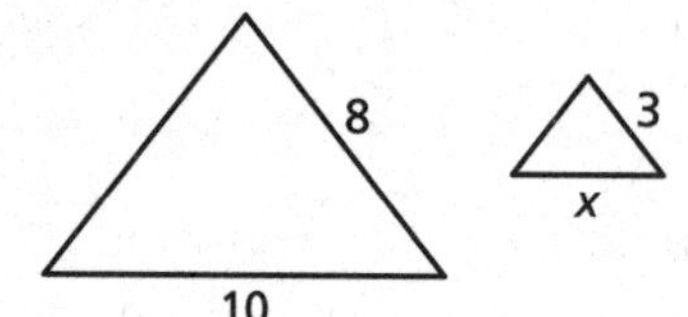

4.

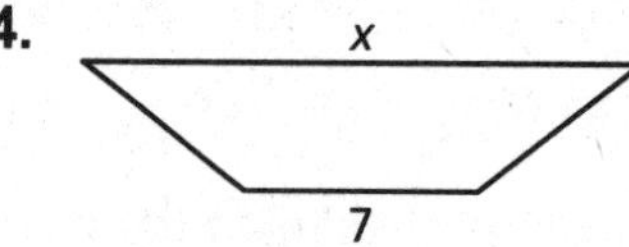

5. You buy two picture frames that are similar. The ratio of the corresponding side lengths is 4 : 5. What is the ratio of the areas?

Name____ Date__________

2.7 Dilations
For use with Activity 2.7

Essential Question How can you enlarge or reduce a figure in the coordinate plane?

1 ACTIVITY: Comparing Triangles in a Coordinate Plane

Work with a partner. Write the coordinates of the vertices of the shaded triangle. Then write the coordinates of the vertices of the nonshaded triangle.

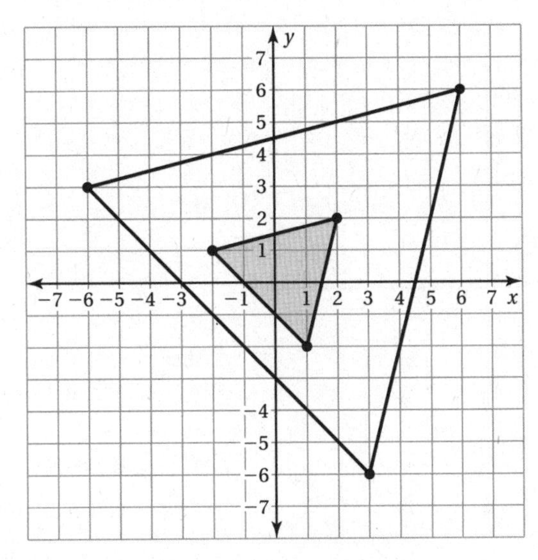

a. How are the two sets of coordinates related?

b. How are the two triangles related? Explain your reasoning.

c. Draw a dashed triangle whose coordinates are twice the values of the corresponding coordinates of the shaded triangle. How are the dashed and shaded triangles related? Explain your reasoning.

Name ________________________________ Date ________

d. How are the coordinates of the nonshaded and dashed triangles related? How are the two triangles related? Explain your reasoning.

2 ACTIVITY: Drawing Triangles in a Coordinate Plane

Work with a partner.

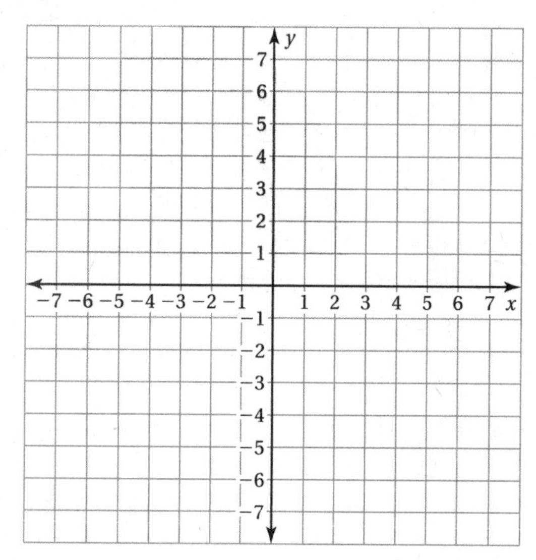

a. Draw the triangle whose vertices are $(0, 2)$, $(-2, 2)$, and $(1, -2)$.

b. Multiply each coordinate of the vertices by 2 to obtain three new vertices. Draw the triangle given by the three new vertices. How are the two triangles related?

c. Repeat part (b) by multiplying by 3 instead of 2.

Name________________________________ Date__________

3 ACTIVITY: Summarizing Transformations

Work with a partner. Make a table that summarizes the relationships between the original figure and its image for the four types of transformations you studied in this chapter.

What Is Your Answer?

4. **IN YOUR OWN WORDS** How can you enlarge or reduce a figure in the coordinate plane?

5. Describe how knowing how to enlarge or reduce figures in a technical drawing is important in a career such as drafting.

Name ______________________________ Date __________

2.7 Practice

For use after Lesson 2.7

Tell whether the shaded figure is a dilation of the nonshaded figure.

1. 

2. 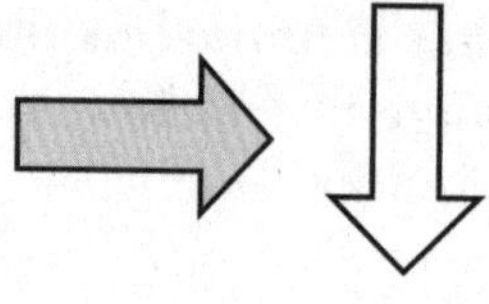

3. 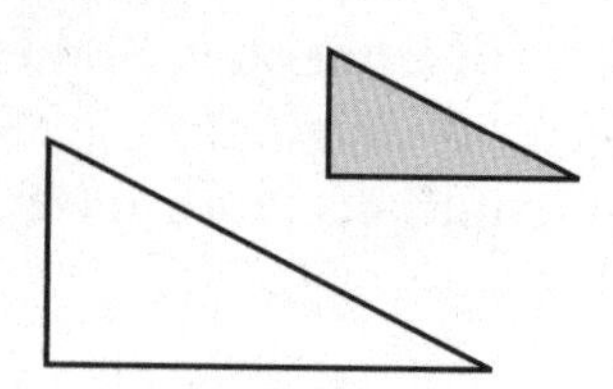

The vertices of a figure are given. Draw the figure and its image after a dilation with the given scale factor. Identify the type of dilation.

4. $A(-2, 2), B(1, 2), C(1, -1); k = 3$

5. $D(4, 2), E(4, 8), F(8, 8), G(8, 2); k = \frac{1}{2}$

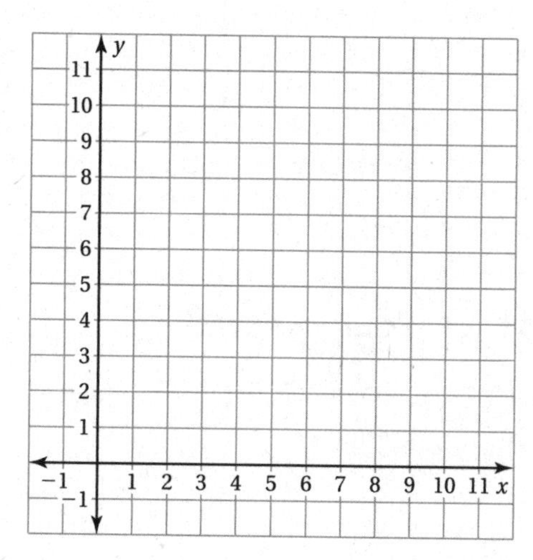

6. A rectangle is dilated using a scale factor of 6. The image is then dilated using a scale factor of $\frac{1}{3}$. What scale factor could you use to dilate the original rectangle to get the final rectangle? Explain.

Name______________________________ Date__________

Chapter 3 Fair Game Review

Tell whether the angles are *adjacent* or *vertical*. Then find the value of *x*.

1.

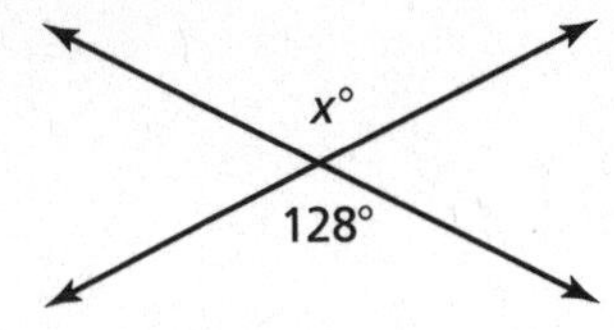

2.

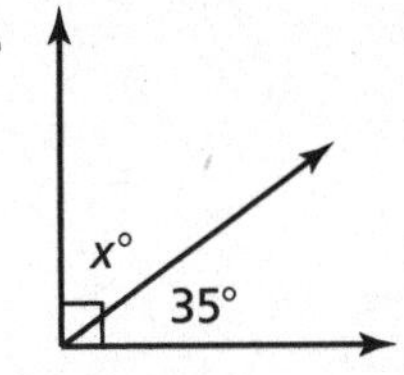

3.

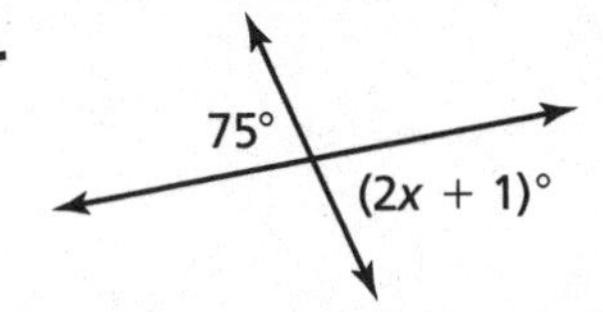

4.

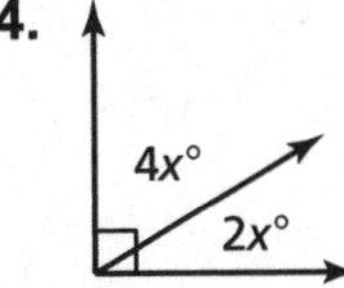

5. The tree is tilted 14°. Find the value of x.

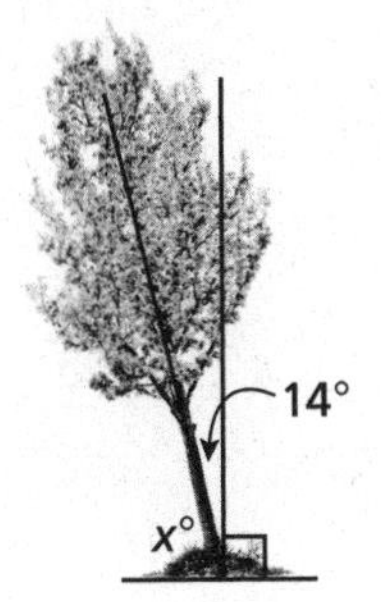

Name ____ Date __________

Fair Game Review (continued)

Tell whether the angles are *complementary* or *supplementary*. Then find the value of *x*.

6.

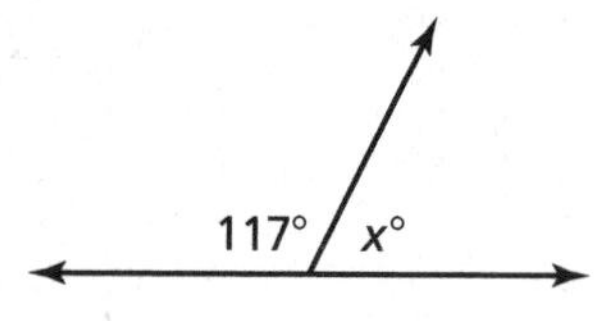

7.

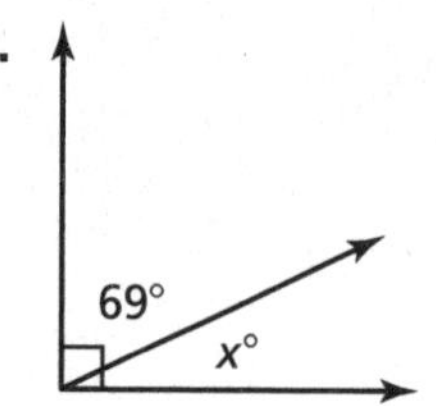

8.

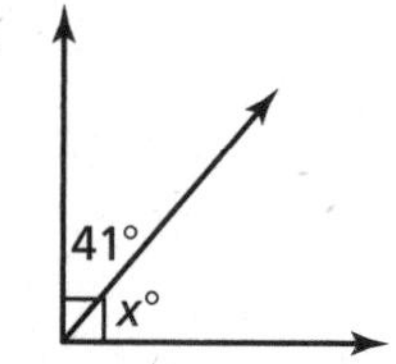

9.

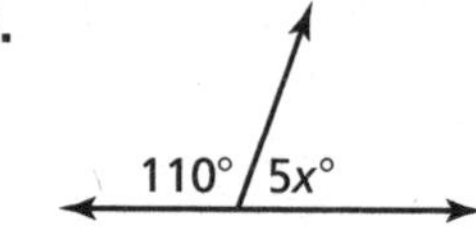

10. A tributary joins a river at an angle. Find the value of x.

Name____ Date__________

3.2 Ang of Triangles

For u ith Activity 3.2

Essential Que on How can you describe the relationships among the angles of a triang

1 ACTIVITY xploring the Interior Angles of a Triangle

Work with rtner.

a. Draw angle. Label the interior angles A, B, and C.

b. Ca y cut out the triangle. Tear off the three corners of the triangle.

c. ge angles A and B so that they share a vertex and are adjacent.

d v can you place the third angle to determine the sum of the measures ne interior angles? What is the sum?

ompare your results with others in your class.

STRUCTURE How does your result in part (d) compare to the rule you wrote in Lesson 1.1, Activity 2?

Name ____ Date __________

3.2 Angles of Triangles (continued)

2 ACTIVITY: Exploring the Interior Angles of a Triangle

Work with a partner.

a. Describe the figure.

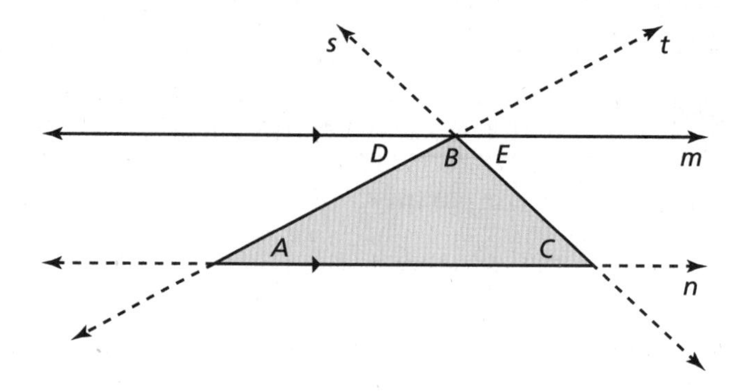

b. LOGIC Use what you know about parallel lines and transversals to justify your result in part (d) of Activity 1.

3 ACTIVITY: Exploring an Exterior Angle of a Triangle

Work with a partner.

a. Draw a triangle. Label the interior angles *A*, *B*, and *C*.

b. Carefully cut out the triangle.

c. Place the triangle on a piece of paper and extend one side to form *exterior angle D*, as shown.

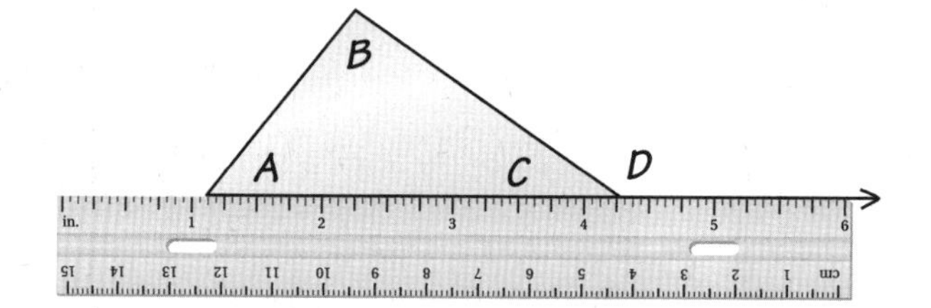

d. Tear off the corners that are not adjacent to the exterior angle. Arrange them to fill the exterior angle, as shown. What does this tell you about the measure of exterior angle *D*?

Name________________________________ Date__________

3.3 Angles of Polygons
For use with Activity 3.3

Essential Question How can you find the sum of the interior angle measures and the sum of the exterior angle measures of a polygon?

1 ACTIVITY: Exploring the Interior Angles of a Polygon

Work with a partner. In parts (a)–(e), identify each polygon and the number of sides *n*. Then find the sum of the interior angle measures of the polygon.

a. Polygon: ______ Number of sides: n = ______

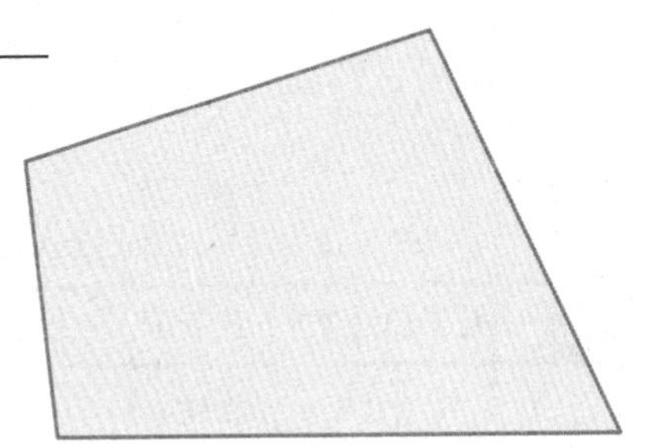

Draw a line segment on the figure that divides it into two triangles. Is there more than one way to do this? Explain.

What is the sum of the interior angle measures of each triangle?

What is the sum of the interior angle measures of the figure?

b.

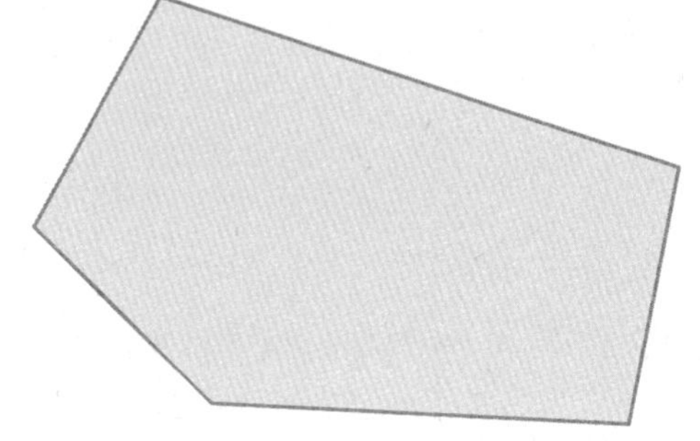

c.

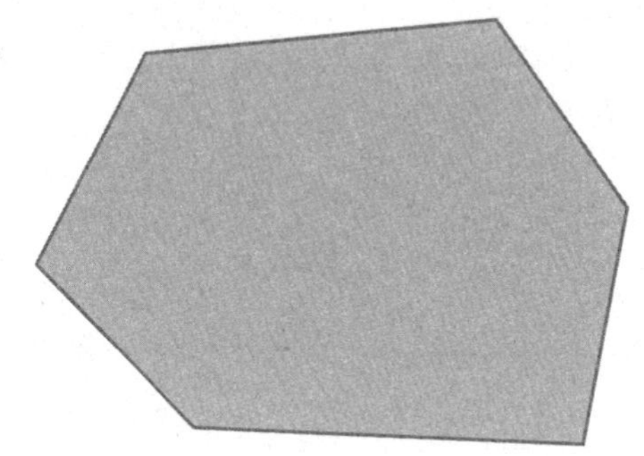

d.

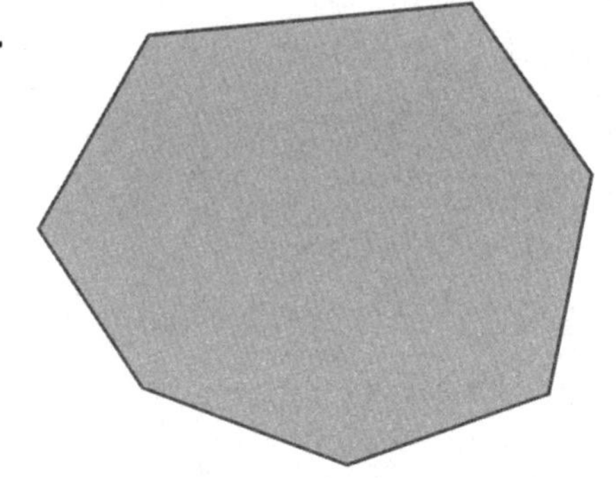

e.

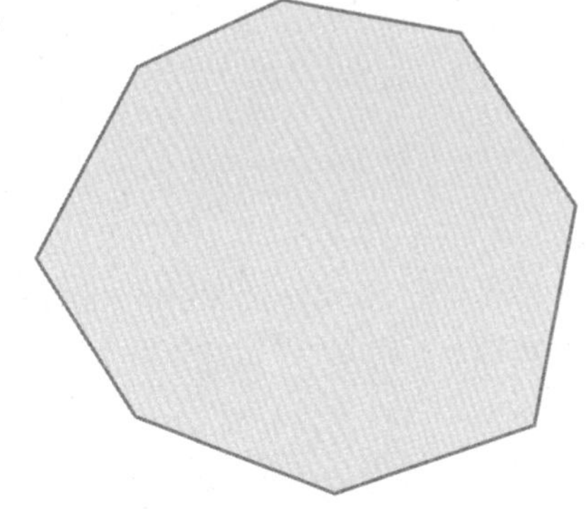

f. REPEATED REASONING Use your results to complete the table. Then find the sum of the interior angle measures of a polygon with 12 sides.

Number of Sides, *n*	3	4	5	6	7	8
Number of Triangles						
Angle Sum, *S*						

2 ACTIVITY: Exploring the Exterior Angles of a Polygon

Work with a partner.

a. Draw a convex pentagon. Extend the sides to form the exterior angles. Label one exterior angle at each vertex A, B, C, D, and E, as shown.

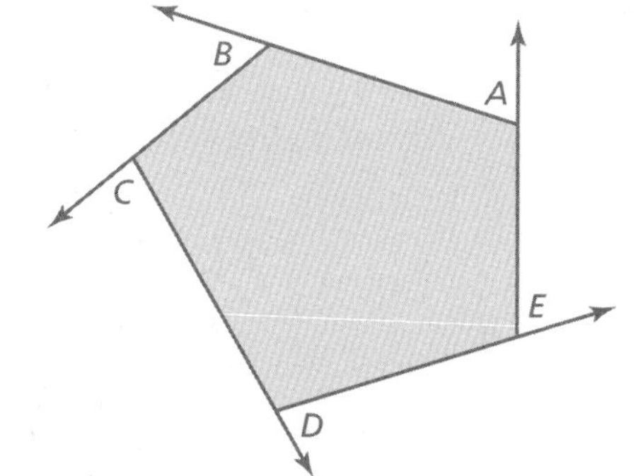

b. Cut out the exterior angles. How can you join the vertices to determine the sum of the angle measures? What do you notice?

c. REPEATED REASONING Repeat the procedure in parts (a) and (b) for each figure below.

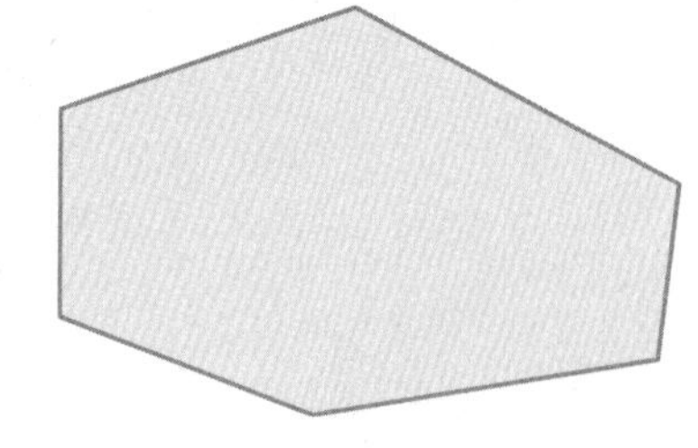

What can you conclude about the sum of the measures of the exterior angles of a convex polygon? Explain.

What Is Your Answer?

3. **STRUCTURE** Use your results from Activity 1 to write an expression that represents the sum of the interior angle measures of a polygon.

4. **IN YOUR OWN WORDS** How can you find the sum of the interior angle measures and the sum of the exterior angle measures of a polygon?

Name ______________________________ Date __________

3.3 Practice

For use after Lesson 3.3

Find the sum of the interior angle measures of the polygon.

1. 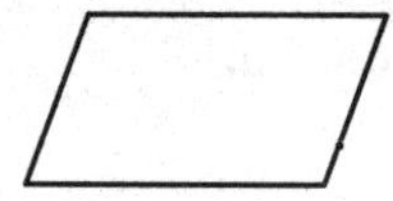

2. 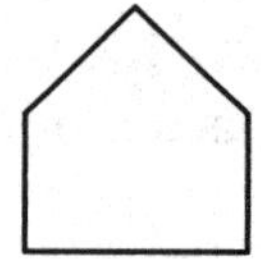

3. 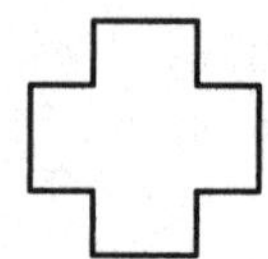

Find the measures of the interior angles.

4. 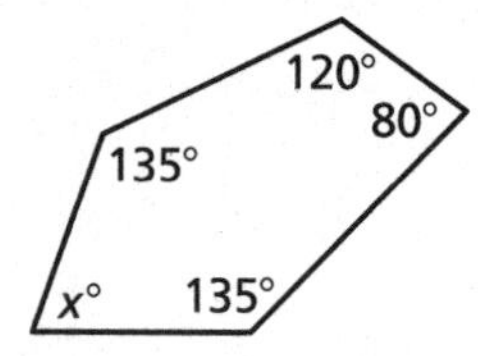

5.

120° 120°
x° x°

Find the measure of each interior angle of the regular polygon.

6. 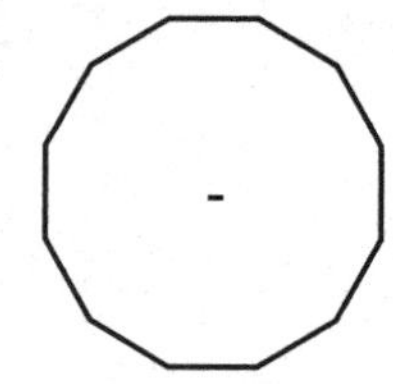

7. 

8. In pottery class, you are making a pot that is shaped as a regular hexagon. What is the measure of each angle in the regular hexagon?

Name______________________________ Date__________

Using Similar Triangles

For use with Activity 3.4

Essential Question How can you use angles to tell whether triangles are similar?

1 ACTIVITY: Constructing Similar Triangles

Work with a partner.

- **Use a straightedge to draw a line segment that is 4 centimeters long.**
- **Then use the line segment and a protractor to draw a triangle that has a 60° and a 40° angle as shown. Label the triangle *ABC*.**

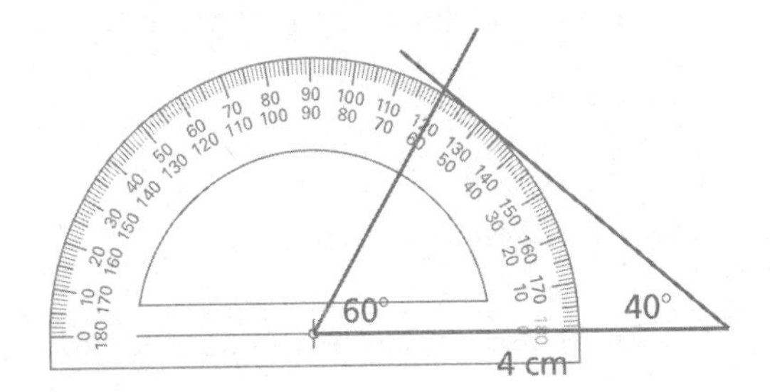

a. Explain how to draw a larger triangle that has the same two angle measures. Label the triangle *JKL*.

b. Explain how to draw a smaller triangle that has the same two angle measures. Label the triangle *PQR*.

c. Are all of the triangles similar? Explain.

Name _______________________________ Date _________

2 ACTIVITY: Using Technology to Explore Triangles

Work with a partner. Use geometry software to draw the triangle shown.

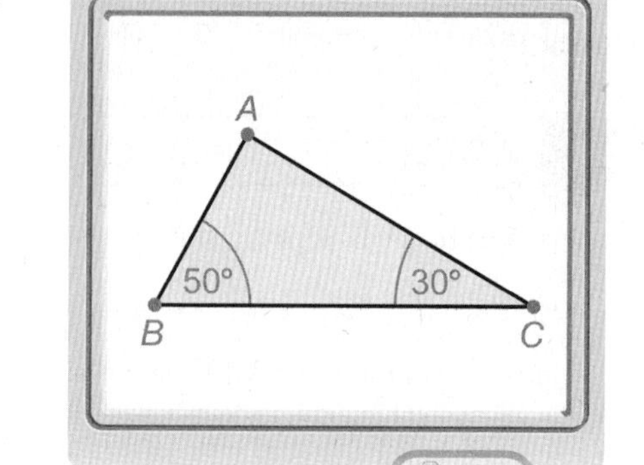

a. Dilate the triangle by the following scale factors.

2 $\qquad \frac{1}{2}$ $\qquad \frac{1}{4}$ $\qquad 2.5$

b. Measure the third angle in each triangle. What do you notice?

c. REASONING When two angles in one triangle are congruent to two angles in another triangle, can you conclude that the triangles are similar? Explain.

3 ACTIVITY: Indirect Measurement

Work with a partner.

a. Use the fact that two rays from the Sun are parallel to explain why $\triangle ABC$ and $\triangle DEF$ are similar.

F

x ft

Sun's ray

C

Sun's ray

5 ft

A 3 ft B

D 36 ft E

b. Explain how to use similar triangles to find the height of the flagpole.

What Is Your Answer?

4. IN YOUR OWN WORDS How can you use angles to tell whether triangles are similar?

5. PROJECT Work with a partner or in a small group.

a. Explain why the process in Activity 3 is called "indirect" measurement.

b. CHOOSE TOOLS Use indirect measurement to measure the height of something outside your school (a tree, a building, a flagpole). Before going outside, decide what materials you need to take with you.

c. MODELING Draw a diagram of the indirect measurement process you used. In the diagram, label the lengths that you actually measured and also the lengths that you calculated.

6. PRECISION Look back at Exercise 17 in Section 2.5. Explain how you can show that the two triangles are similar.

Name ________________________________ Date __________

3.4 Practice
For use after Lesson 3.4

Tell whether the triangles are similar. Explain.

1.

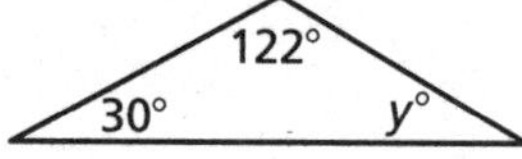

2.

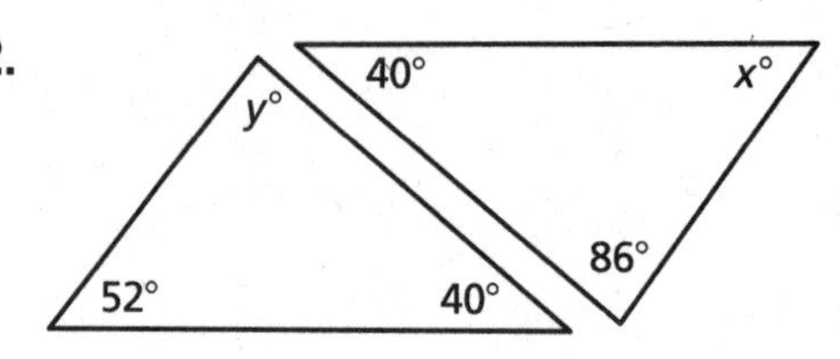

3.

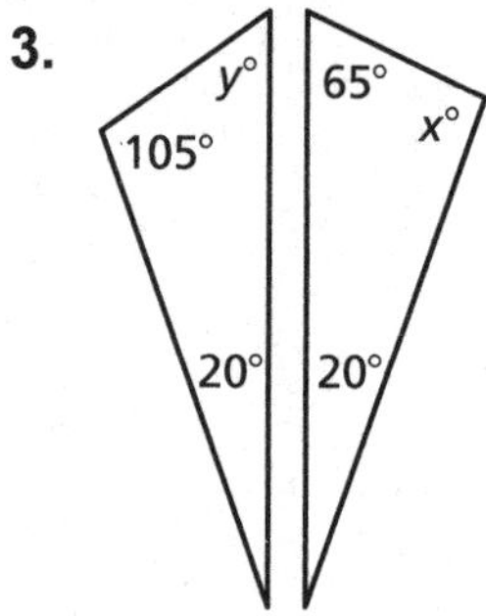

4.

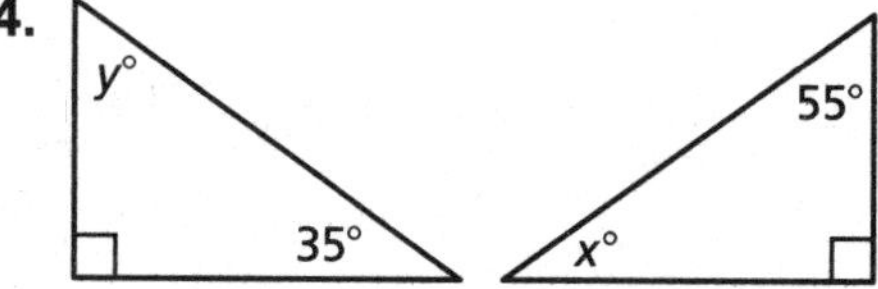

5. You can use similar triangles to find the height of a tree. Triangle ABC is similar to triangle DEC. What is the height of the tree?

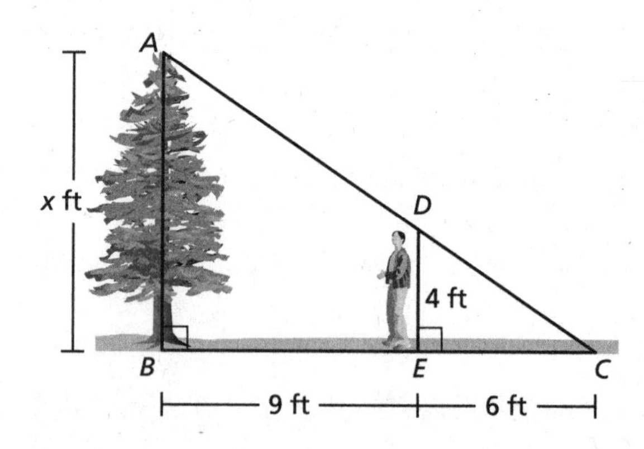

Name Roselynn Sim Date 05/11/21

Fair Game Review

Evaluate the expression when $x = \frac{1}{2}$ and $y = -5$.

1. $-2xy$

   $-2(\frac{1}{2})(-5)$

   $\boxed{5}$

2. $4x^2 - 3y$

   $4(\frac{1}{2})^2 - 3(-5)$

   $1-(-15) = \boxed{16}$

3. $\dfrac{10y}{12x + 4}$

   $\dfrac{10(-5)}{12(\frac{1}{2})+4}$

   $\dfrac{-50}{6+4}$

   $\boxed{-5}$

4. $11x - 8(x - y)$

   $11(\frac{1}{2}) - 8[\frac{1}{2} - (-5)]$

   $5.5 - 8(5.5)$

   $5.5 - 44 = \boxed{-38.5}$

Evaluate the expression when $a = -9$ and $b = -4$.

5. $3ab$

   $3(-9)(-4)$

   $\boxed{108}$

6. $a^2 - 2(b + 12)$

   $(-9)^2 - 2(-4+12)$

   $81 - 16 = \boxed{65}$

7. $\dfrac{4b^2}{3b - 7}$

   $\dfrac{4(-4)^2}{3(-4)-7}$

   64

   -19

   $\boxed{3.37}$

8. $7b^2 + 5(ab - 6)$

   $7(-4)^2 + 5[(-9)(-4) - 6]$

   $7(16) + 5(30)$

   $112 + 150$

   $\boxed{262}$

9. You go to the movies with five friends. You and one of your friends each buy a ticket and a bag of popcorn. The rest of your friends buy just one ticket each. The expression $4x + 2(x + y)$ represents the situation. Evaluate the expression when tickets cost \$7.25 and a bag of popcorn costs \$3.25.

   $x = 7.25$

   $y = 3.25$

   $4(7.25) + 2(7.25 + 3.25) =$

   $29 + 21 = \boxed{50}$

Name Roselynn Sim Date ________

Chapter 4 Fair Game Review (continued)

Use the graph to answer the question.

10. Write the ordered pair that corresponds to Point D.

(-5, 1)

11. Write the ordered pair that corresponds to Point H.

(3, -5)

12. Which point is located at $(-2, 4)$?

F

13. Which point is located at $(0, 3)$?

G

14. Which point(s) are located in Quadrant IV?

B + H

15. Which point(s) are located in Quadrant III?

D, E, C

Plot the point.

16. $(3, -1)$

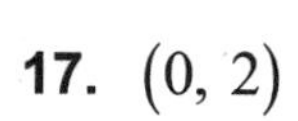

17. $(0, 2)$

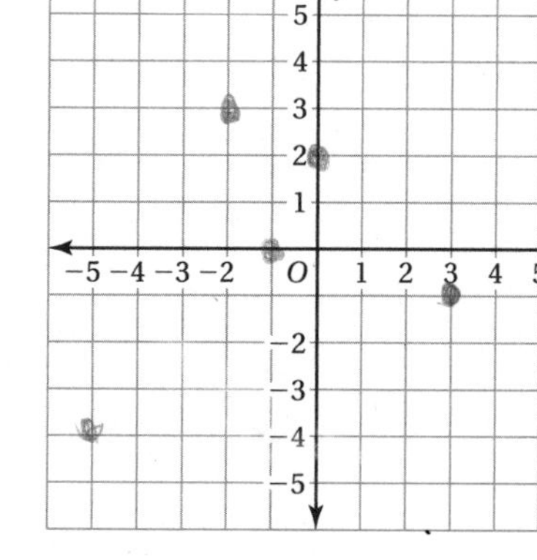

18. $(-5, -4)$

19. $(-1, 0)$

20. $(-2, 3)$

Name______________________________ Date__________

4.1 Graphing Linear Equations
For use with Activity 4.1

Essential Question How can you recognize a linear equation? How can you draw its graph?

1 ACTIVITY: Graphing a Linear Equation

Work with a partner.

a. Use the equation $y = \frac{1}{2}x + 1$ to complete the table. (Choose any two x-values and find the y-values).

	Solution Points	
x		
$y = \frac{1}{2}x + 1$		

b. Write the two ordered pairs given by the table. These are called *solution points* of the equation.

c. PRECISION Plot the two solution points. Draw a line *exactly* through the two points.

d. Find a different point on the line. Check that this point is a solution point of the equation $y = \frac{1}{2}x + 1$.

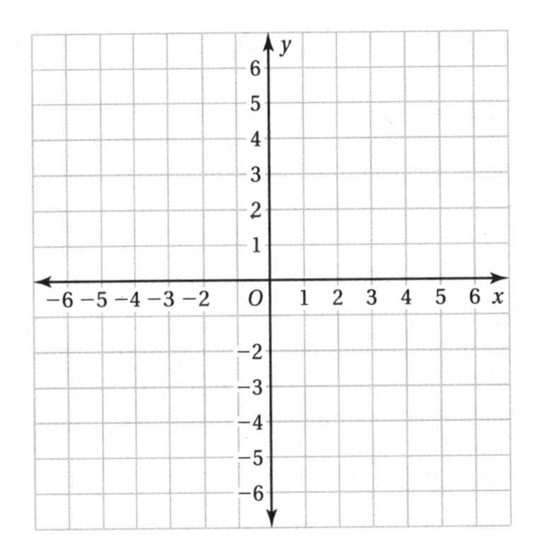

e. LOGIC Do you think it is true that *any* point on the line is a solution point of the equation $y = \frac{1}{2}x + 1$? Explain.

f. Choose five additional x-values for the table. (Choose positive and negative x-values.) Plot the five corresponding solution points on the previous page. Does each point lie on the line?

	Solution Points				
x					
$y = \frac{1}{2}x + 1$					

g. LOGIC Do you think it is true that *any* solution point of the equation $y = \frac{1}{2}x + 1$ is a point on the line? Explain.

h. Why do you think $y = ax + b$ is called a *linear equation*?

2 ACTIVITY: Using a Graphing Calculator

Use a graphing calculator to graph $y = 2x + 5$.

a. Enter the equation $y = 2x + 5$ into your calculator.

b. Check the settings of the *viewing window.* The boundaries of the graph are set by the minimum and maximum x- and y-values. The numbers of units between the tick marks are set by the x- and y-scales.

WINDOW
Xmin=-10
Xmax=10
Xscl=1
Ymin=-10
Ymax=10
Yscl=1
Xres=1

This is the standard viewing window.

Name____ Date__________

c. Graph $y = 2x + 5$ on your calculator.

d. Change the settings of the viewing window to match those shown. Compare the two graphs.

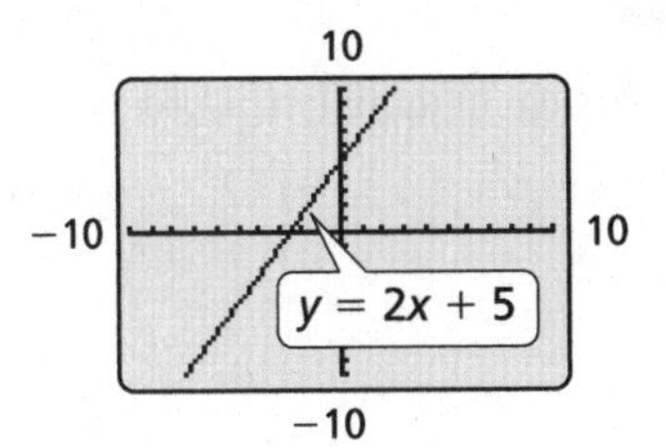

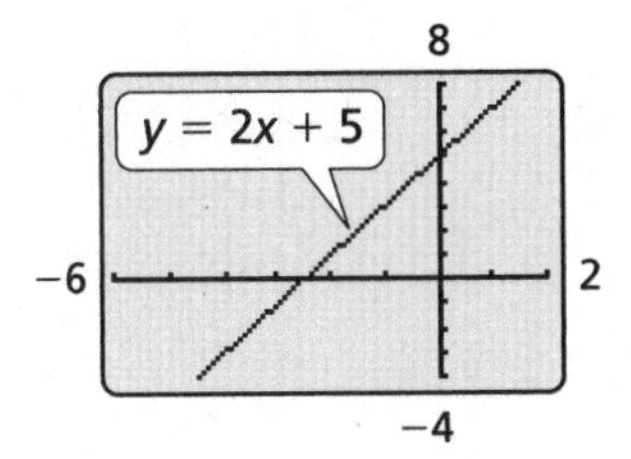

What Is Your Answer?

3. IN YOUR OWN WORDS How can you recognize a linear equation? How can you draw its graph? Write an equation that is linear. Write an equation that is *not* linear.

4. Use a graphing calculator to graph $y = 5x - 12$ in the standard viewing window.

a. Can you tell where the line crosses the x-axis? Can you tell where the line crosses the y-axis?

b. How can you adjust the viewing window so that you can determine where the line crosses the x- and y-axes?

5. CHOOSE TOOLS You want to graph $y = 2.5x - 3.8$. Would you graph it by hand or by using a graphing calculator? Why?

Name ______________________________ Date __________

4.1 Practice

For use after Lesson 4.1

Graph the linear equation. Use a graphing calculator to check your graph, if possible.

1. $y = 4$

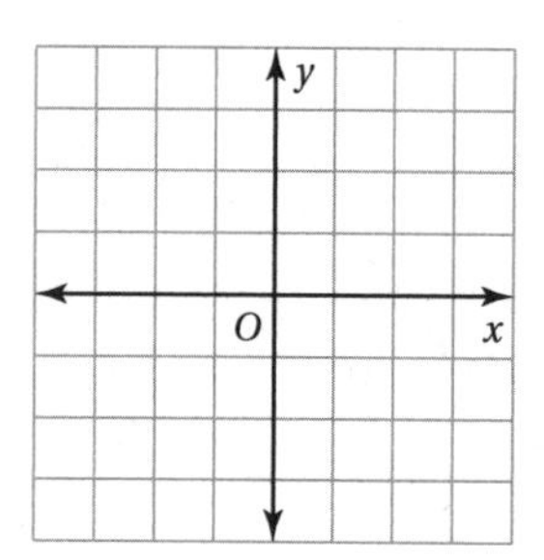

2. $y = -\frac{1}{3}x$

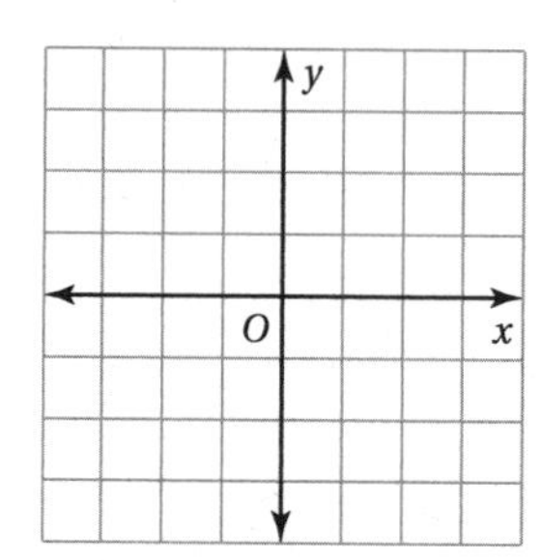

Solve for *y*. Then graph the equation. Use a graphing calculator to check your graph.

3. $y + 2x = 3$

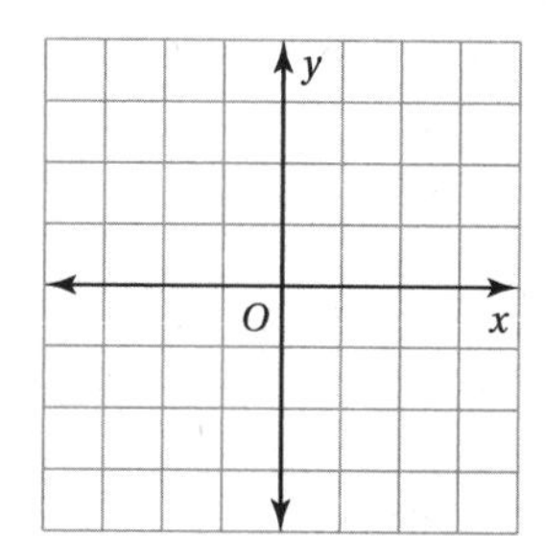

4. $2y - 3x = 1$

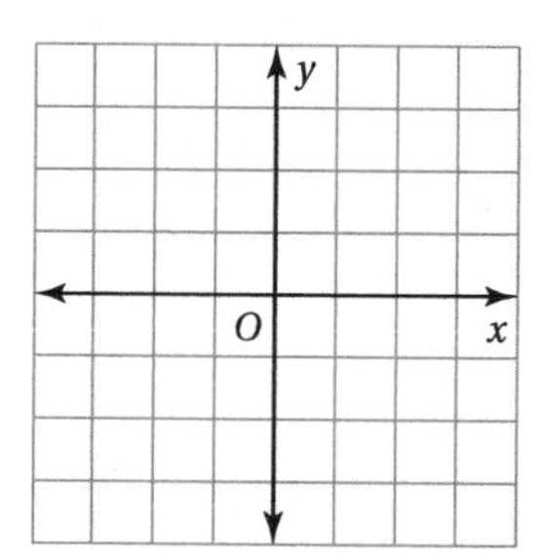

5. The equation $y = 2x + 4$ represents the cost y (in dollars) of renting a movie after x days of late charges.

a. Graph the equation.

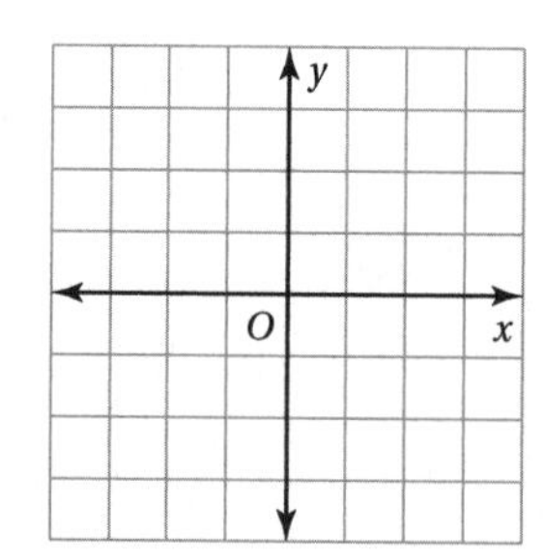

b. Use the graph to determine how much it costs after 3 days of late charges.

Name ________________________________ Date __________

4.2 Slope of a Line
For use with Activity 4.2

Essential Question How can the slope of a line be used to describe the line?

Slope is the rate of change between any two points on a line. It is the measure of the *steepness* of the line.

To find the slope of a line, find the ratio of the change in y (vertical change) to the change in x (horizontal change).

$$\text{slope} = \frac{\text{change in } y}{\text{change in } x}$$

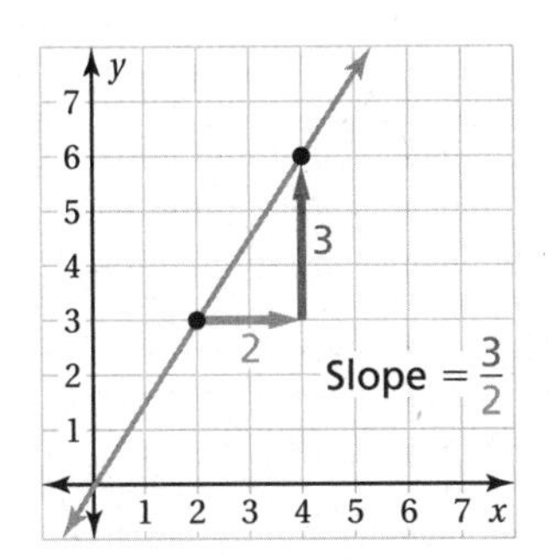

1 ACTIVITY: Finding the Slope of a Line

Work with a partner. Find the slope of each line using two methods.

Method 1: Use the two black points.

Method 2: Use the two gray points.

Do you get the same slope using each method? Why do you think this happens?

a.

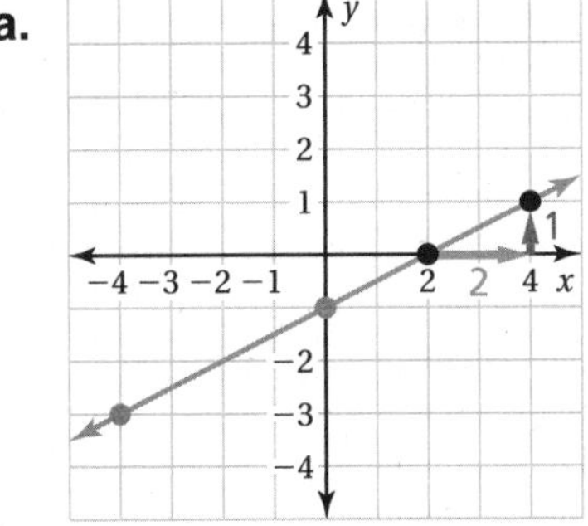

b.

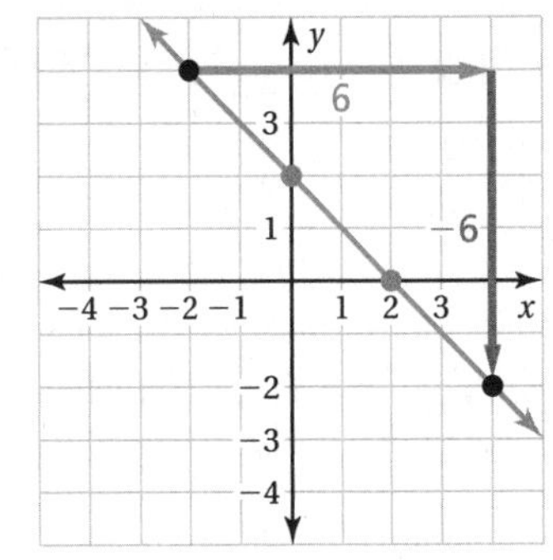

c.

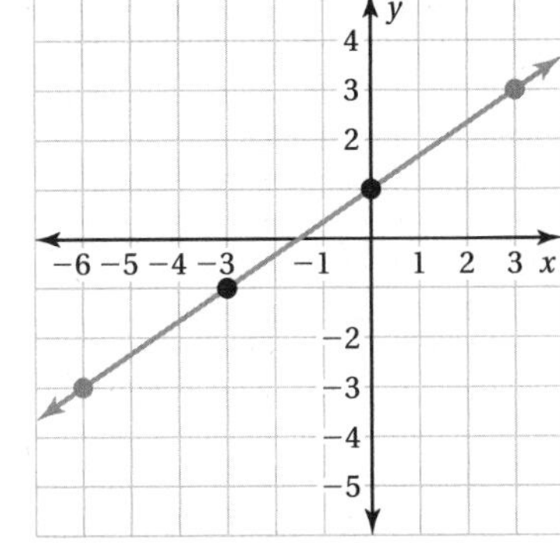

d.

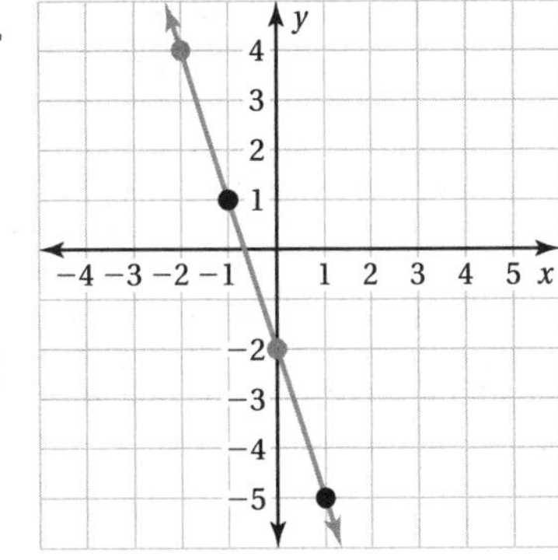

Name ______________________________ Date __________

2 ACTIVITY: Using Similar Triangles

Work with a partner. Use the figure shown.

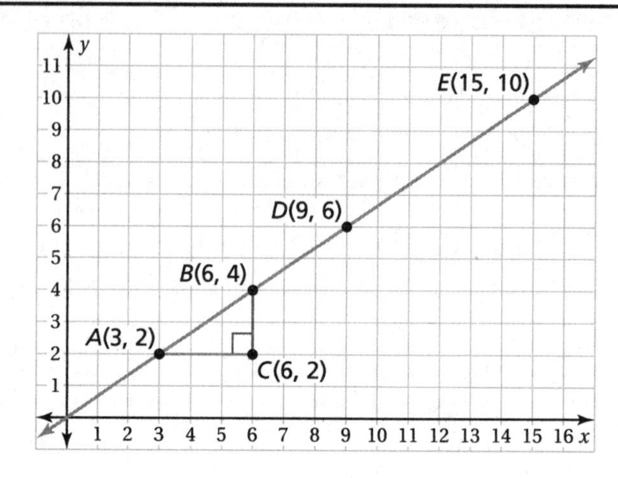

a. $\triangle ABC$ is a right triangle formed by drawing a horizontal line segment from point A and a vertical line segment from point B. Use this method to draw another right triangle, $\triangle DEF$.

b. What can you conclude about $\triangle ABC$ and $\triangle DEF$? Justify your conclusion.

c. For each triangle, find the ratio of the length of the vertical side to the length of the horizontal side. What do these ratios represent?

d. What can you conclude about the slope between any two points on the line?

3 ACTIVITY: Drawing Lines with Given Slopes

Work with a partner.

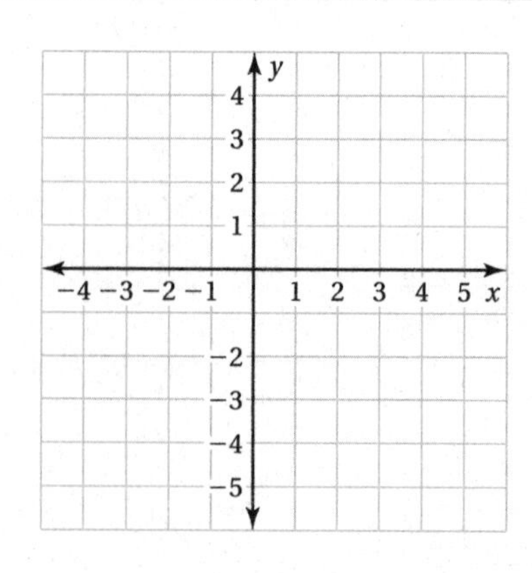

a. Draw two lines with slope $\frac{3}{4}$. One line passes through $(-4, 1)$, and the other line passes through $(4, 0)$. What do you notice about the two lines?

Name____ Date__________

b. Draw two lines with slope $-\frac{4}{3}$. One line passes through $(2, 1)$, and the other line passes through $(-1, -1)$. What do you notice about the two lines?

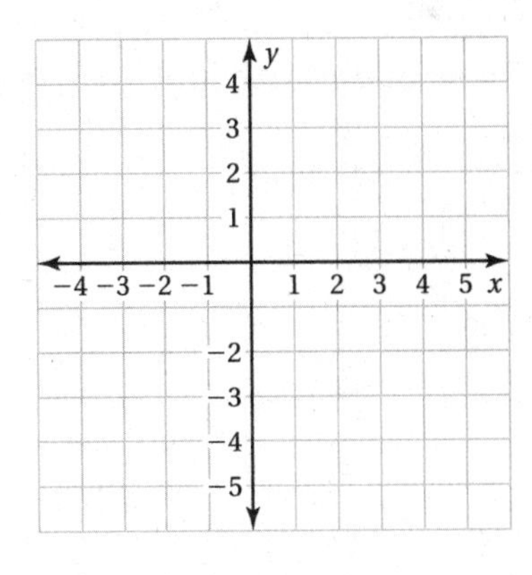

c. CONJECTURE Make a conjecture about two different nonvertical lines in the same plane that have the same slope.

d. Graph one line from part (a) and one line from part (b) in the same coordinate plane. Describe the angle formed by the two lines. What do you notice about the product of the slopes of the two lines?

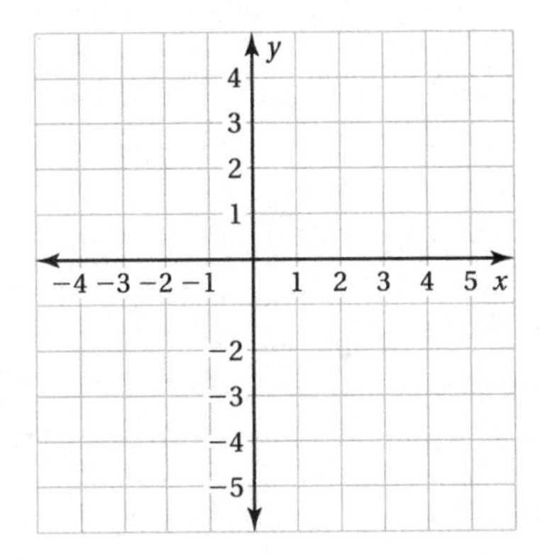

e. REPEATED REASONING Repeat part (d) for the two lines you did *not* choose. Based on your results, make a conjecture about two lines in the same plane whose slopes have a product of -1.

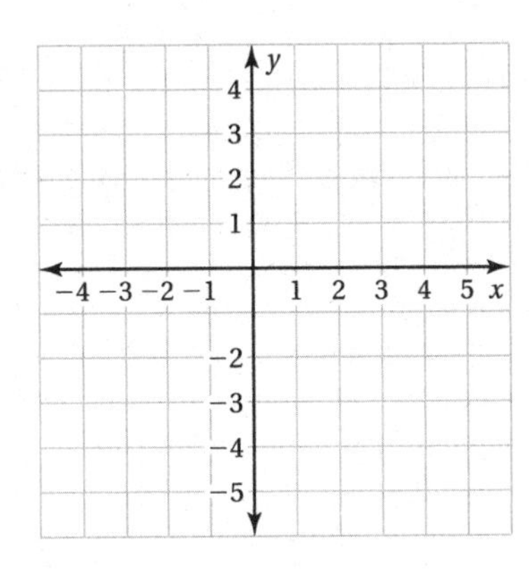

What Is Your Answer?

4. IN YOUR OWN WORDS How can you use the slope of a line to describe the line?

Name ______________________ Date ________

4.2 Practice

For use after Lesson 4.2

Find the slope of the line.

1.

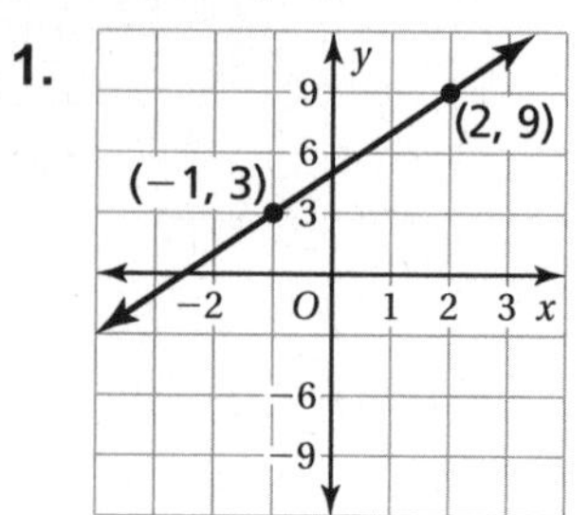

2.

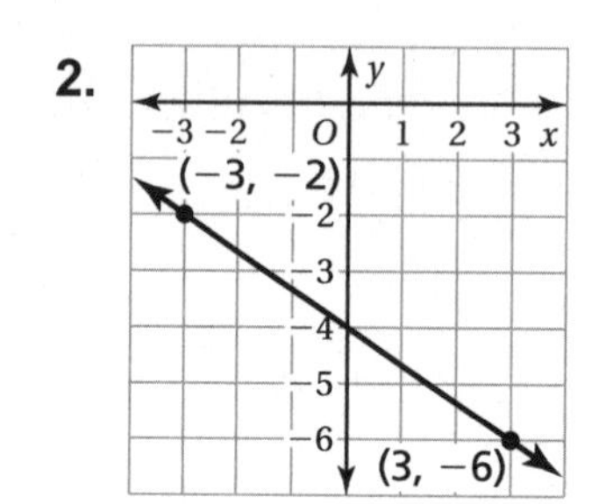

3.

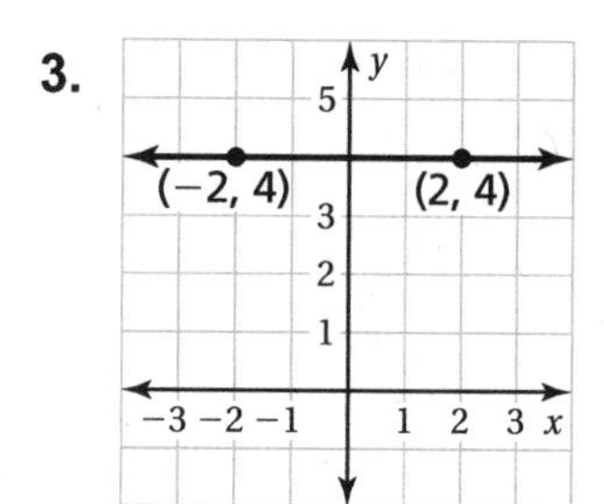

4.

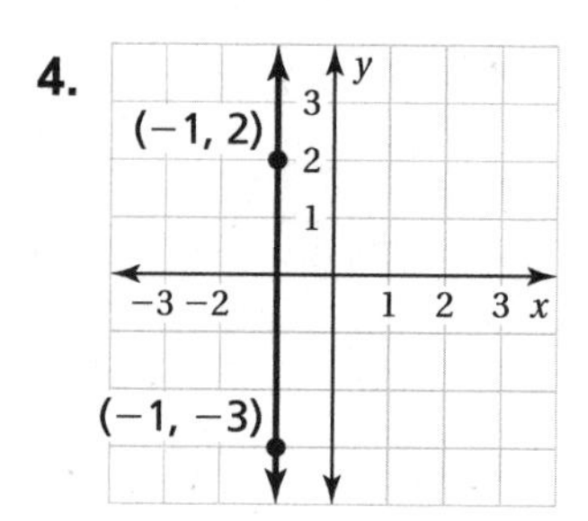

5. Which set of stairs is more difficult to climb? Explain.

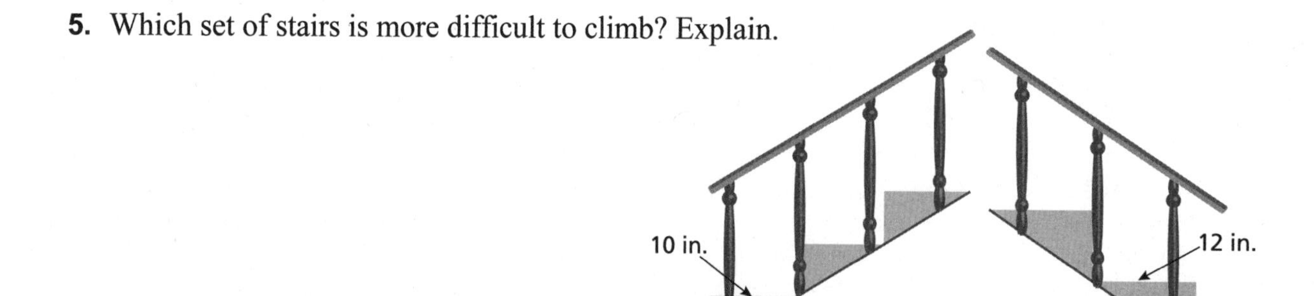

Name____ Date__________

Extension 4.2 Practice

For use after Extension 4.2

Which lines are parallel? How do you know?

1.

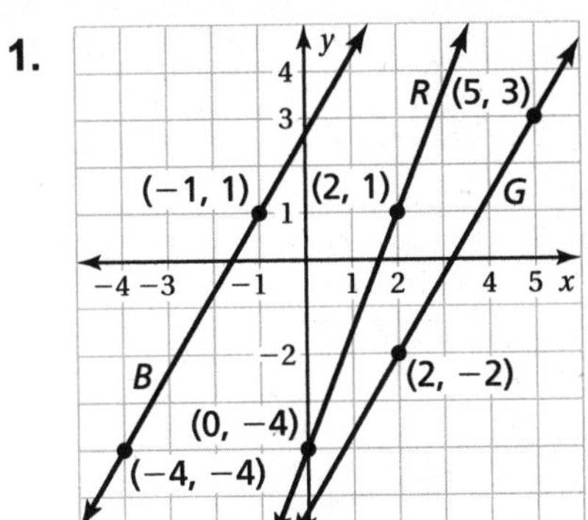

2.

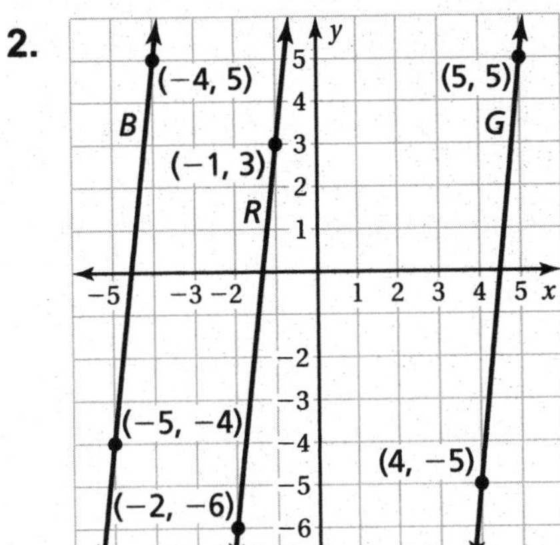

Are the given lines parallel? Explain your reasoning.

3. $y = 2, y = -4$

4. $x = 3, y = -3$

5. Is the quadrilateral a parallelogram? Justify your answer.

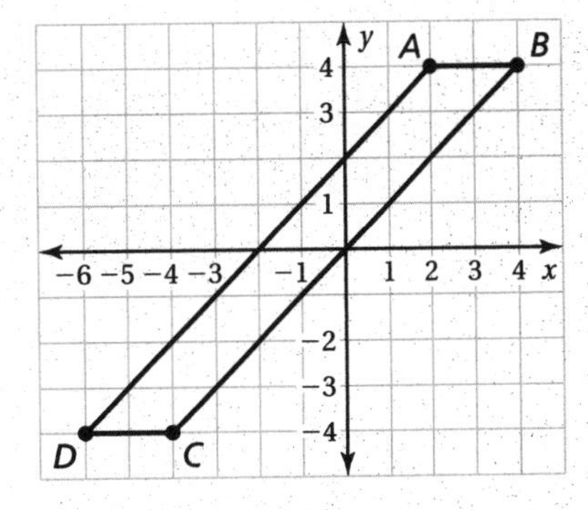

Name ______________________________ Date __________

Extension 4.2 **Practice (continued)**

Which lines are perpendicular? How do you know?

6.

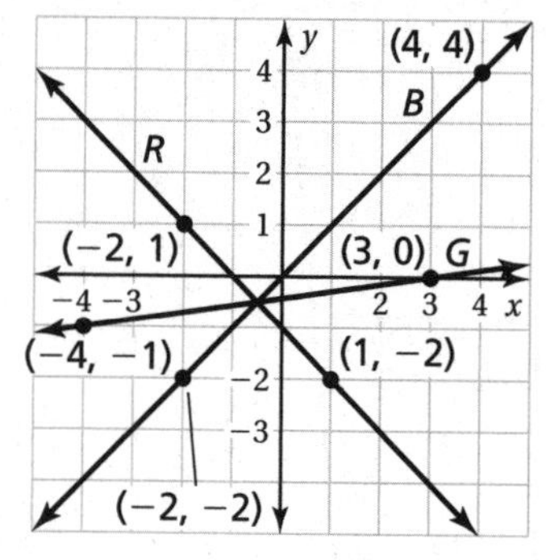

7.

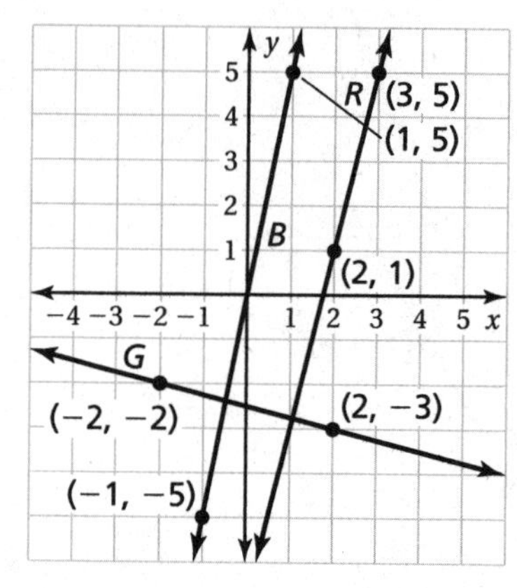

Are the given lines perpendicular? Explain your reasoning.

8. $x = 0, y = 3$

9. $y = 2, y = -\frac{1}{2}$

10. Is the parallelogram a rectangle? Justify your answer.

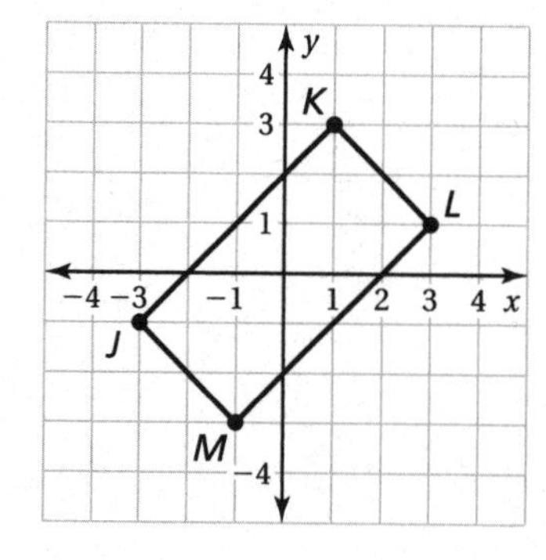

Name____ Date__________

4.3 Graphing Proportional Relationships
For use with Activity 4.3

Essential Question How can you describe the graph of the equation $y = mx$?

1 ACTIVITY: Identifying Proportional Relationships

Work with a partner. Tell whether x and y are in a proportional relationship. Explain your reasoning.

a.

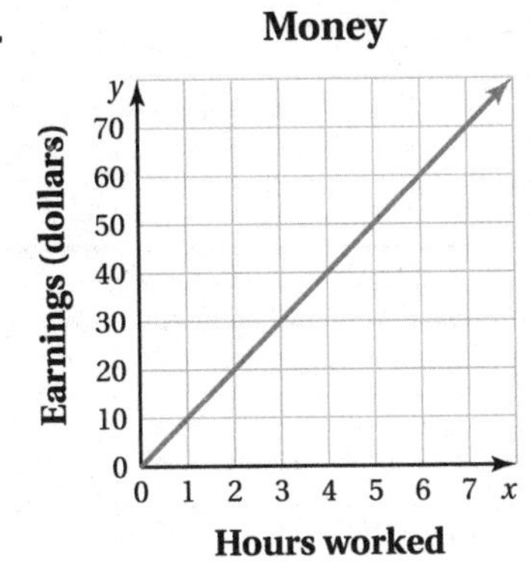

b.

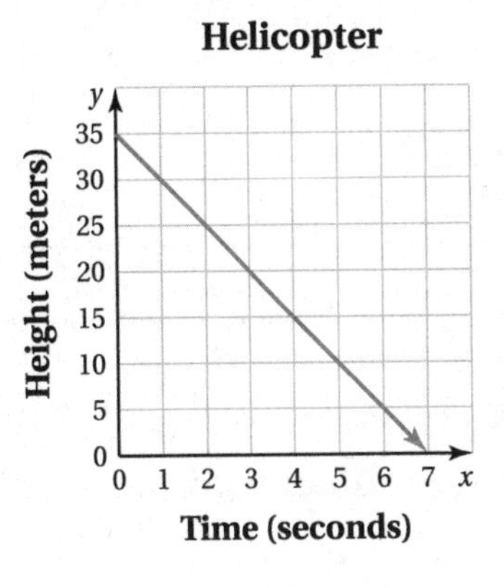

c.

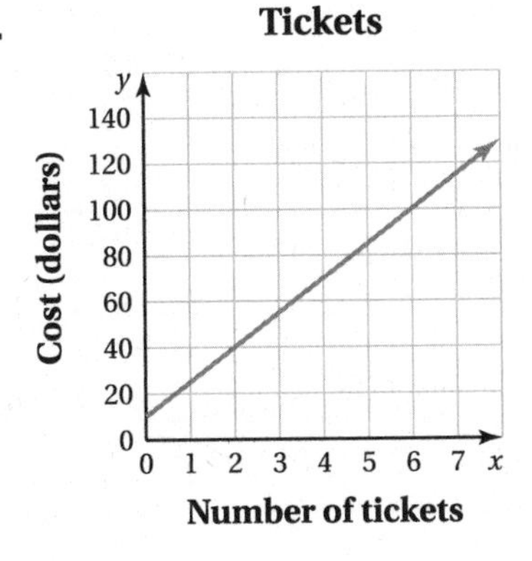

d.

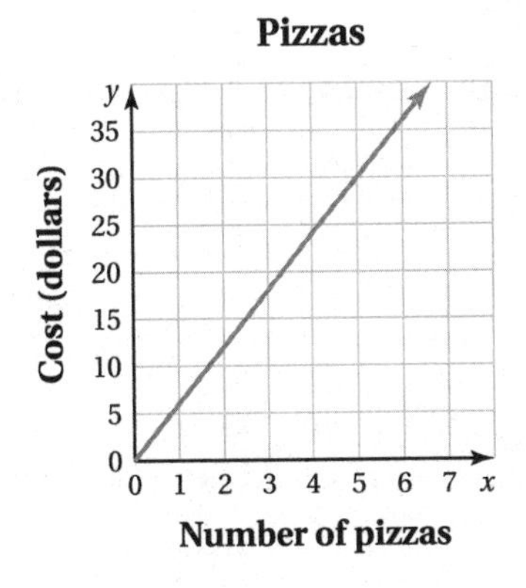

e.

Laps, x	1	2	3	4
Time (seconds), y	90	200	325	480

f.

Cups of Sugar, x	$\frac{1}{2}$	1	$1\frac{1}{2}$	2
Cups of Flour, y	1	2	3	4

Name ____________________ Date ________

2 ACTIVITY: Analyzing Proportional Relationships

Work with a partner. Use only the proportional relationships in Activity 1 to do the following.

- **Find the slope of the line.**
- **Find the value of y for the ordered pair $(1, y)$.**

What do you notice? What does the value of y represent?

3 ACTIVITY: Deriving an Equation

Work with a partner. Let (x, y) represent any point on the graph of a proportional relationship.

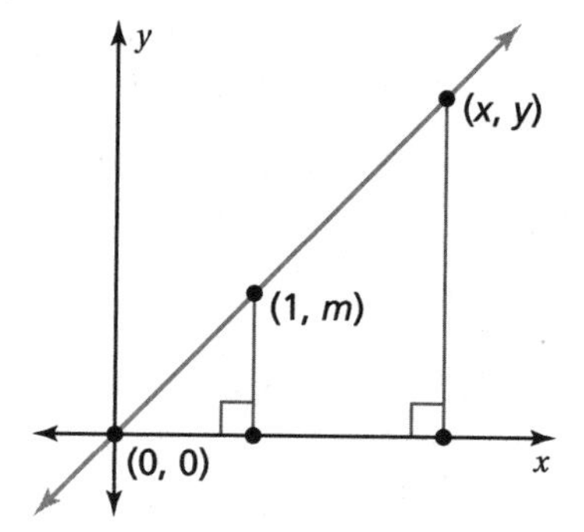

a. Explain why the two triangles are similar.

b. Because the triangles are similar, the corresponding side lengths are proportional. Use the vertical and horizontal side lengths to complete the steps below.

$$\frac{\square}{\square} = \frac{m}{1}$$ Ratios of side lengths

$$\frac{\square}{\square} = m$$ Simplify.

$$\square = m \bullet \square$$ Multiplication Property of Equality

What does the final equation represent?

c. Use your result in part (b) to write an equation that represents each proportional relationship in Activity 1.

What Is Your Answer?

4. IN YOUR OWN WORDS How can you describe the graph of the equation $y = mx$? How does the value of m affect the graph of the equation?

5. Give a real-life example of two quantities that are in a proportional relationship. Write an equation that represents the relationship and sketch its graph.

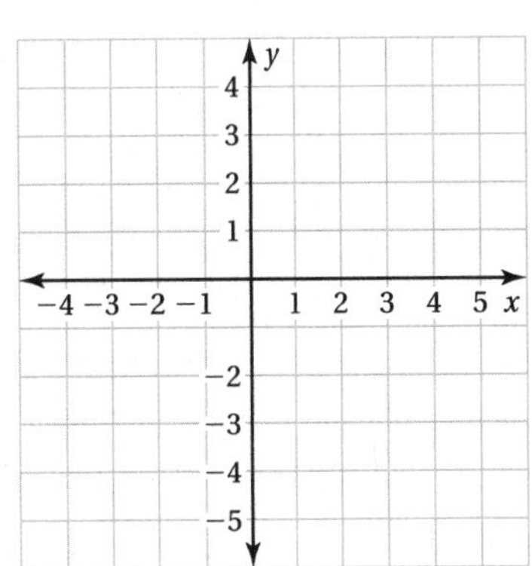

Name ______________________________ Date __________

4.3 Practice

For use after Lesson 4.3

1. The amount p (in dollars) that you earn by working h hours is represented by the equation $p = 9h$. Graph the equation and interpret the slope.

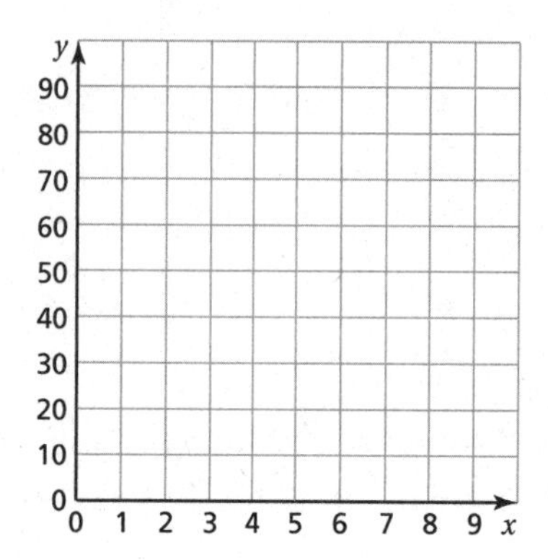

2. The cost c (in dollars) to rent a bicycle is proportional to the number h of hours that you rent the bicycle. It costs \$20 to rent the bicycle for 4 hours.

   a. Write an equation that represents the situation.

   b. Interpret the slope.

   c. How much does it cost to rent the bicycle for 6 hours?

Name ____ Date __________

4.4 Graphing Linear Equations in Slope-Intercept Form

For use with Activity 4.4

Essential Question How can you describe the graph of the equation $y = mx + b$?

1 ACTIVITY: Analyzing Graphs of Lines

Work with a partner.

- **Graph each equation.**
- **Find the slope of each line.**
- **Find the point where each line crosses the *y*-axis.**
- **Complete the table.**

Equation	Slope of Graph	Point of Intersection with *y*-axis
a. $y = -\frac{1}{2}x + 1$		
b. $y = -x + 2$		
c. $y = -x - 2$		
d. $y = \frac{1}{2}x + 1$		
e. $y = x + 2$		
f. $y = x - 2$		
g. $y = \frac{1}{2}x - 1$		
h. $y = -\frac{1}{2}x - 1$		
i. $y = 3x + 2$		

4.4 Graphing Linear Equations in Slope-Intercept Form (continued)

Equation	Slope of Graph	Point of Intersection with *y*-axis
j. $y = 3x - 2$		

k. Do you notice any relationship between the slope of the graph and its equation? Between the point of intersection with the *y*-axis and its equation? Compare the results with those of other students in your class.

2 ACTIVITY: Deriving an Equation

Work with a partner.

a. Look at the graph of each equation in Activity 1. Do any of the graphs represent a proportional relationship? Explain.

b. For a nonproportional linear relationship, the graph crosses the *y*-axis at some point $(0, b)$, where b does not equal 0.

Let (x, y) represent any other point on the graph. You can use the formula for slope to write the equation for a nonproportional linear relationship.

Use the graph to complete the steps.

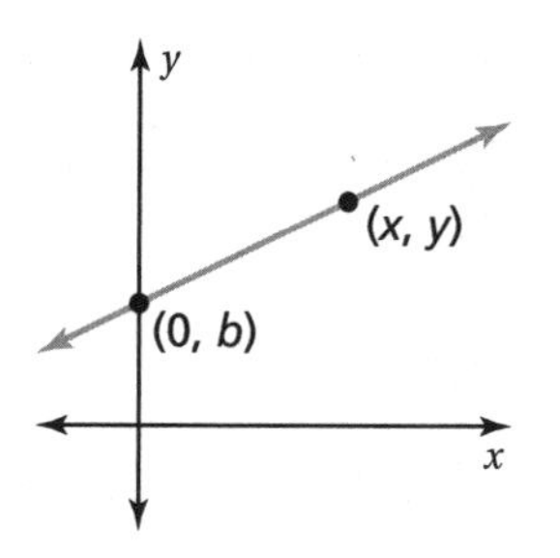

$\dfrac{y_2 - y_1}{x_2 - x_1} = m$ **Slope formula**

$\dfrac{y - \square}{x - \square} = m$ **Substitute values.**

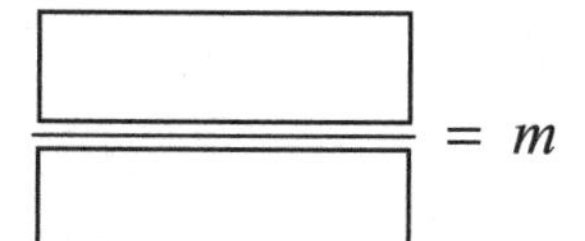
$= m$ **Simplify.**

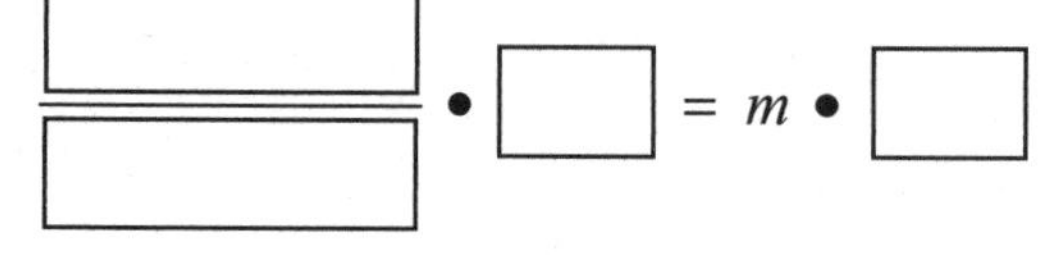
Multiplication Property of Equality

$y - \square = m \bullet \square$ **Simplify.**

$y = m\square + \square$ **Addition Property of Equality**

c. What do m and b represent in the equation?

What Is Your Answer?

3. IN YOUR OWN WORDS How can you describe the graph of the equation $y = mx + b$?

a. How does the value of m affect the graph of the equation?

b. How does the value of b affect the graph of the equation?

c. Check your answers to parts (a) and (b) with three equations that are not in Activity 1.

4. LOGIC Why do you think $y = mx + b$ is called the *slope-intercept form* of the equation of a line? Use drawings or diagrams to support your answer.

Name ______________________________ Date ________

4.4 Practice

For use after Lesson 4.4

Find the slope and *y*-intercept of the graph of the linear equation.

1. $y = -3x + 9$

2. $y = 4 - \frac{2}{5}x$

3. $6 + y = 8x$

Graph the linear equation. Identify the *x*-intercept. Use a graphing calculator to check your answer.

4. $y = \frac{2}{3}x + 6$

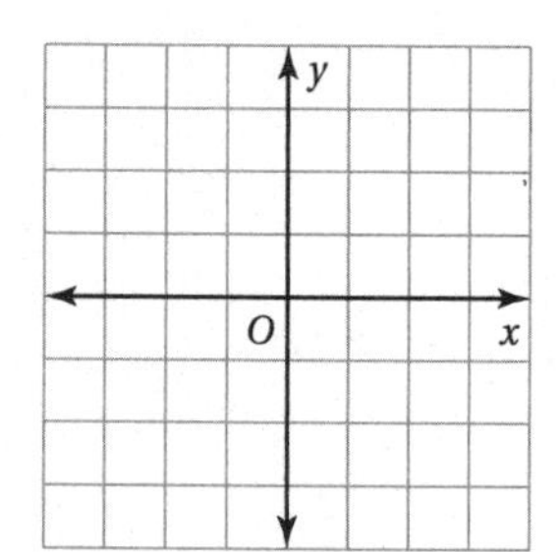

5. $y - 10 = -5x$

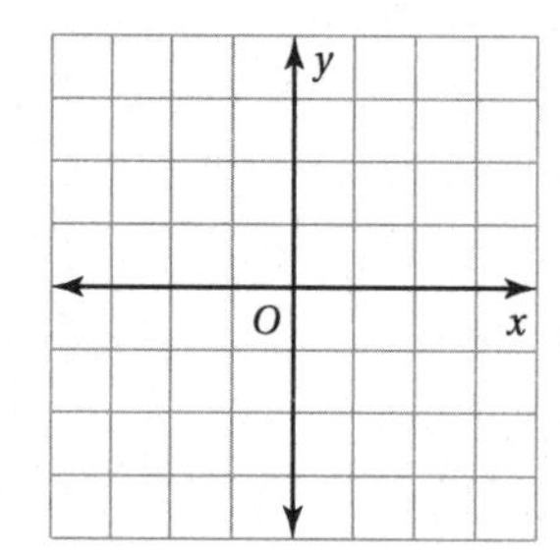

6. The equation $y = -90x + 1440$ represents the time (in minutes) left after x games of a tournament.

a. Graph the equation.

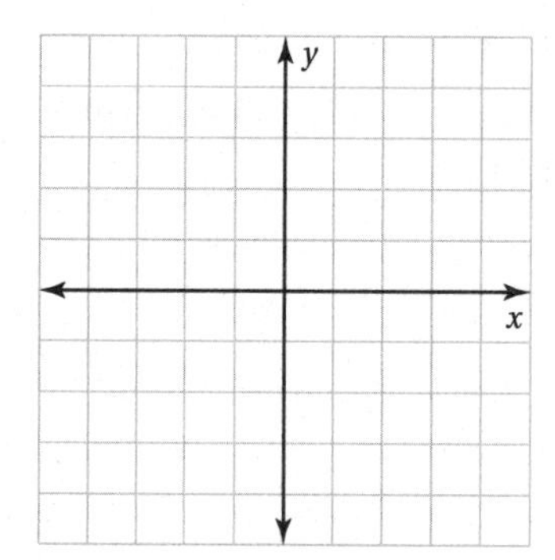

b. Interpret the x-intercept and slope.

Name____ Date__________

4.5 Graphing Linear Equations in Standard Form

For use with Activity 4.5

Essential Question How can you describe the graph of the equation $ax + by = c$?

1 ACTIVITY: Using a Table to Plot Points

Work with a partner. You sold a total of $16 worth of tickets to a school concert. You lost track of how many of each type of ticket you sold.

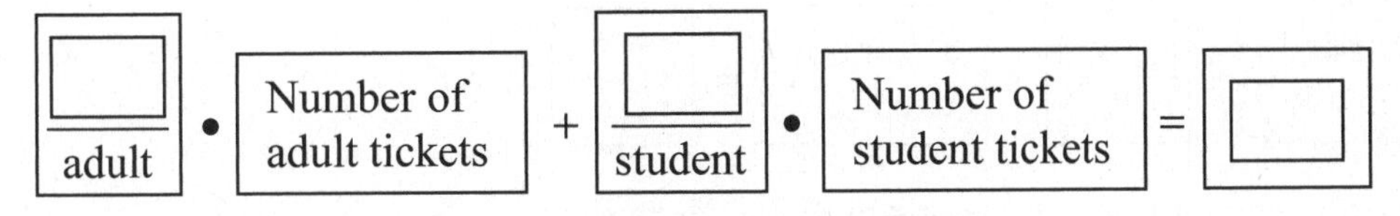

a. Let x represent the number of adult tickets.
Let y represent the number of student tickets.
Write an equation that relates x and y.

b. Complete the table showing the different combinations of tickets you might have sold.

Number of Adult Tickets, *x*					
Number of Student Tickets, *y*					

c. Plot the points from the table. Describe the pattern formed by the points.

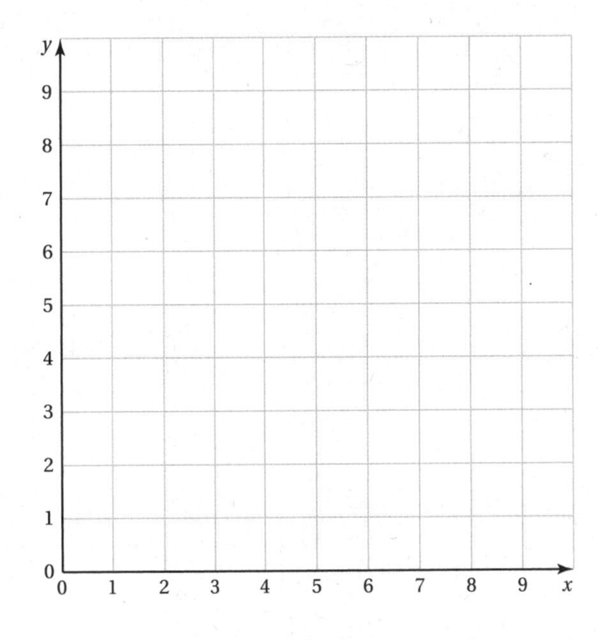

d. If you remember how many adult tickets you sold, can you determine how many student tickets you sold? Explain your reasoning.

Name ______________________________ Date __________

2 ACTIVITY: Rewriting an Equation

Work with a partner. You sold a total of $16 worth of cheese. You forgot how many pounds of each type of cheese you sold.

$\frac{\square}{\text{lb}}$ • Pounds of swiss + $\frac{\square}{\text{lb}}$ • Pound of cheddar = $\square$

a. Let x represent the number of pounds of swiss cheese.
Let y represent the number of pounds of cheddar cheese.
Write an equation that relates x and y.

b. Rewrite the equation in slope-intercept form. Then graph the equation.

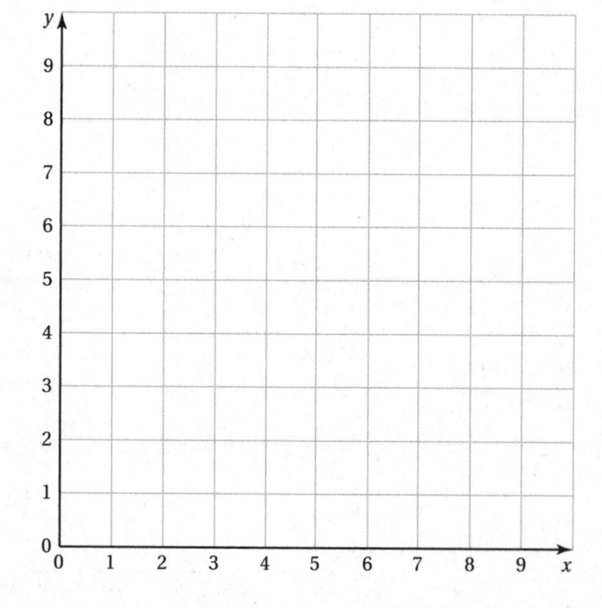

c. You sold 2 pounds of cheddar cheese. How many pounds of swiss cheese did you sell?

d. Does the value $x = 2.5$ make sense in the context of the problem? Explain.

What Is Your Answer?

3. IN YOUR OWN WORDS How can you describe the graph of the equation $ax + by = c$?

4. Activities 1 and 2 show two different methods for graphing $ax + by = c$. Describe the two methods. Which method do you prefer? Explain.

5. Write a real-life problem that is similar to those shown in Activities 1 and 2.

6. Why do you think it might be easier to graph $x + y = 10$ without rewriting it in slope-intercept form and then graphing?

Name ______________________________ Date __________

4.5 Practice
For use after Lesson 4.5

Write the linear equation in slope-intercept form.

1. $2x - y = 7$

2. $\frac{1}{4}x + y = -\frac{2}{7}$

3. $3x - 5y = -20$

Graph the linear equation using intercepts. Use a graphing calculator to check your graph.

4. $2x - 3y = 12$

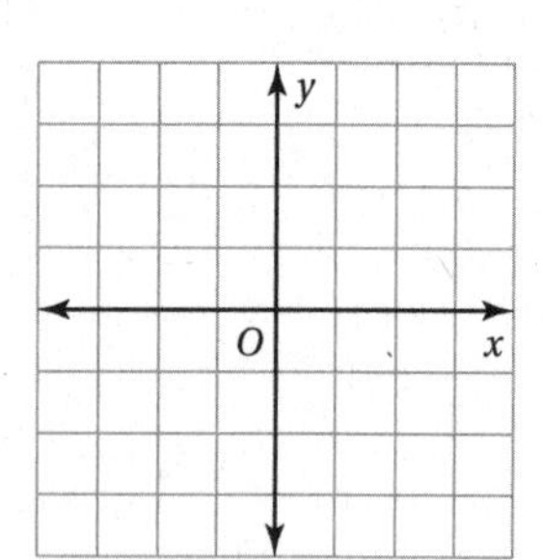

5. $x + 9y = -27$

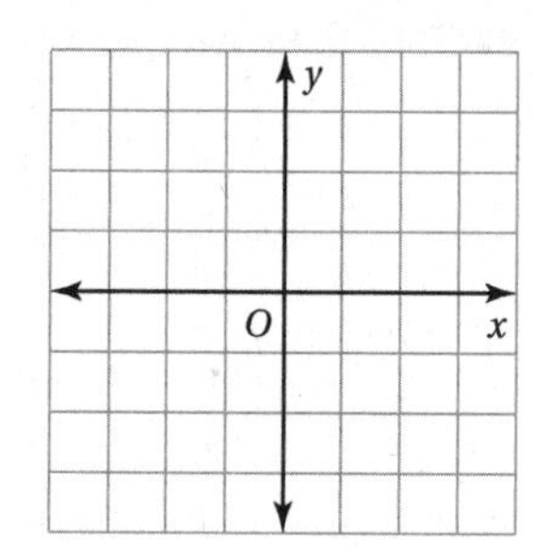

6. You go shopping and buy x shirts for \$12 and y jeans for \$28. The total spent is \$84.

   a. Write an equation in standard form that models how much money you spent.

   b. Graph the equation and interpret the intercepts.

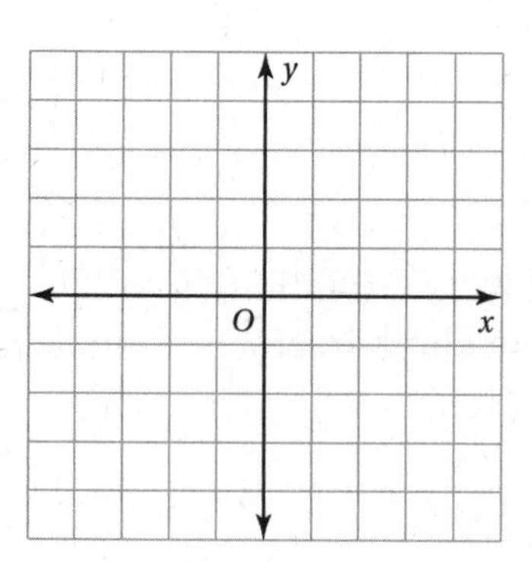

Name________________________ Date__________

4.6 Writing Equations in Slope-Intercept Form

For use with Activity 4.6

Essential Question How can you write an equation of a line when you are given the slope and *y*-intercept of the line?

1 ACTIVITY: Writing Equations of Lines

Work with a partner.

- **Find the slope of each line.**
- **Find the *y*-intercept of each line.**
- **Write an equation for each line.**
- **What do the three lines have in common?**

a.

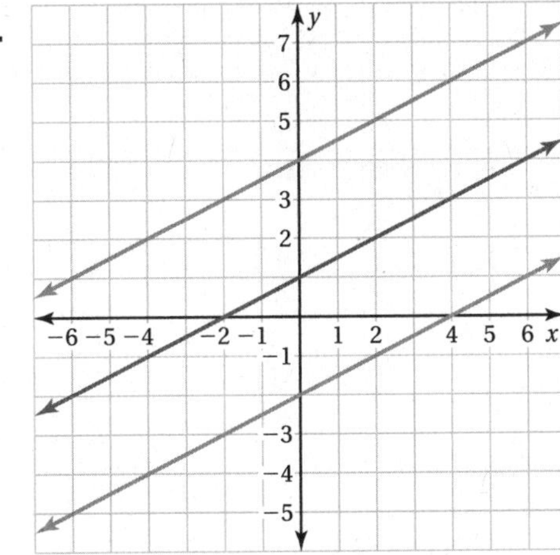

b.

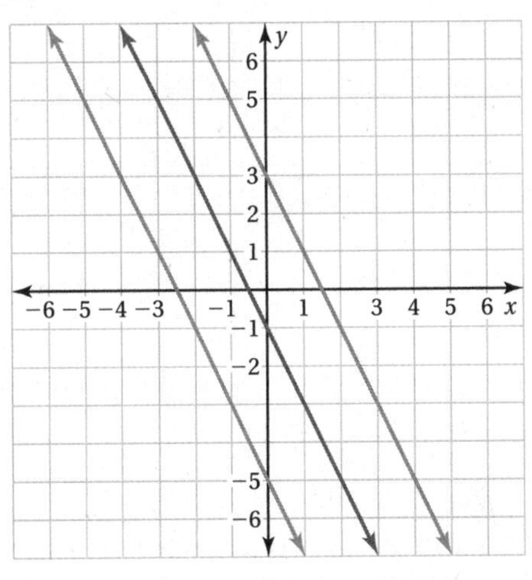

c.

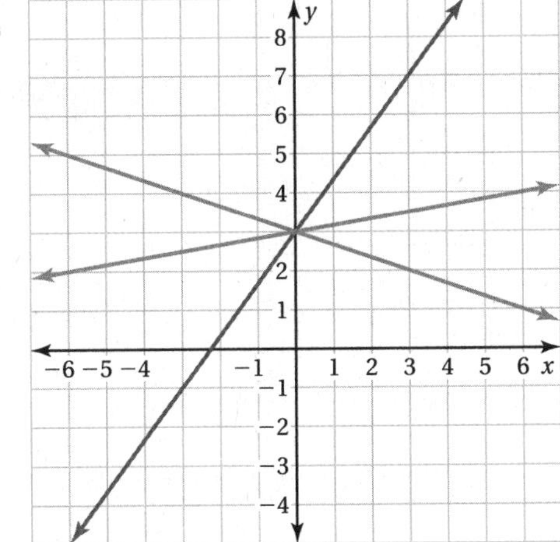

d.

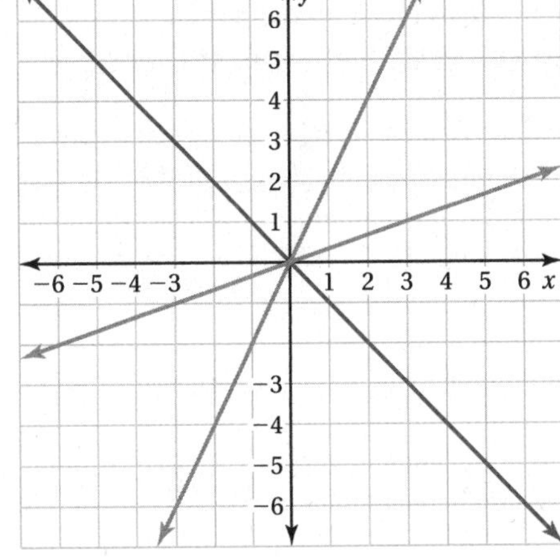

Name ______________________________ Date __________

2 ACTIVITY: Describing a Parallelogram

Work with a partner.

- **Find the area of each parallelogram.**

- **Write an equation that represents each side of each parallelogram.**

a.

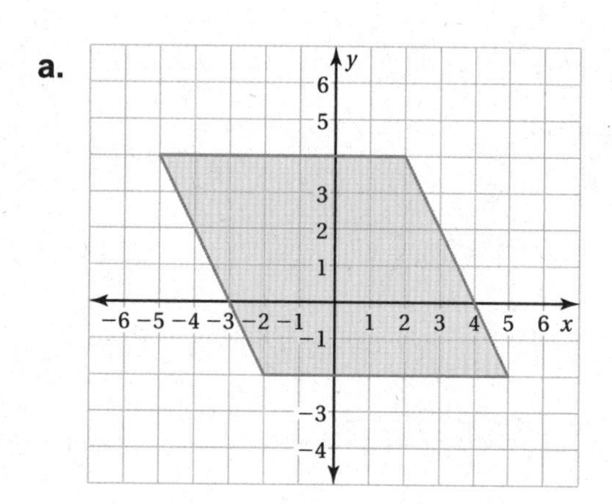

b.

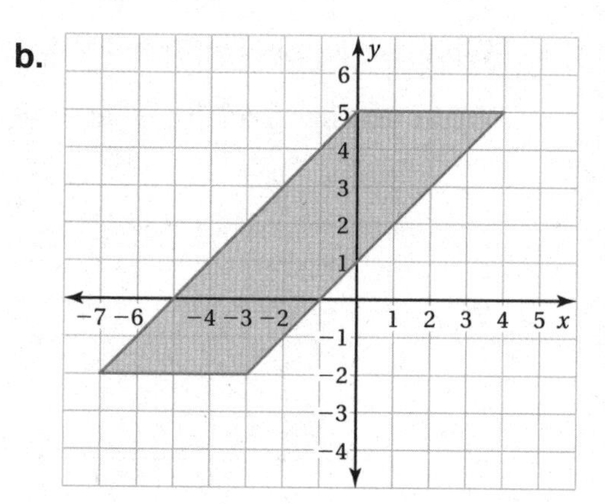

3 ACTIVITY: Interpreting the Slope and the *y*-Intercept

Work with a partner. The graph shows a trip taken by a car, where *t* is the time (in hours) and *y* is the distance (in miles) from Phoenix.

a. Find the *y*-intercept of the graph. What does it represent?

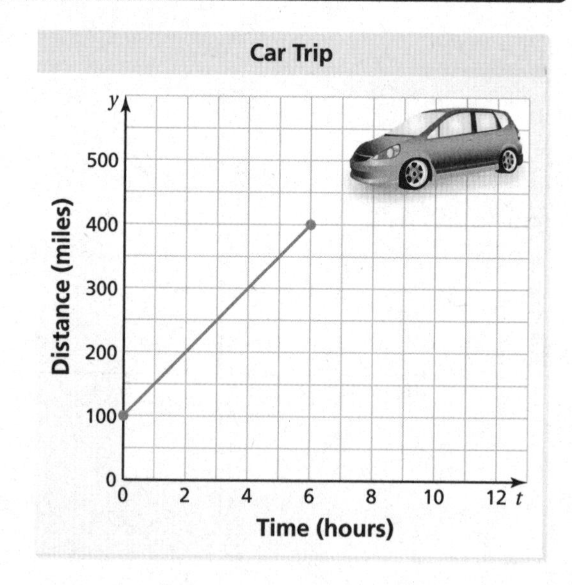

b. Find the slope of the graph. What does it represent?

c. How long did the trip last?

d. How far from Phoenix was the car at the end of the trip?

e. Write an equation that represents the graph.

What Is Your Answer?

4. IN YOUR OWN WORDS How can you write an equation of a line when you are given the slope and the y-intercept of the line? Give an example that is different from those in Activities 1, 2, and 3.

5. Two sides of a parallelogram are represented by the equations $y = 2x + 1$ and $y = -x + 3$. Give two equations that can represent the other two sides.

Name ______________________________ Date __________

4.6 Practice

For use after Lesson 4.6

Write an equation of the line in slope-intercept form.

1.

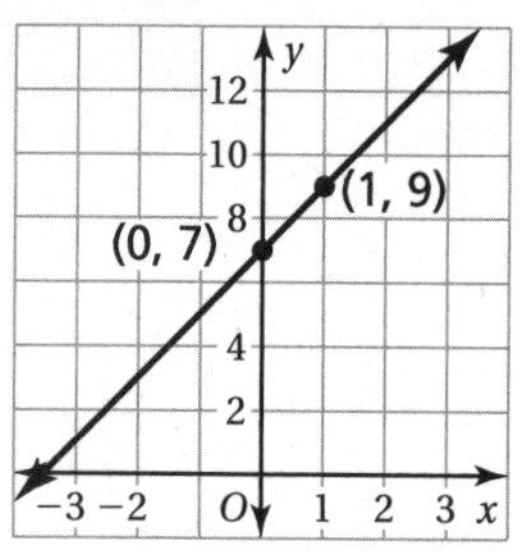

2.

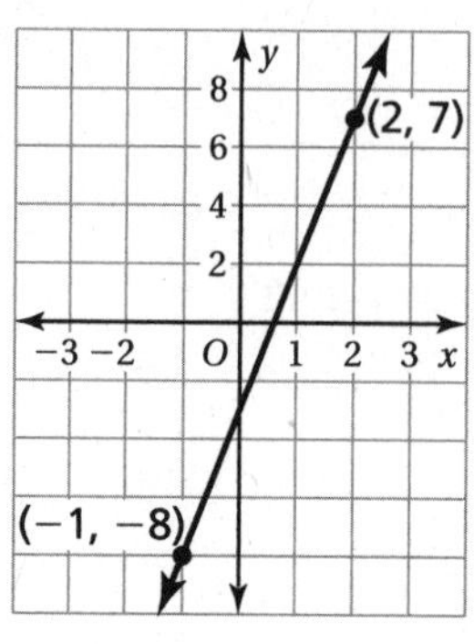

3.

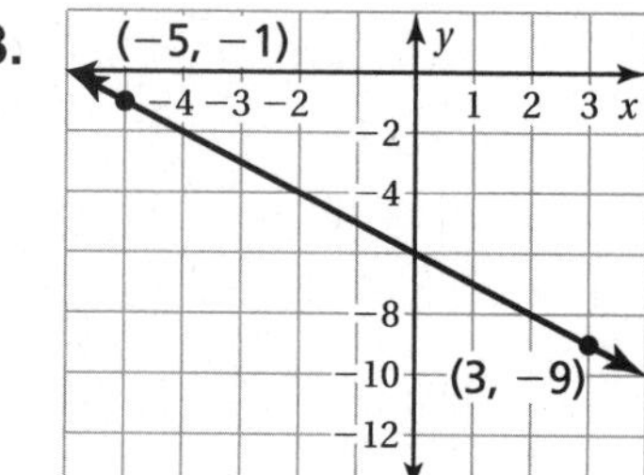

4.

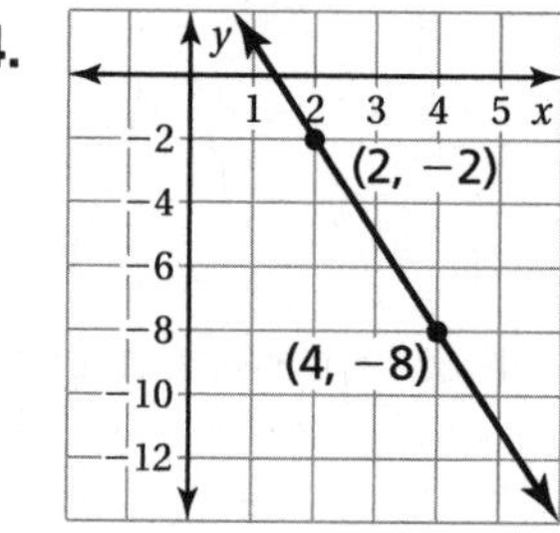

Write an equation of the line that passes through the points.

5. $(3, 8), (-2, 8)$

6. $(4, 3), (6, -3)$

7. $(-1, 0), (-5, 0)$

8. You organize a garage sale. You have \$30 at the beginning of the sale. You earn an average of \$20 per hour. Write an equation that represents the amount of money y you have after x hours.

Name________________________________ Date__________

4.7 Writing Equations in Point-Slope Form

For use with Activity 4.7

Essential Question How can you write an equation of a line when you are given the slope and a point on the line?

1 ACTIVITY: Writing Equations of Lines

Work with a partner.

- **Sketch the line that has the given slope and passes through the given point.**
- **Find the *y*-intercept of the line.**
- **Write an equation of the line.**

a. $m = -2$

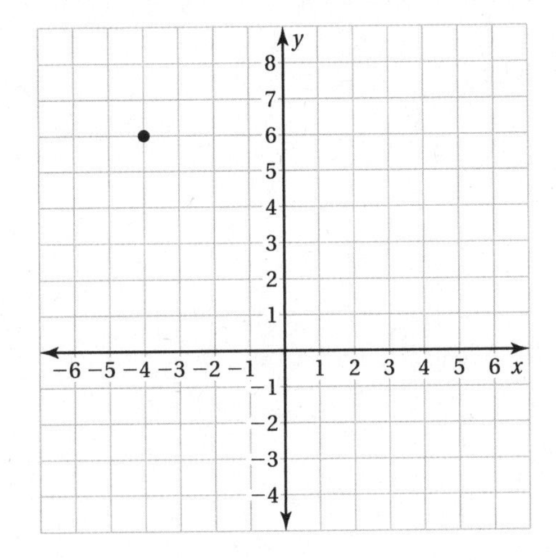

b. $m = \frac{1}{3}$

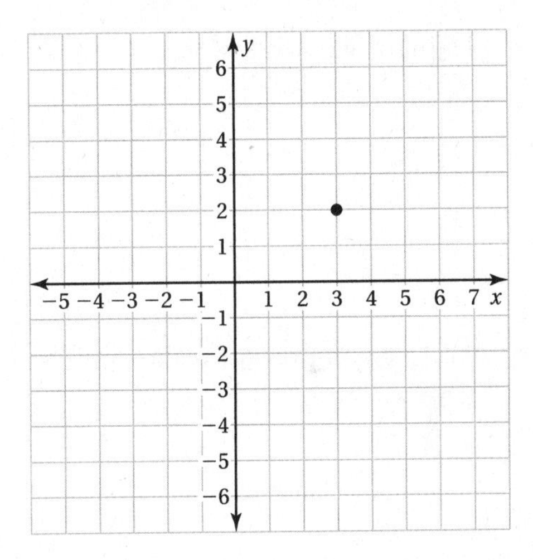

c. $m = -\frac{2}{3}$

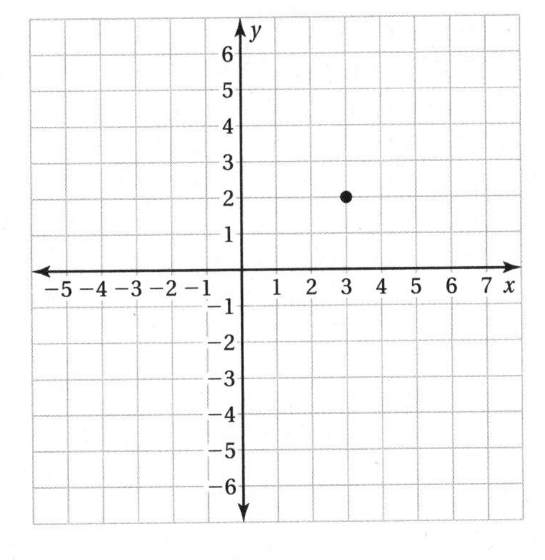

d. $m = \frac{5}{2}$

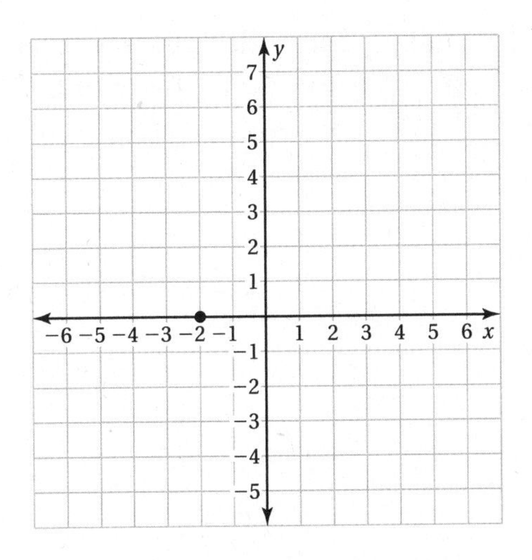

2 ACTIVITY: Deriving an Equation

Work with a partner.

a. Draw a nonvertical line that passes through the point (x_1, y_1).

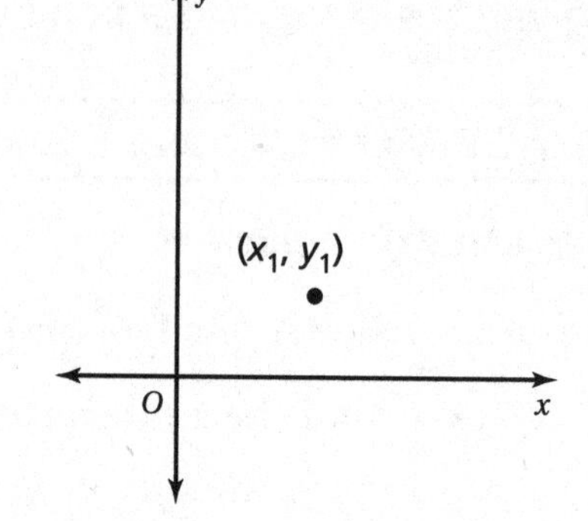

b. Plot another point on your line. Label this point as (x, y). This point represents any other point on the line.

c. Label the rise and run of the line through the points (x_1, y_1) and (x, y).

d. The rise can be written as $y - y_1$. The run can be written as $x - x_1$. Explain why this is true.

e. Write an equation for the slope m of the line using the expressions from part (d).

f. Multiply each side of the equation by the expression in the denominator. Write your result. What does this result represent?

Name____ Date__________

3 ACTIVITY: Writing an Equation

Work with a partner.

For 4 months, you saved \$25 a month. You now have \$175 in your savings account.

- Draw a graph that shows the balance in your account after *t* months.
- Use your result from Activity 2 to write an equation that represents the balance *A* after *t* months.

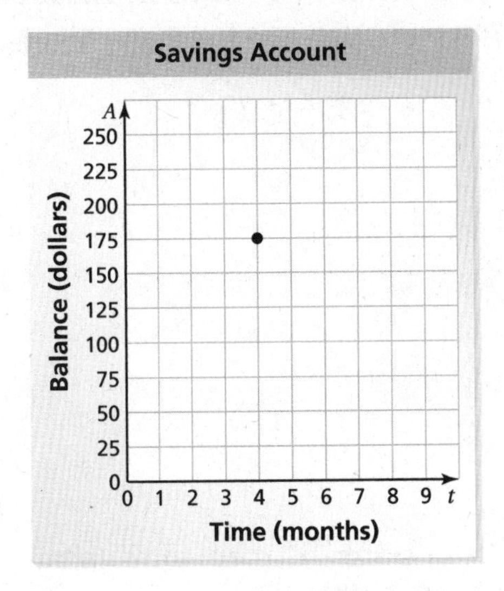

What Is Your Answer?

4. Redo Activity 1 using the equation you found in Activity 2. Compare the results. What do you notice?

5. Why do you think $y - y_1 = m(x - x_1)$ is called the *point-slope form* of the equation of a line? Why do you think this is important?

6. **IN YOUR OWN WORDS** How can you write an equation of a line when you are given the slope and a point on the line? Give an example that is different from those in Activity 1.

Name ____________________ Date __________

4.7 Practice

For use after Lesson 4.7

Write in point-slope form an equation of the line that passes through the given point that has the given slope.

1. $m = -3; (-4, 6)$

2. $m = -\frac{4}{3}; (3, -1)$

Write in slope-intercept form an equation of the line that passes through the given points.

3. $(-3, 0), (-2, 3)$

4. $(-6, 10), (6, -10)$

5. The total cost for bowling includes the fee for shoe rental plus a fee per game. The cost of each game increases the price by $4. After 3 games, the total cost with shoe rental is $14.

a. Write an equation to represent the total cost y to rent shoes and bowl x games.

b. How much is shoe rental? How is this represented in the equation?

Name____ Date__________

Fair Game Review

Simplify the expression.

1. $2x + 5 - x$

2. $4 + 2d - 4d$

3. $7y - 8 + 6y - 3$

4. $5 + 4z - 3 + 3z$

5. $4(s + 2) + s - 1$

6. $2(4x - 5) - 3$

7. The width of a garden is $(4x - 1)$ feet and the length is $2x$ feet. Find the perimeter of the garden.

Name ______________________________ Date __________

Chapter 5 Fair Game Review (continued)

Solve the equation. Check your solution.

8. $8y - 3 = 13$

9. $4a + 11 - a = 2$

10. $9 = 4(3k - 4) - 7k$

11. $-12 - 5(6 - 2m) = 18$

12. $15 - t + 8t = -13$

13. $5h - 2\left(\frac{3}{2}h + 4\right) = 10$

14. The profit P (in dollars) from selling x calculators is $P = 25x - (10x + 250)$. How many calculators are sold when the profit is \$425?

Name______________________________ Date__________

5.1 Solving Systems of Linear Equations by Graphing

For use with Activity 5.1

Essential Question How can you solve a system of linear equations?

1 ACTIVITY: Writing a System of Linear Equations

Work with a partner.

Your family starts a bed-and-breakfast. It spends $500 fixing up a bedroom to rent. The cost for food and utilities is $10 per night. Your family charges $60 per night to rent the bedroom.

a. Write an equation that represents the costs.

Cost, C (in dollars)	=	$10 per night	•	Number of nights, x	+	$500

b. Write an equation that represents the revenue (income).

Revenue, R (in dollars)	=	$60 per night	•	Number of nights, x

c. A set of two (or more) linear equations is called a **system of linear equations**. Write the system of linear equations for this problem.

Name ______________________________ Date __________

5.1 Solving Systems of Linear Equations by Graphing (continued)

2 ACTIVITY: Using a Table to Solve a System

Work with a partner. Use the cost and revenue equations from Activity 1 to find how many nights your family needs to rent the bedroom before recovering the cost of fixing up the bedroom. This is the *break-even point*.

a. Complete the table.

x	0	1	2	3	4	5	6	7	8	9	10	11
C												
R												

b. How many nights does your family need to rent the bedroom before breaking even?

3 ACTIVITY: Using a Graph to Solve a System

Work with a partner.

a. Graph the cost equation from Activity 1.

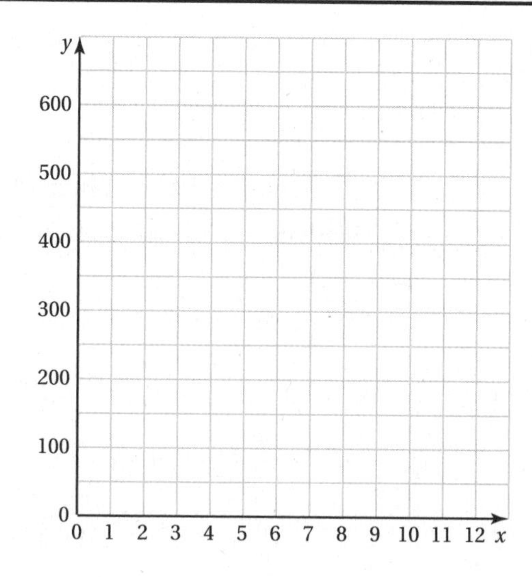

b. In the same coordinate plane, graph the revenue equation from Activity 1.

c. Find the point of intersection of the two graphs. What does this point represent? How does this compare to the break-even point in Activity 2? Explain.

Name________________________________ Date__________

5.2 Solving Systems of Linear Equations by Substitution

For use with Activity 5.2

Essential Question How can you use substitution to solve a system of linear equations?

1 ACTIVITY: Using Substitution to Solve a System

Work with a partner. Solve each system of linear equations using two methods.

Method 1: Solve for x first.

Solve for x in one of the equations. Use the expression for x to find the solution of the system. Explain how you did it.

Method 2: Solve for y first.

Solve for y in one of the equations. Use the expression for y to find the solution of the system. Explain how you did it.

Is the solution the same using both methods?

a. $6x - y = 11$
$2x + 3y = 7$

b. $2x - 3y = -1$
$x - y = 1$

c. $3x + y = 5$
$5x - 4y = -3$

d. $5x - y = 2$
$3x - 6y = 12$

e. $x + y = -1$
$5x + y = -13$

f. $2x - 6y = -6$
$7x - 8y = 5$

Name ____________________ Date __________

5.2 Solving Systems of Linear Equations by Substitution (continued)

2 ACTIVITY: Writing and Solving a System of Equations

Work with a partner.

a. Roll a pair of number cubes that have different colors. Then write the ordered pair shown by the number cubes. The ordered pair at the right is (3, 4).

b. Write a system of linear equations that has this ordered pair as its solution.

c. Exchange systems with your partner and use one of the methods from Activity 1 to solve the system.

3 ACTIVITY: Solving a Secret Code

Work with a partner. Decode the quote by Archimedes.

___ ___ ___ ___   ___ ___   ___   ___ ___ ___ ___ ___   ___ ___   ___ ___ ___ ___ ___ ,
−8 −7 7 −5   −4 −5   −3   −2 −1 −3 0 −5   1 2   3 1 −3 4 5

___ ___ ___   ___   ___ ___ ___ ___   ___ ___ ___ ___   ___ ___ ___   ___ ___ ___ ___ ___ .
−3 4 5   −7   6 −7 −1 −1   −4 2 7 −5   1 8 −5   −5 −3 9 1 8

5.2 Solving Systems of Linear Equations by Substitution (continued)

(A, C) $x + y = -3$
$x - y = -3$

(D, E) $x + y = 0$
$x - y = 10$

(G, H) $x + y = 0$
$x - y = -16$

(I, L) $x + 2y = -9$
$2x - y = -13$

(M, N) $x + 2y = 4$
$2x - y = -12$

(O, P) $x + 2y = -2$
$2x - y = 6$

(R, S) $2x + y = 21$
$x - y = 6$

(T, U) $2x + y = -7$
$x - y = 10$

(V, W) $2x + y = 20$
$x - y = 1$

What Is Your Answer?

4. **IN YOUR OWN WORDS** How can you use substitution to solve a system of linear equations?

Name ____ Date __________

5.2 Practice

For use after Lesson 5.2

Solve the system of linear equations by substitution. Check your solution.

1. $y = -2x + 4$

$-x + 3y = -9$

2. $\frac{3}{4}x - 5y = 7$

$x = -4y + 12$

3. $5x - y = 4$

$2x + 2y = 16$

4. $2x + 3y = 0$

$8x + 9y = 18$

5. A gas station sells a total of 4500 gallons of regular gas and premium gas in one day. The ratio of gallons of regular gas sold to gallons of premium gas sold is 7 : 2.

a. Write a system of linear equations that represents this situation.

b. How many gallons sold were regular gas? premium gas?

Name____ Date__________

5.3 Solving Systems of Linear Equations by Elimination

For use with Activity 5.3

Essential Question How can you use elimination to solve a system of linear equations?

1 ACTIVITY: Using Elimination to Solve a System

Work with a partner. Solve each system of linear equations using two methods.

Method 1: Subtract.

Subtract Equation 2 from Equation 1. What is the result? Explain how you can use the result to solve the system of equations.

Method 2: Add.

Add the two equations. What is the result? Explain how you can use the result to solve the system of equations.

Is the solution the same using both methods?

a. $2x + y = 4$
$2x - y = 0$

b. $3x - y = 4$
$3x + y = 2$

c. $x + 2y = 7$
$x - 2y = -5$

2 ACTIVITY: Using Elimination to Solve a System

Work with a partner.

$2x + y = 2$ Equation 1

$x + 5y = 1$ Equation 2

a. Can you add or subtract the equations to solve the system of linear equations? Explain.

Name ______________________________ Date __________

b. Explain what property you can apply to Equation 1 in the system so that the y coefficients are the same.

c. Explain what property you can apply to Equation 2 in the system so that the x coefficients are the same.

d. You solve the system in part (b). Your partner solves the system in part (c). Compare your solutions.

e. Use a graphing calculator to check your solution.

3 ACTIVITY: Solving a Secret Code

Work with a partner. Solve the puzzle to find the name of a famous mathematician who lived in Egypt around 350 A.D.

4	B	W	R	M	F	Y	K	N
3	O	J	A	S	I	D	X	Z
2	Q	P	C	E	G	B	T	J
1	M	R	C	Z	N	O	U	W
0	K	X	U	H	L	Y	S	Q
−1	F	E	A	S	W	K	R	M
−2	G	J	Z	N	H	V	D	G
−3	E	L	X	L	F	Q	O	B
	−3	−2	−1	0	1	2	3	4

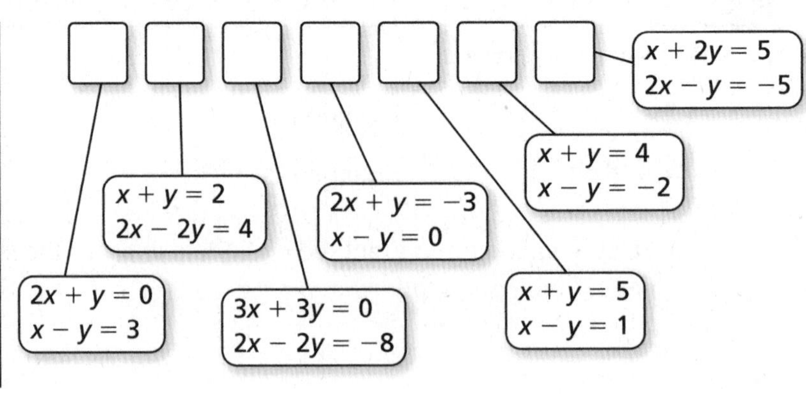

What Is Your Answer?

4. **IN YOUR OWN WORDS** How can you use elimination to solve a system of linear equations?

5. **STRUCTURE** When can you add or subtract equations in a system to solve the system? When do you have to multiply first? Justify your answers with examples.

6. **LOGIC** In Activity 2, why can you multiply the equations in the system by a constant and not change the solution of the system? Explain your reasoning.

Name ____________________ Date __________

5.3 Practice

For use after Lesson 5.3

Solve the system of linear equations by elimination. Check your solution.

1. $x + y = 7$

$3x - y = 1$

2. $-2x - 5y = -8$

$-2x + y = 16$

3. $8x - 9y = 7$

$2x - 3y = -5$

4. $-5x + 3y = -6$

$9x - 4y = 1$

5. A high school has a total of 850 students. There are 60 more female students than there are male students.

a. Write a system of linear equations that represents this situation.

b. How many students are female? male?

Name____ Date__________

5.4 Solving Special Systems of Linear Equations
For use with Activity 5.4

Essential Question Can a system of linear equations have no solution? Can a system of linear equations have many solutions?

1 ACTIVITY: Writing a System of Linear Equations

Work with a partner. Your cousin is 3 years older than you. Your ages can be represented by two linear equations.

$y = t$ Your age

$y = t + 3$ Your cousin's age

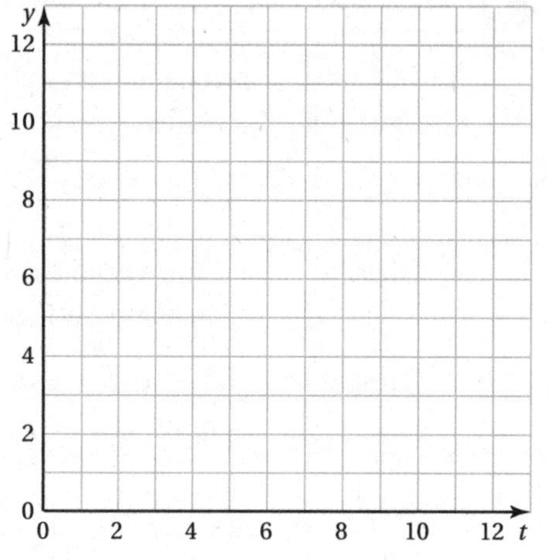

a. Graph both equations in the same coordinate plane.

b. What is the vertical distance between the two graphs? What does this distance represent?

c. Do the two graphs intersect? Explain what this means in terms of your age and your cousin's age.

2 ACTIVITY: Using a Table to Solve a System

Work with a partner. You invest \$500 for equipment to make dog backpacks. Each backpack costs you \$15 for materials. You sell each backpack for \$15.

a. Complete the table for your cost C and your revenue R.

x	0	1	2	3	4	5	6	7	8	9	10
C											
R											

b. When will you break even? What is wrong?

3 ACTIVITY: Using a Graph to Solve a Puzzle

Work with a partner. Let x and y be two numbers. Here are two clues about the values of x and y.

	Words	Equation
Clue 1:	The value of y is 4 more than twice the value of x.	$y = 2x + 4$
Clue 2:	The difference of $3y$ and $6x$ is 12.	$3y - 6x = 12$

a. Graph both equations in the same coordinate plane.

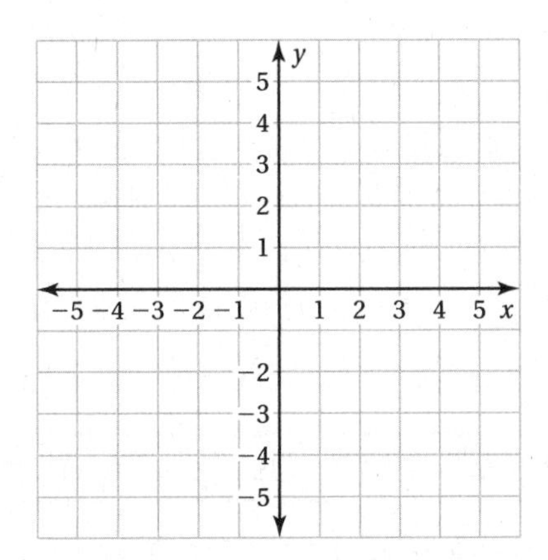

b. Do the two lines intersect? Explain.

c. What is the solution of the puzzle?

d. Use the equation $y = 2x + 4$ to complete the table.

x	0	1	2	3	4	5	6	7	8	9	10
y											

e. Does each solution in the table satisfy *both* clues?

f. What can you conclude? How many solutions does the puzzle have? How can you describe them?

What Is Your Answer?

4. IN YOUR OWN WORDS Can a system of linear equations have no solution? Can a system of linear equations have many solutions? Give examples to support your answers.

Name ____________________ Date ________

5.4 Practice

For use after Lesson 5.4

Solve the system of linear equations. Check your solution.

1. $y = 2x - 5$

$y = 2x + 7$

2. $3x + 4y = -10$

$y = -\frac{3}{4}x - \frac{5}{2}$

3. $x - y = 8$

$2y = 2x - 16$

4. $3y = -6x + 4$

$2x + y = 9$

5. You start reading a book for your literature class two days before your friend. You both read 10 pages per night. A system of linear equations that represents this situation is $y = 10x + 20$ and $y = 10x$. Will your friend finish the book before you? Justify your answer.

Name______________________________ Date__________

Extension 5.4 Practice

For use after Extension 5.4

Use a graph to solve the equation. Check your solution.

1. $3x - 4 = -x$

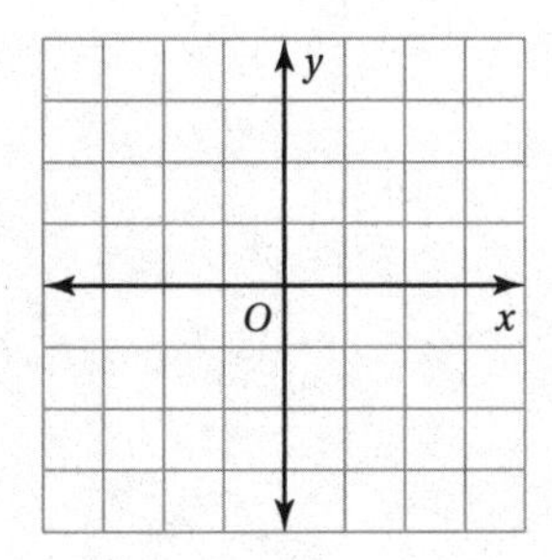

2. $\frac{1}{3}x + 3 = 4x - 8$

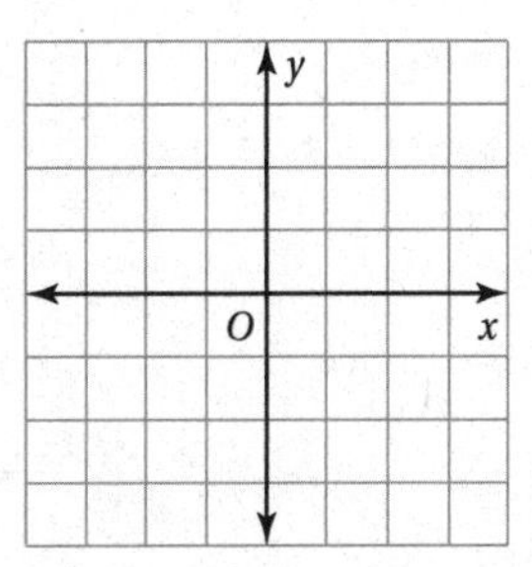

3. $\frac{1}{2}x + 4 = -x - 11$

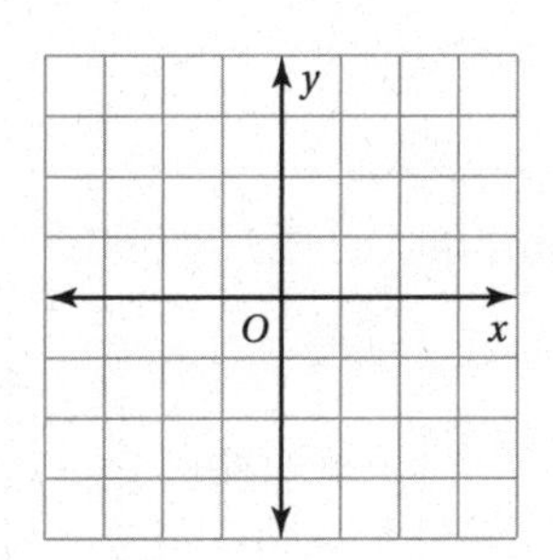

4. $-x + 1 = -\frac{1}{4}x - \frac{1}{2}$

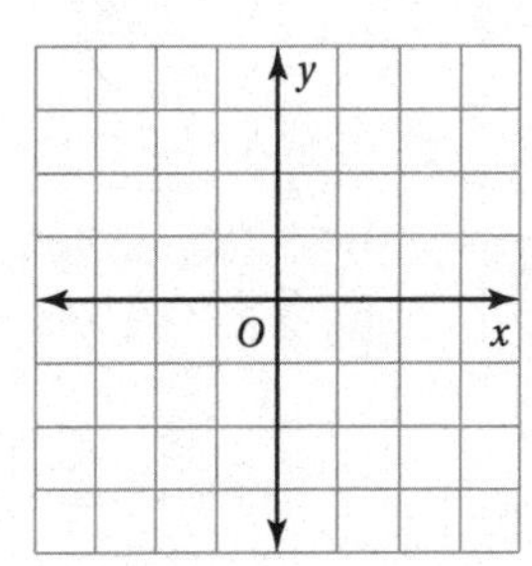

5. On the first day of your garage sale, you earned $12x + 9$ dollars. The next day you earned $22x$ dollars. Is it possible that you earned the same amount each day? Explain.

Name ______________________________ Date __________

Extension 5.4 **Practice (continued)**

6. You hike uphill at a rate of 200 feet per minute. Your friend hikes downhill on the same trail at a rate of 250 feet per minute. How long will it be until you meet?

7. Two savings accounts earn simple interest. Account A has a beginning balance of \$500 and grows by \$25 per year. Account B has a beginning balance of \$750 and grows by \$15 per year.

Growth rate	•	Years, x	+	Beginning balance	=	Growth rate	•	Years, x	+	Beginning balance

a. Use the model to write an equation.

b. After how many years x do the accounts have the same balance?

Name____ Date__________

Fair Game Review

Find the missing value in the table.

1.

x	y
1	5
3	7
5	9
7	

2.

x	y
2	6
4	12
8	24
12	

3.

x	y
6	11
14	19
26	31
41	

4.

x	y
8	4
18	9
28	14
38	

5.

x	y
4	2.5
11	9.5
15	13.5
21	

6.

x	y
6	5.8
15	14.8
22.8	22.6
31.4	

Name ____ Date __________

Chapter 6 Fair Game Review (continued)

Evaluate the expression when $x = 2$, $y = 3$, and $z = -4$.

7. $3x - 2$

8. $-6 - 2y$

9. $2z^2$

10. $3y - 3z$

11. $\frac{8}{x} - 1$

12. $-1 + \frac{z}{2}$

Name____ Date__________

6.1 Relations and Functions

For use with Activity 6.1

Essential Question How can you use a mapping diagram to show the relationship between two data sets?

ACTIVITY: Constructing Mapping Diagrams

Work with a partner. Complete the mapping diagram.

a. Area A

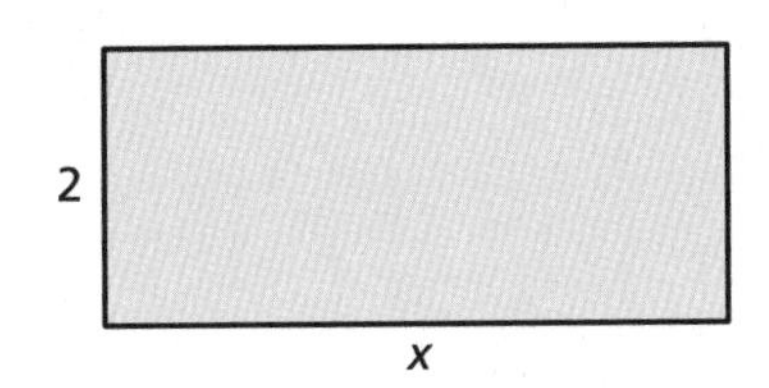

Input, x **Output, A**

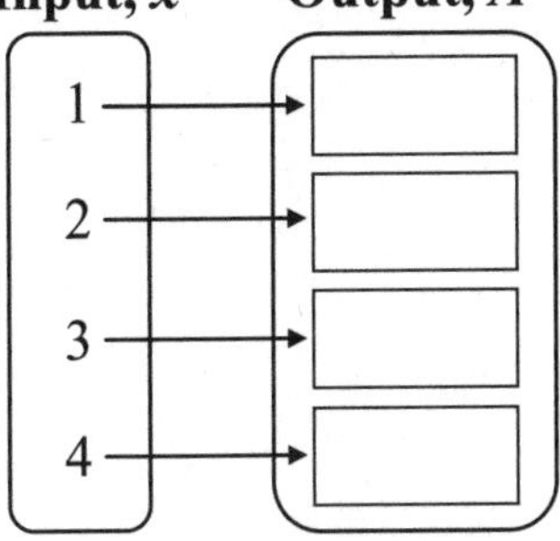

b. Perimeter P

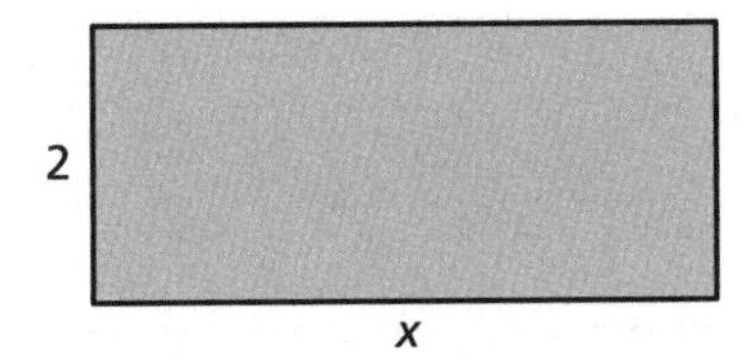

Input, x **Output, P**

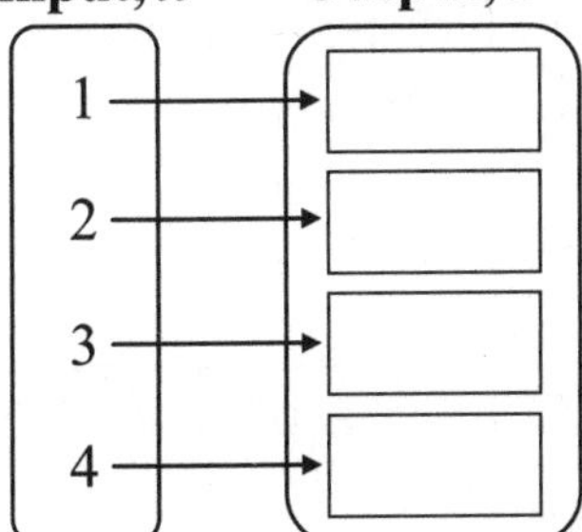

c. Circumference C

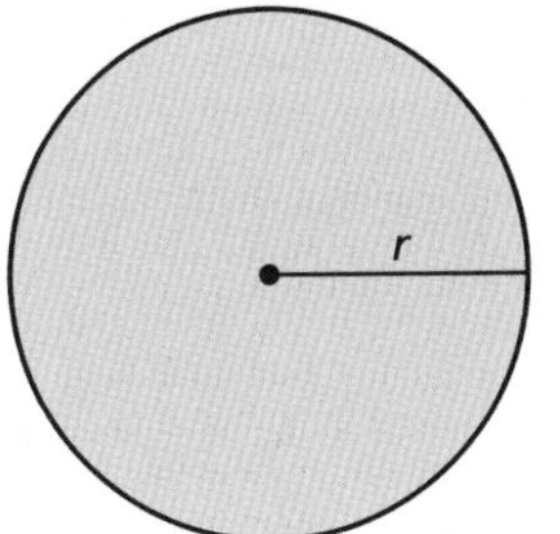
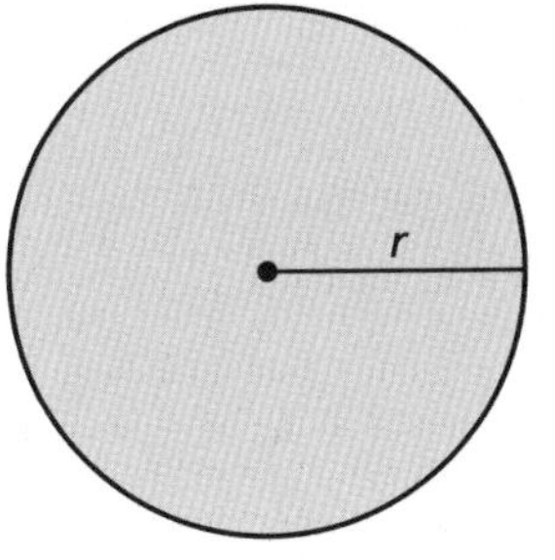

Input, r **Output, C**

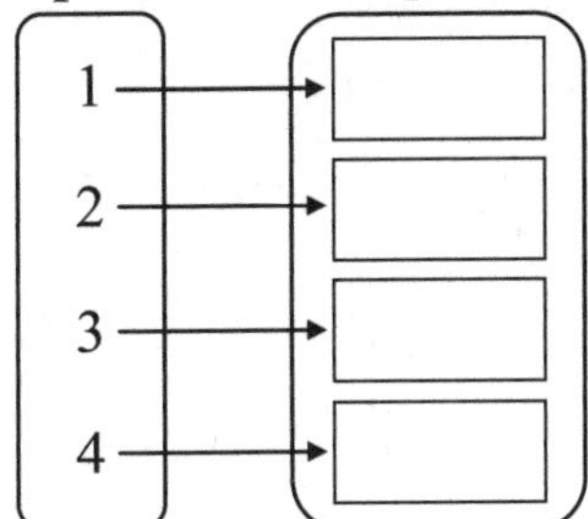

d. Volume V

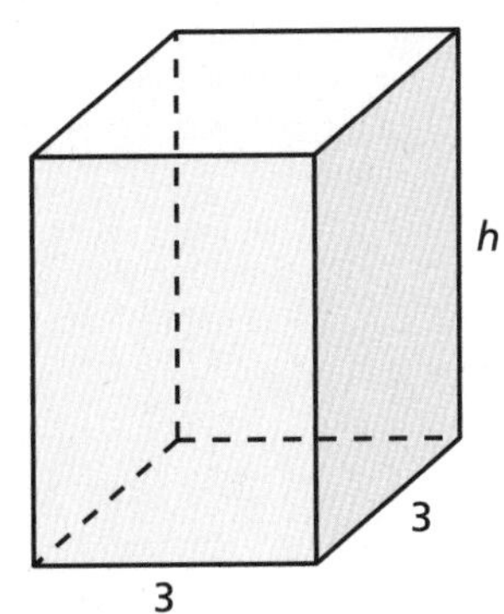

Input, h **Output, V**

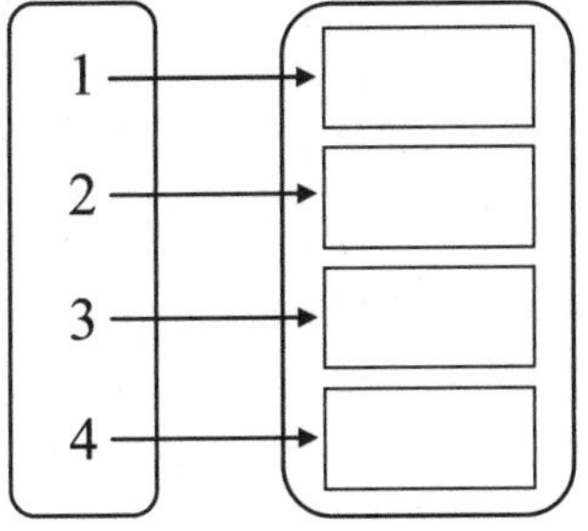

Name _______________ Date _______

2 ACTIVITY: Describing Situations

Work with a partner. How many outputs are assigned to each input? Describe a possible situation for each mapping diagram.

a. **Input, x** **Output, y**

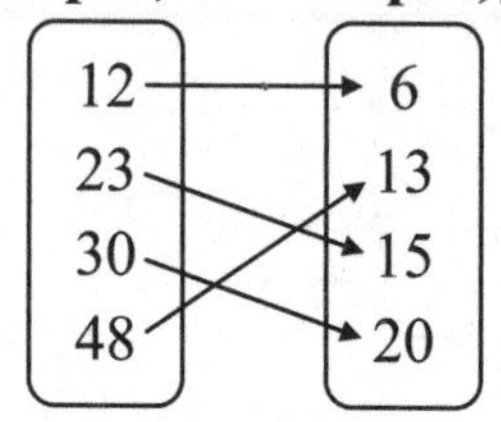

b. **Input, x** **Output, y**

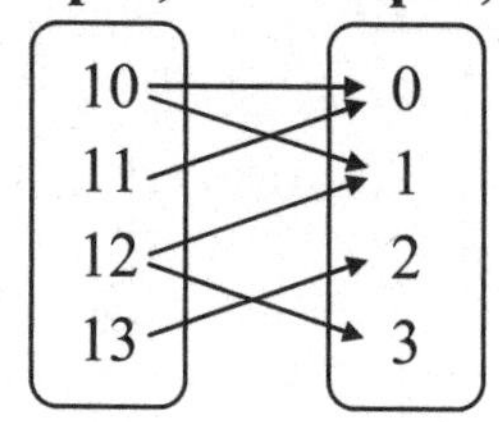

3 ACTIVITY: Interpreting Mapping Diagrams

Work with a partner. Describe the pattern in the mapping diagram. Complete the diagram.

a. **Input, t** **Output, M**

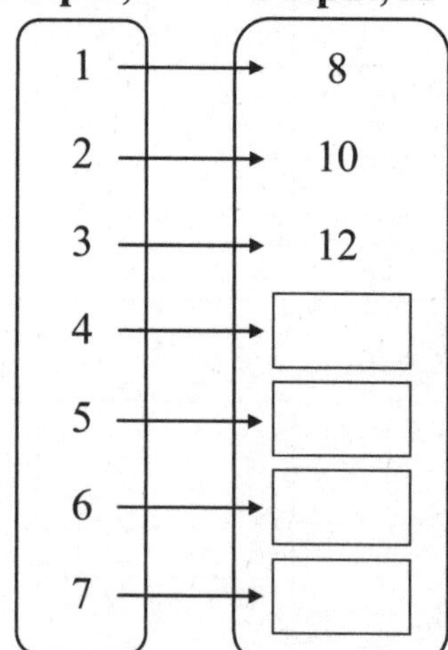

Name________________________________ Date__________

b.

Input, x	Output, A
1	4/3
2	5/3
3	2
4	
5	
6	
7	

What Is Your Answer?

4. **IN YOUR OWN WORDS** How can you use a mapping diagram to show the relationship between two data sets?

"I made a mapping diagram."

"It shows how I feel about my skateboard with each passing day."

Name ______________________________ Date __________

6.1 Practice

For use after Lesson 6.1

List the ordered pairs shown in the mapping diagram.

1. 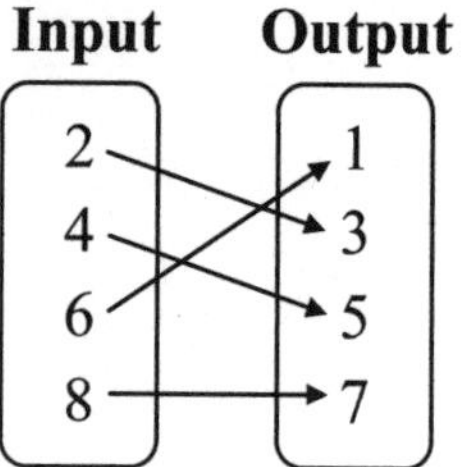

2. 

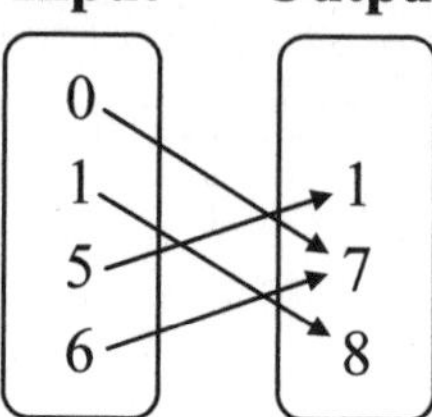

3. Draw a mapping diagram for the graph. Then describe the pattern of inputs and outputs.

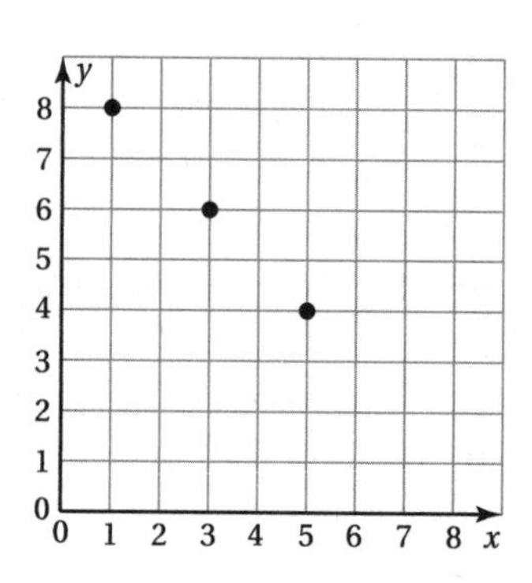

4. The table shows the number of beads needed to make a bracelet. Use the table to draw a mapping diagram.

Bracelet Length (in.)	Number of Beads
6	12
7	14
8	16
9	18

Name____ Date__________

6.2 Representations of Functions

For use with Activity 6.2

Essential Question How can you represent a function in different ways?

1 ACTIVITY: Describing a Function

Work with a partner. Complete the mapping diagram for the area of the figure. Then write an equation that describes the function.

a.

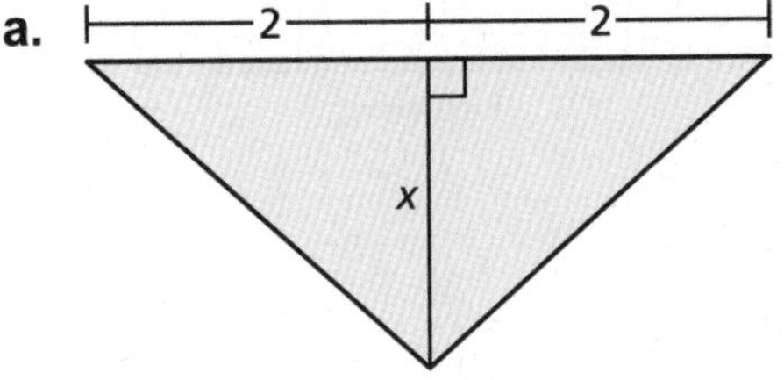

b.

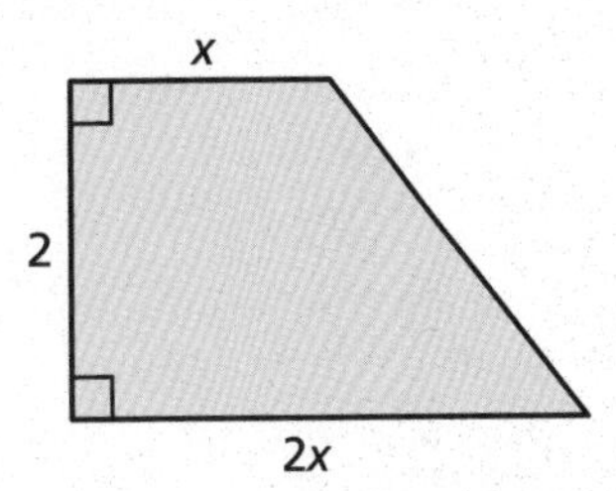

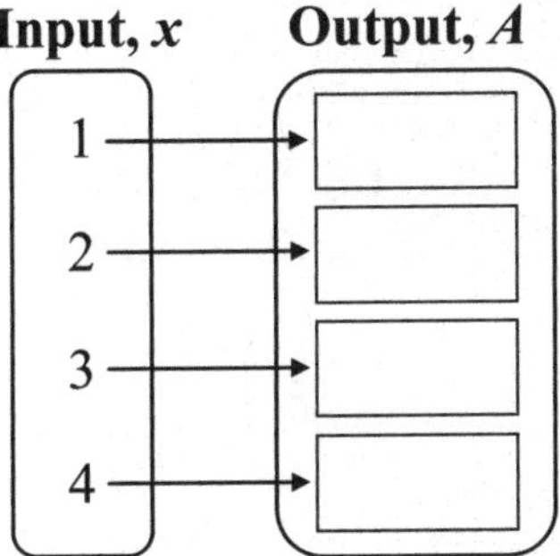

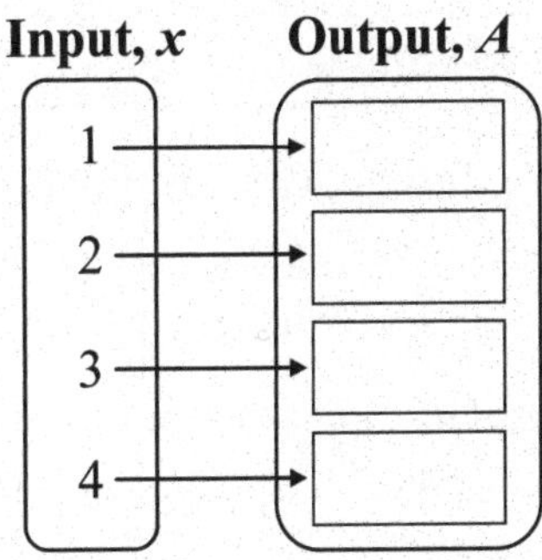

2 ACTIVITY: Using a Table

Work with a partner. Make a table that shows the pattern for the area, where the input is the figure number x and the output is the area A. Write an equation that describes the function. Then use your equation to find which figure has an area of 81 when the pattern continues.

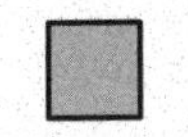

1 square unit

a.

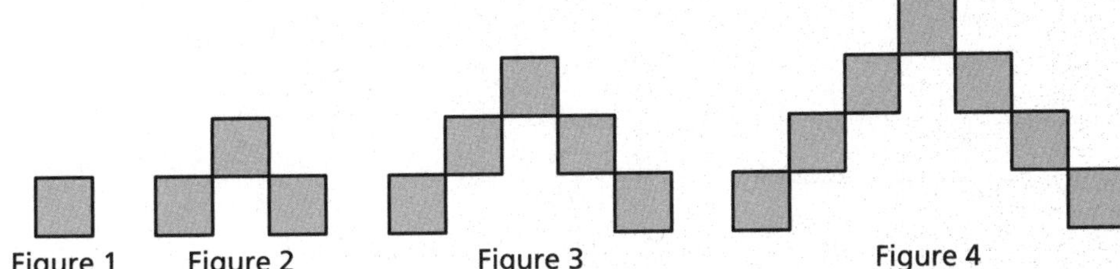

b.

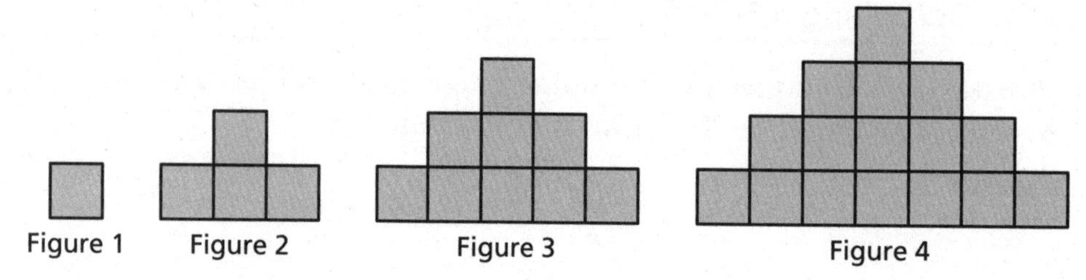

3 ACTIVITY: Using a Graph

Work with a partner. Graph the data. Use the graph to test the truth of each statement. If the statement is true, write an equation that shows how to obtain one measurement from the other measurement.

a. "You can find the horsepower of a race car engine if you know its volume in cubic inches."

Volume (cubic inches), x	200	350	350	500
Horsepower, y	375	650	250	600

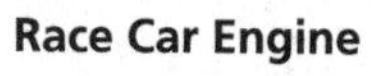

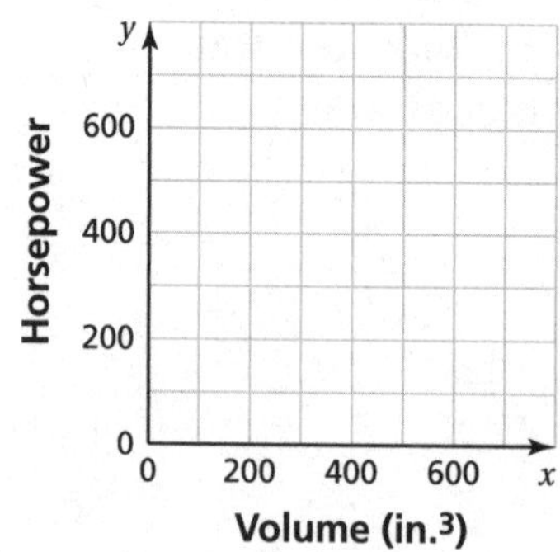

Name____ Date__________

b. "You can find the volume of a race car engine in cubic centimeters if you know its volume in cubic inches."

Volume (cubic inches), x	100	200	300
Volume (cubic centimeters), y	1640	3280	4920

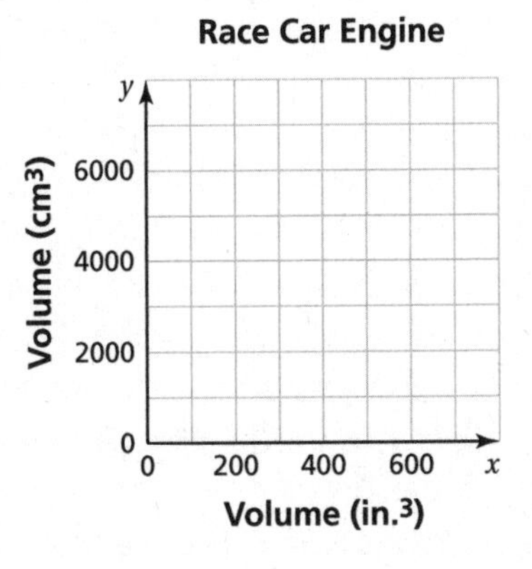

4 ACTIVITY: Interpreting a Graph

Work with a partner. The table shows the average speeds of the winners of the Daytona 500. Graph the data. Can you use the graph to predict future winning speeds? Explain why or why not.

Year	2004	2005	2006	2007	2008	2009	2010	2011	2012
Speed (mi/h)	156	135	143	149	153	133	137	130	140

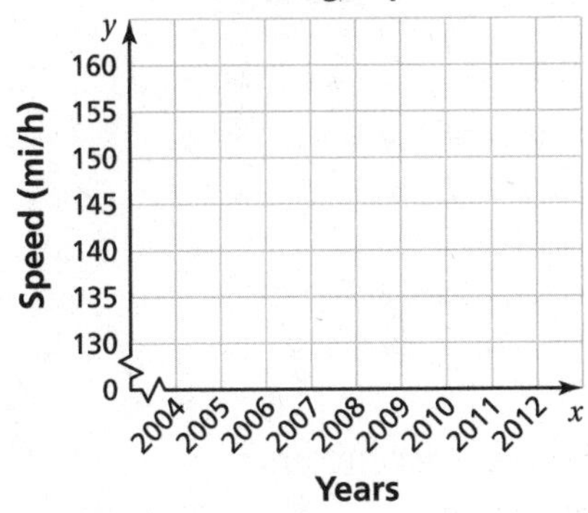

What Is Your Answer?

5. **IN YOUR OWN WORDS** How can you represent a function in different ways?

Name ______________________ Date __________

6.2 Practice

For use after Lesson 6.2

Write a function rule for the statement.

1. The output is four times the input.

2. The output is eight less than the input.

Find the value of *y* for the given value of *x*.

3. $y = \frac{x}{3}; x = 12$

4. $y = 5x + 9; x = 2$

5. You set up a hot chocolate stand at a football game. The cost of your supplies is \$75. You charge \$0.50 for each cup of hot chocolate.

a. Write a function that represents the profit P for selling c cups of hot chocolate.

b. You will *break even* when the cost of your supplies equals your income. How many cups of hot chocolate must you sell to break even?

Name______________________________ Date__________

6.3 Linear Functions
For use with Activity 6.3

Essential Question How can you use a function to describe a linear pattern?

1 ACTIVITY: Finding Linear Patterns

Work with a partner.

- **Plot the points from the table in a coordinate plane.**
- **Write a linear equation for the function.**

a.

x	0	2	4	6	8
y	150	125	100	75	50

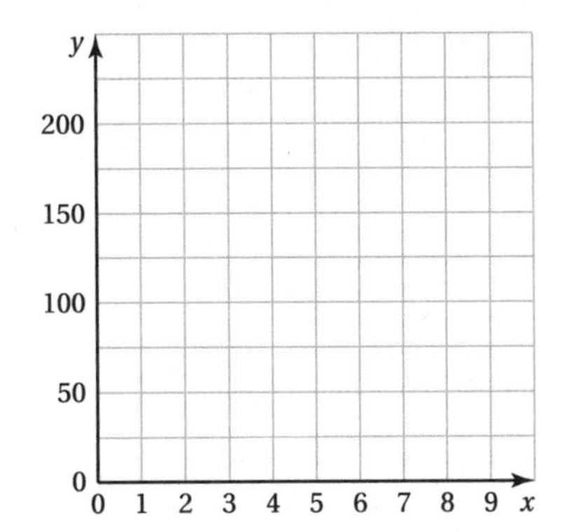

b.

x	4	6	8	10	12
y	15	20	25	30	35

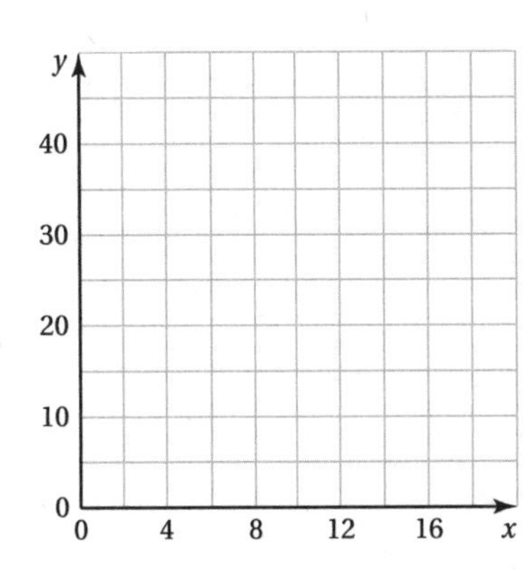

c.

x	−4	−2	0	2	4
y	4	6	8	10	12

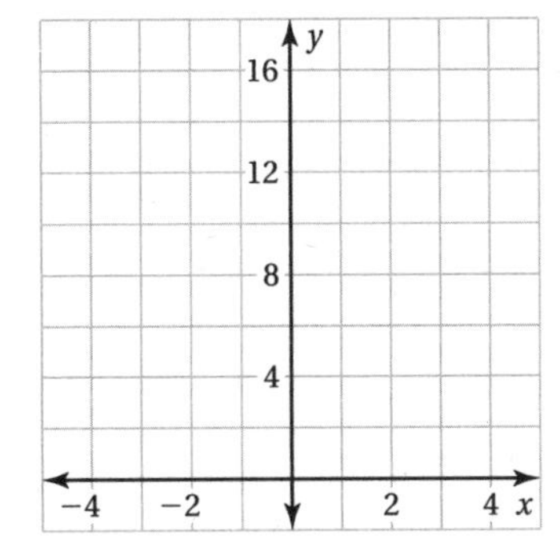

d.

x	−4	−2	0	2	4
y	1	0	−1	−2	−3

6.3 Linear Functions (continued)

2 ACTIVITY: Finding Linear Patterns

Work with a partner. The table shows a familiar linear pattern from geometry.

- **Write a function that relates y to x.**
- **What do the variables x and y represent?**
- **Graph the function.**

a.

x	1	2	3	4	5
y	2π	4π	6π	8π	10π

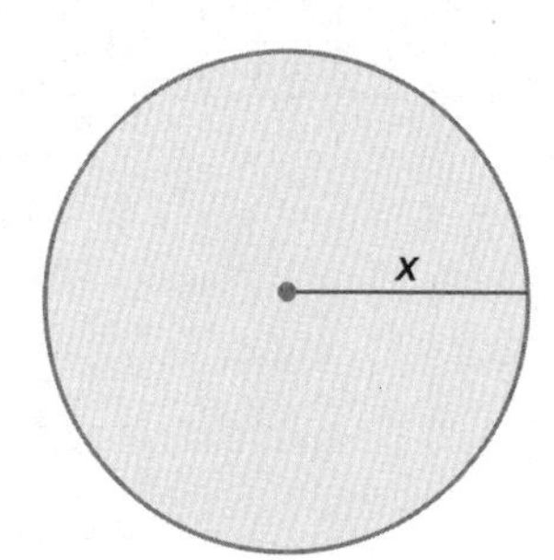

b.

x	1	2	3	4	5
y	10	12	14	16	18

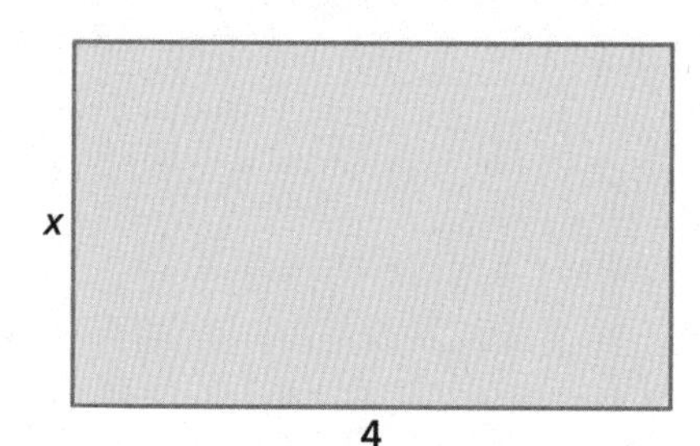

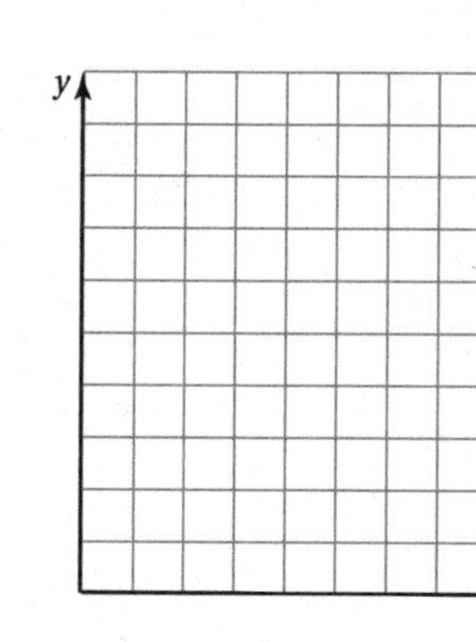

Name________________________________ Date__________

c.

x	1	2	3	4	5
y	5	6	7	8	9

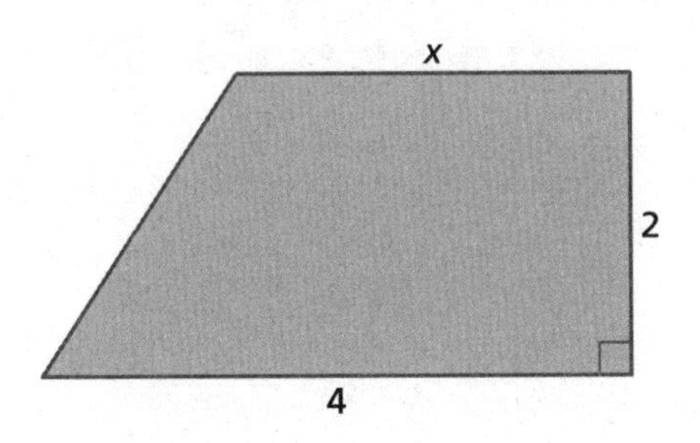

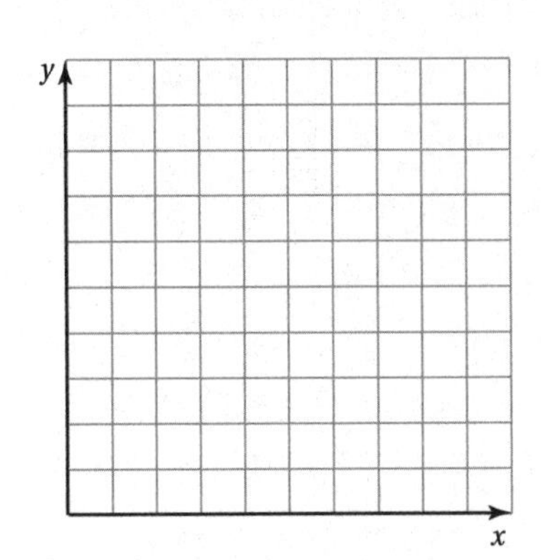

d.

x	1	2	3	4	5
y	28	40	52	64	76

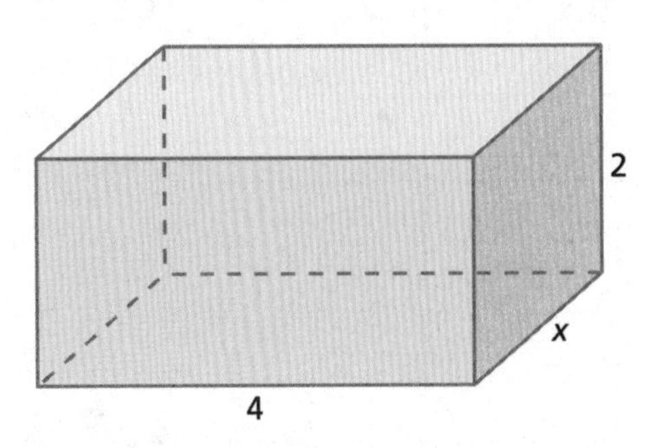

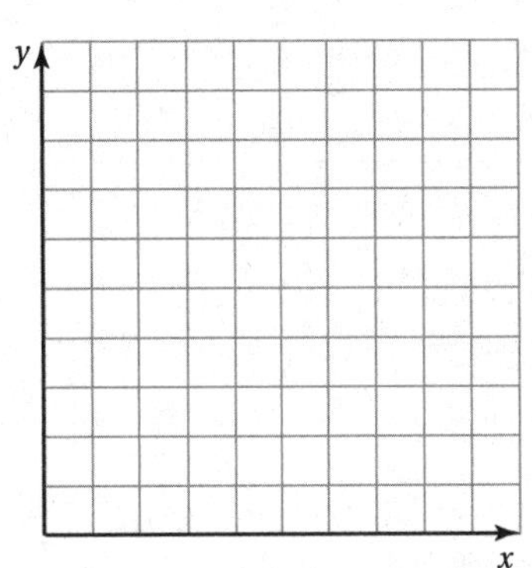

What Is Your Answer?

3. **IN YOUR OWN WORDS** How can you use a function to describe a linear pattern?

4. Describe the strategy you used to find the functions in Activities 1 and 2.

Name ______________________________ Date __________

6.3 Practice

For use after Lesson 6.3

Use the graph or the table to write a linear function that relates *y* to *x*.

1.

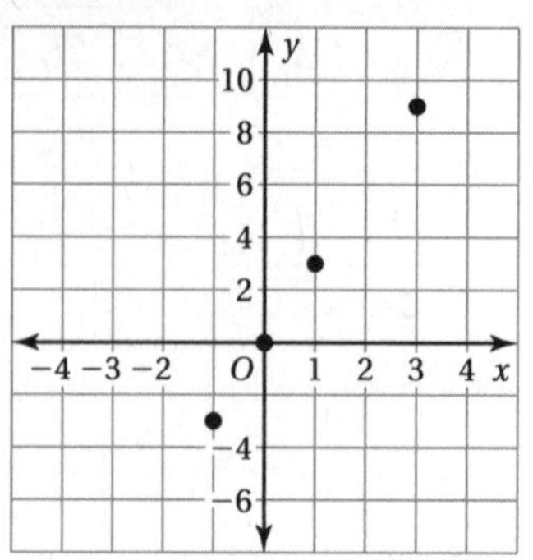

2.

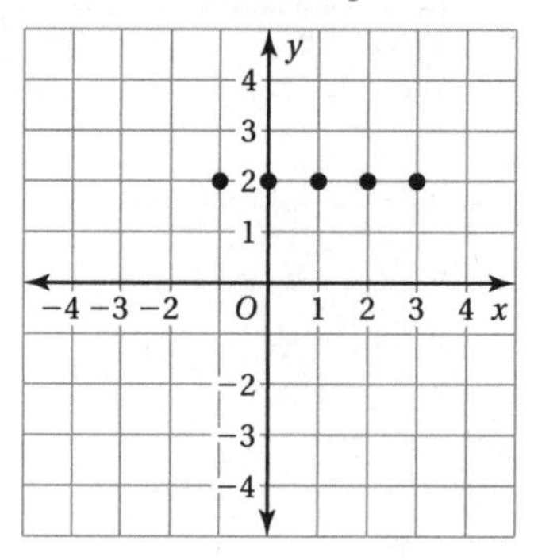

3.

x	0	1	2	3
y	5	7	9	11

4.

x	–2	0	2	4
y	–1	–2	–3	–4

5. The table shows the distance traveled y (in miles) after x hours.

x	0	2	4	6
y	0	120	240	360

a. Write a linear function that relates y to x.

b. Graph the linear function.

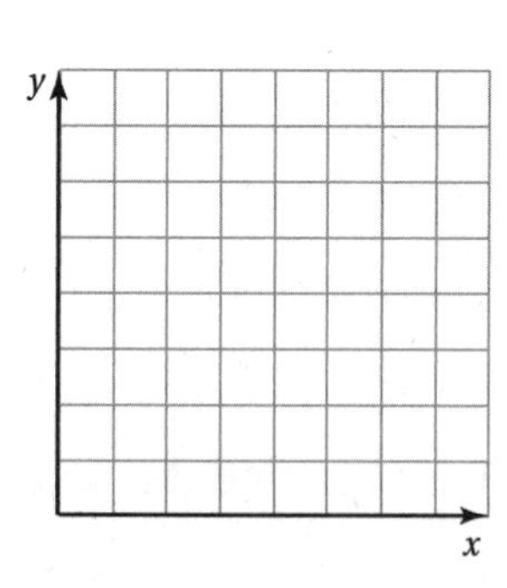

c. What is the distance traveled after three hours?

Name______________________________ Date__________

6.4 Comparing Linear and Nonlinear Functions
For use with Activity 6.4

Essential Question How can you recognize when a pattern in real life is linear or nonlinear?

1 ACTIVITY: Finding Patterns for Similar Figures

Work with a partner. Complete each table for the sequence of similar rectangles. Graph the data in each table. Decide whether each pattern is linear or nonlinear.

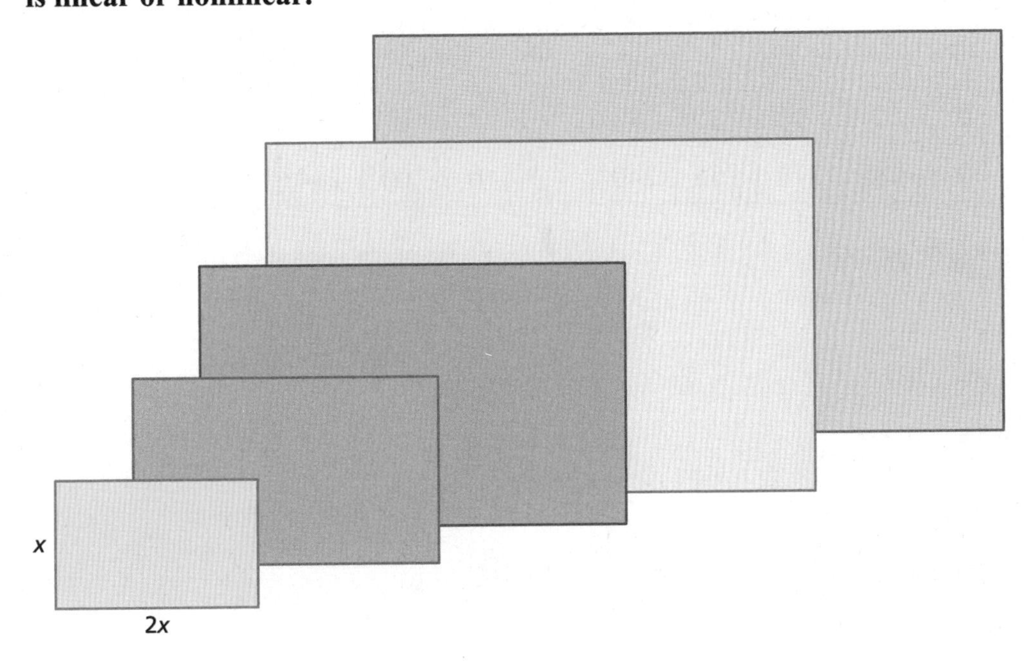

a. Perimeters of similar rectangles

x	1	2	3	4	5
P					

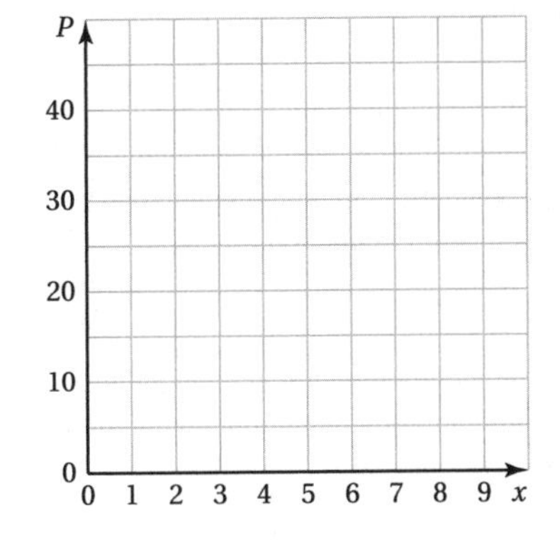

b. Areas of similar rectangles

x	1	2	3	4	5
A					

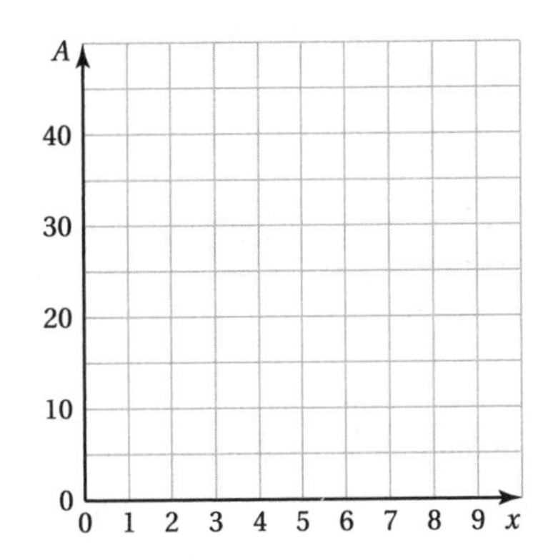

Name ______________________________ Date __________

2 ACTIVITY: Comparing Linear and Nonlinear Functions

Work with a partner. Each table shows the height *h* (in feet) of a falling object at *t* seconds.

- **Graph the data in each table.**
- **Decide whether each graph is linear or nonlinear.**
- **Compare the two falling objects. Which one has an increasing speed?**

a. Falling parachute jumper

t	0	1	2	3	4
h	300	285	270	255	240

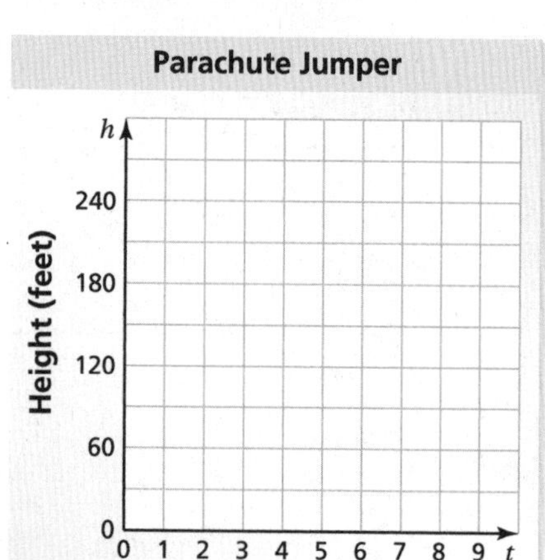

b. Falling bowling ball

t	0	1	2	3	4
h	300	284	236	156	44

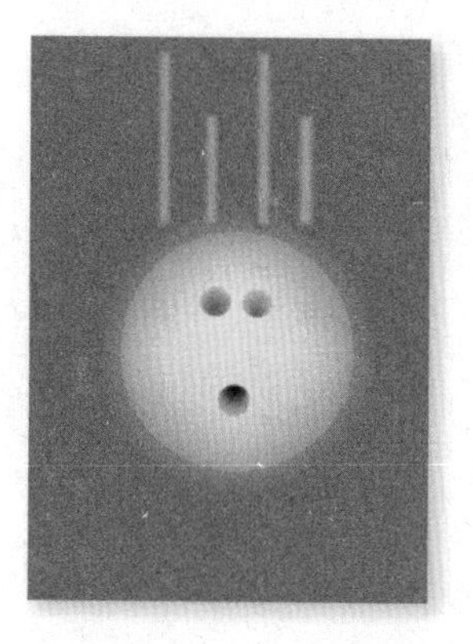

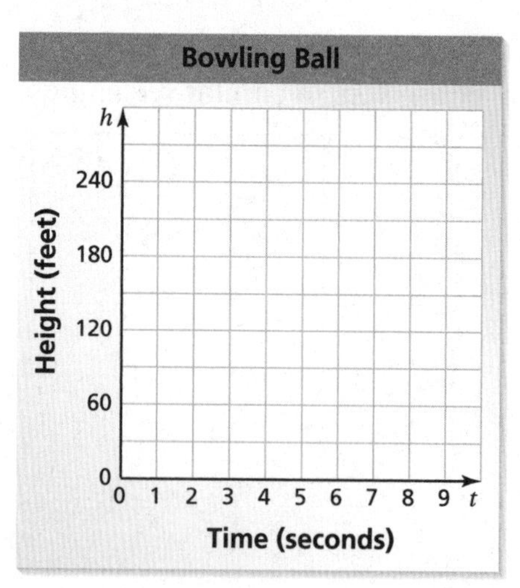

What Is Your Answer?

3. **IN YOUR OWN WORDS** How can you recognize when a pattern in real life is linear or nonlinear? Describe two real-life patterns: one that is linear and one that is nonlinear. Use patterns that are different from those described in Activities 1 and 2.

Name ______________________________ Date __________

6.4 Practice
For use after Lesson 6.4

Graph the data in the table. Decide whether the graph is *linear* or *nonlinear*.

1.

x	–2	0	2	4
y	4	0	4	16

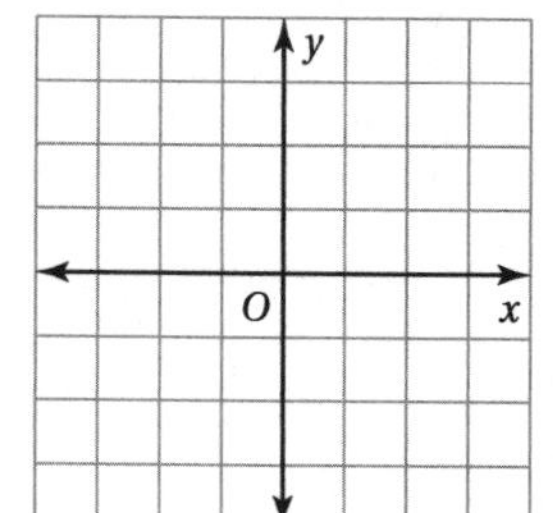

2.

x	–1	0	1	2
y	–1	1	3	5

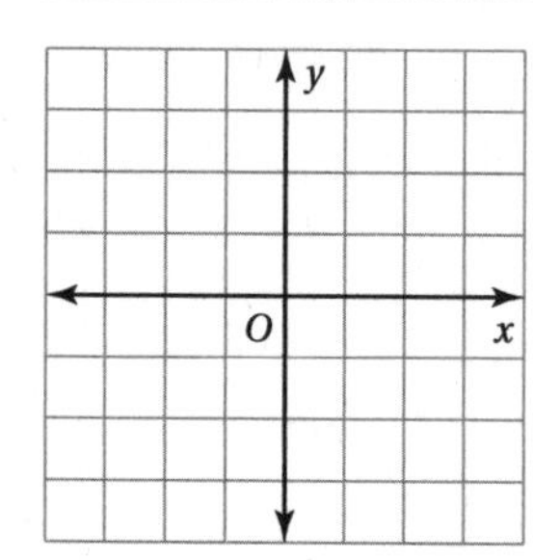

Does the graph represent a *linear* or nonlinear *function*? Explain.

3.

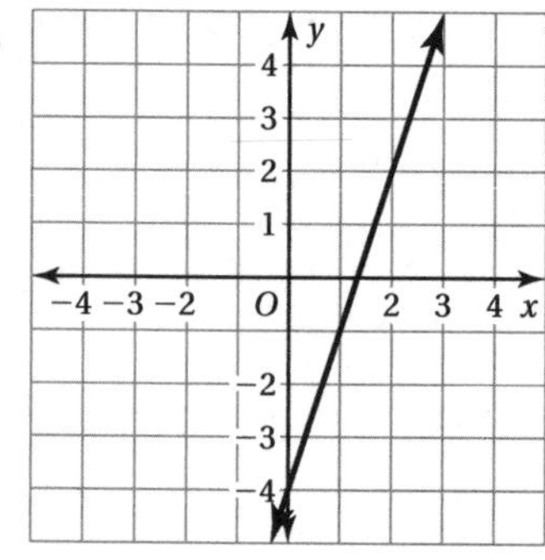

4.

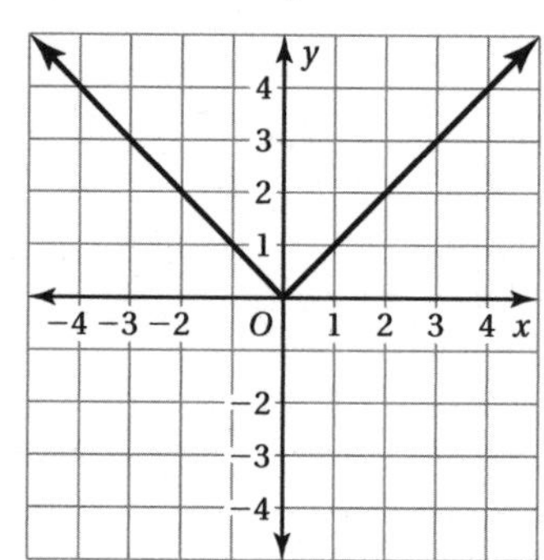

5. The table shows the area of a square with side length *x* inches. Does the table represent a linear or nonlinear function? Explain.

Side Length, *x*	1	2	3	4
Area, *A*	1	4	9	16

Name____ Date__________

6.5 Analyzing and Sketching Graphs

For use with Activity 6.5

Essential Question How can you use a graph to represent relationships between quantities without using numbers?

1 ACTIVITY: Interpreting a Graph

Work with a partner. Use the graph shown.

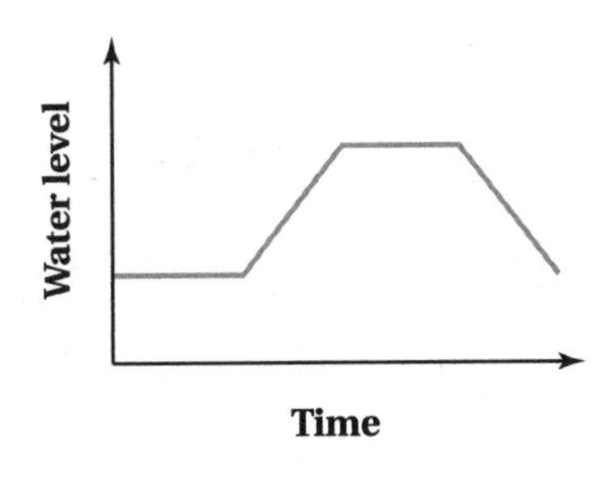

a. How is this graph different from the other graphs you have studied?

b. Write a short paragraph that describes how the water level changes over time.

c. What situation can this graph represent?

2 ACTIVITY: Matching Situations to Graphs

Work with a partner. You are riding your bike. Match each situation with the appropriate graph. Explain your reasoning.

A.

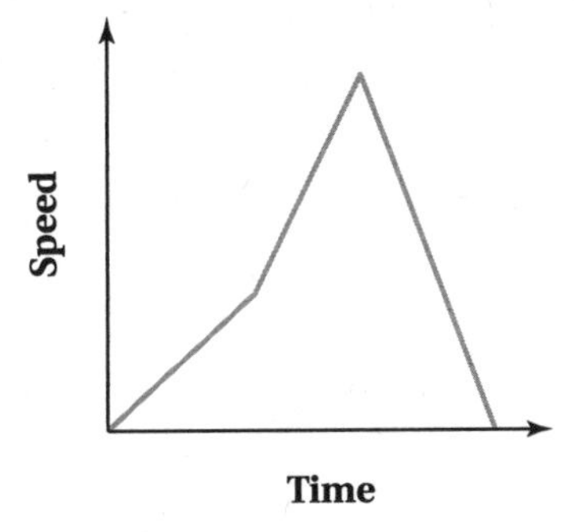

B.

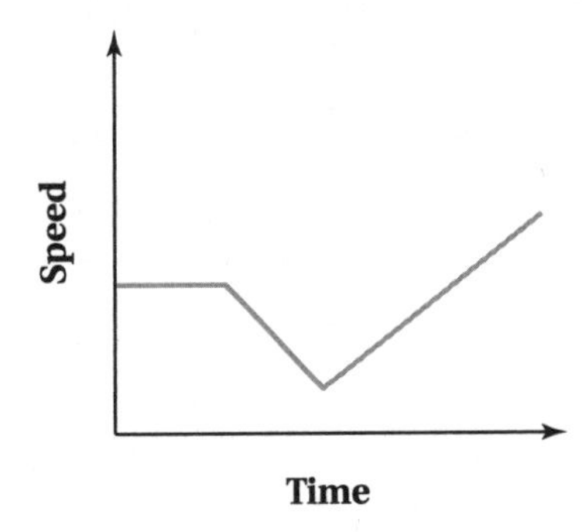

6.5 Analyzing and Sketching Graphs (continued)

C.

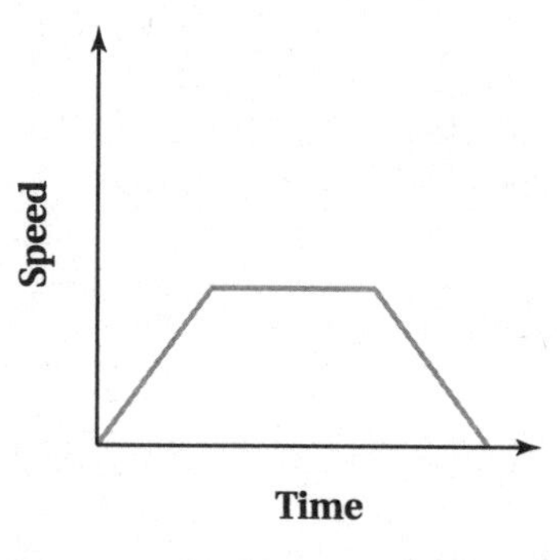

D.

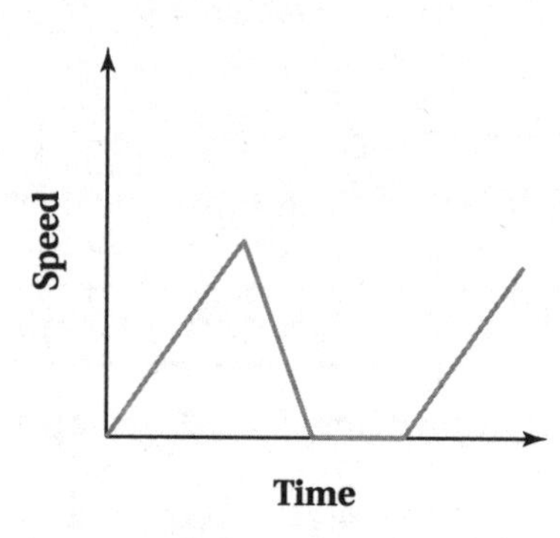

a. You gradually increase your speed, then ride at a constant speed along a bike path. You then slow down until you reach your friend's house.

b. You gradually increase your speed, then go down a hill. You then quickly come to a stop at an intersection.

c. You gradually increase your speed, then stop at a convenience store for a couple of minutes. You then continue to ride, gradually increasing your speed.

d. You ride at a constant speed, then go up a hill. Once on top of the hill, you gradually increase your speed.

3 ACTIVITY: Comparing Graphs

Work with a partner. The graphs represent the heights of a rocket and a weather balloon after they are launched.

Graph A

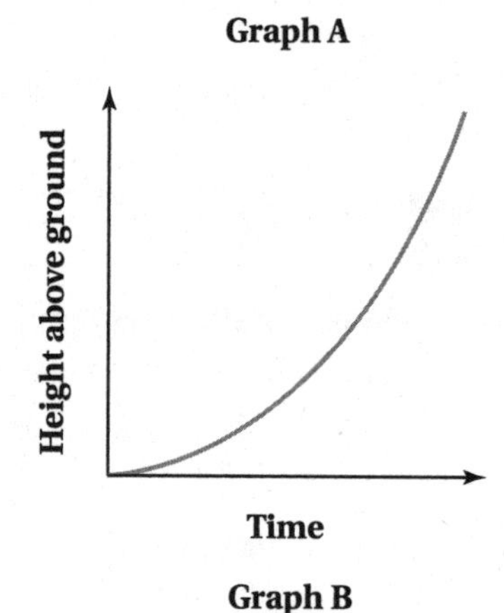

Graph B

Height above ground

Time

a. How are the graphs similar? How are they different? Explain.

b. Compare the steepness of each graph.

c. Which graph do you think represents the height of the rocket? Explain.

Name____ Date__________

4 ACTIVITY: Comparing Graphs

Work with a partner. The graphs represent the speeds of two cars. One car is approaching a stop sign. The other car is approaching a yield sign.

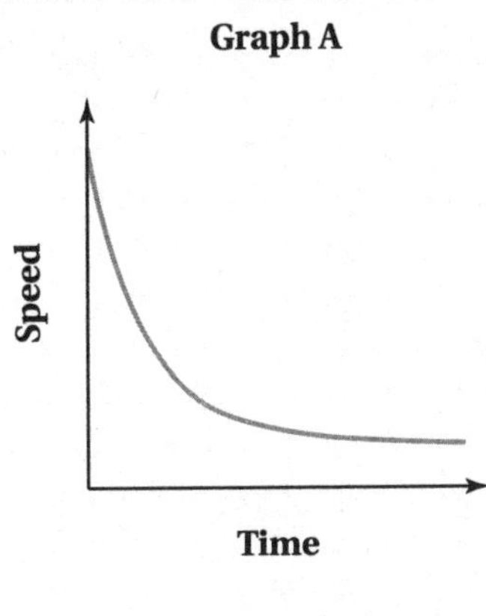

a. How are the graphs similar? How are they different? Explain.

b. Compare the steepness of each graph.

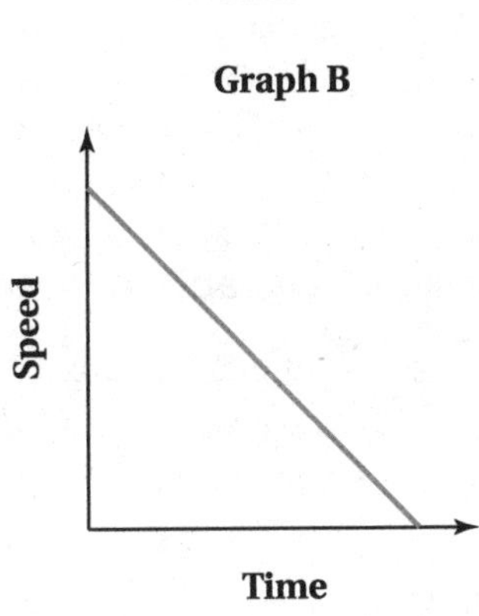

c. Which graph do you think represents the car approaching a stop sign? Explain.

What Is Your Answer?

5. IN YOUR OWN WORDS How can you use a graph to represent relationships between quantities without using numbers?

6. Describe a possible situation represented by the graph shown.

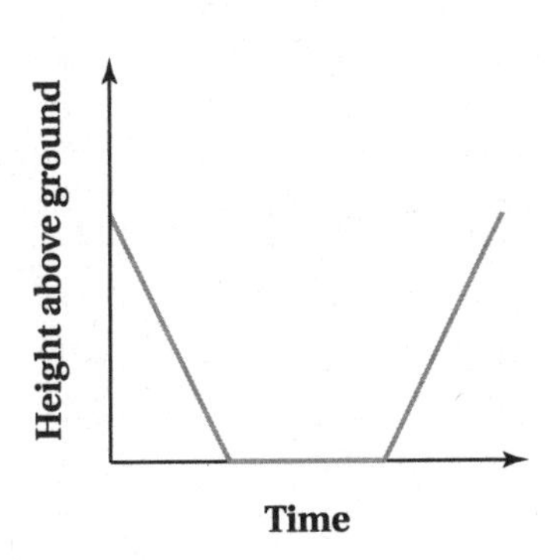

7. Sketch a graph similar to the graphs in Activities 1 and 2. Exchange graphs with a classmate and describe a possible situation represented by the graph. Discuss the results.

Name ______________________________ Date __________

6.5 Practice
For use after Lesson 6.5

Describe the relationship between the two quantities.

1. 

2. 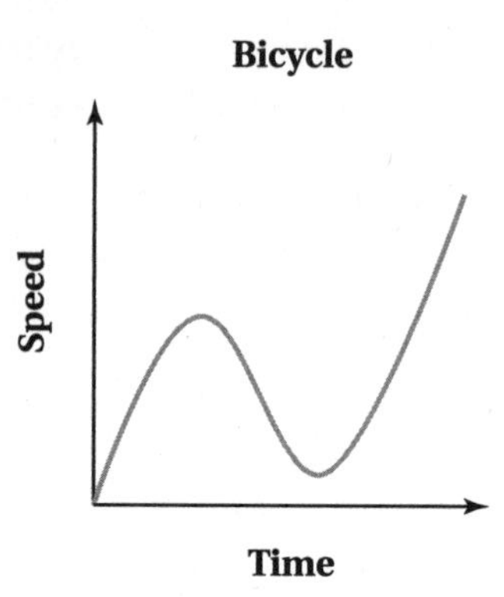

Sketch a graph that represents the situation.

3. You are texting a friend at a constant rate. You send the message then wait for a response. Once you receive a response, you begin texting a reply at a constant rate.

4. You cut your fingernails, let the nails grow back, and then cut them again.

Name____ Date__________

Fair Game Review

Complete the number sentence with <, >, or =.

1. 3.4 ______ 3.45

2. −6.01 ______ −6.1

3. 3.50 ______ 3.5

4. −0.84 ______ −0.91

Find three decimals that make the number sentence true.

5. $-5.2 \geq$ ______

6. $2.65 >$ ______

7. $-3.18 \leq$ ______

8. $0.03 <$ ______

9. The table shows the times of a 100-meter dash. Order the runners from first place to fifth place.

Runner	Time (seconds)
A	12.60
B	12.55
C	12.49
D	12.63
E	12.495

Name ______________________________ Date __________

Chapter 7 Fair Game Review (continued)

Evaluate the expression.

10. $10^2 - 48 \div 6 + 25 \bullet 3$

11. $8\left(\frac{16}{4}\right) + 2^2 - 11 \bullet 3$

12. $\left(\frac{6}{3} + 4\right)^2 \div 4 \bullet 7$

13. $5(9 - 4)^2 - 3^2$

14. $5^2 - 2^2 \bullet 4^2 - 12$

15. $\left(\frac{50}{5^2}\right)^2 \div 4$

16. The table shows the numbers of students in 4 classes. The teachers are combining the classes and dividing the students in half to form two groups for a project. Write an expression to represent this situation. How many students are in each group?

Class	Students
1	24
2	32
3	30
4	28

Name______________________________ Date__________

7.1 Finding Square Roots

For use with Activity 7.1

Essential Question How can you find the dimensions of a square or a circle when you are given its area?

When you multiply a number by itself, you square the number.

Symbol for squaring is the exponent 2. → $4^2 = 4 \bullet 4$

$= 16$

4 squared is 16.

To "undo" this, take the *square root* of the number.

Symbol for square root is a *radical sign*, $\sqrt{\ }$. → $\sqrt{16} = \sqrt{4^2} = 4$

The square root of 16 is 4.

1 ACTIVITY: Finding Square Roots

Work with a partner. Use a square root symbol to write the side length of the square. Then find the square root. Check your answer by multiplying.

a. Sample: $s = \sqrt{121} =$ **Check:**

Area = 121 ft^2

s

s

The length of each side of the square is ____________________.

b. Area = 81 yd^2

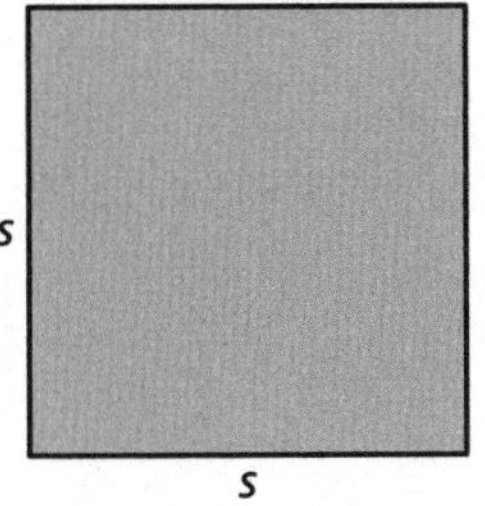

c. Area = 324 cm^2

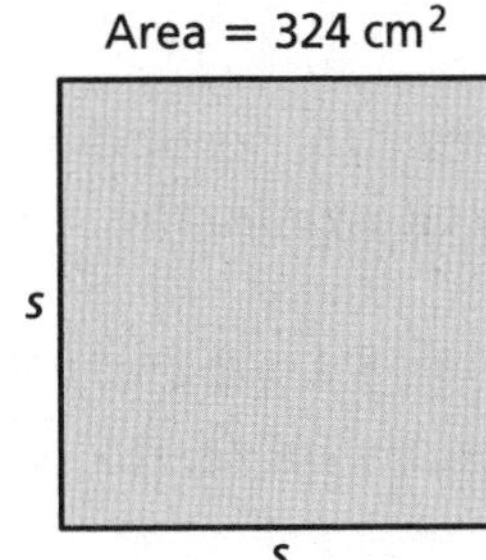

d. Area = 361 mi^2

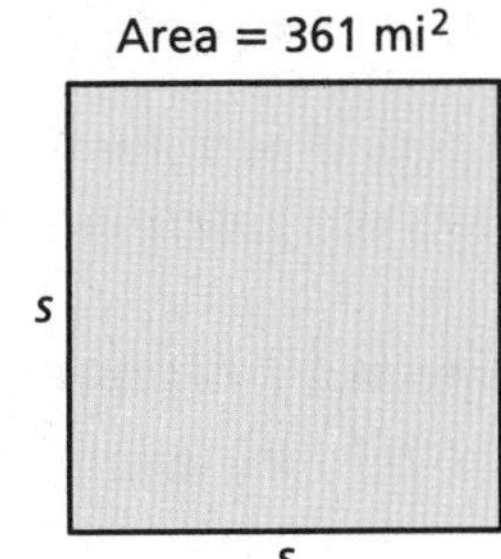

Name ______________________ Date ________

e. Area = 225 mi^2

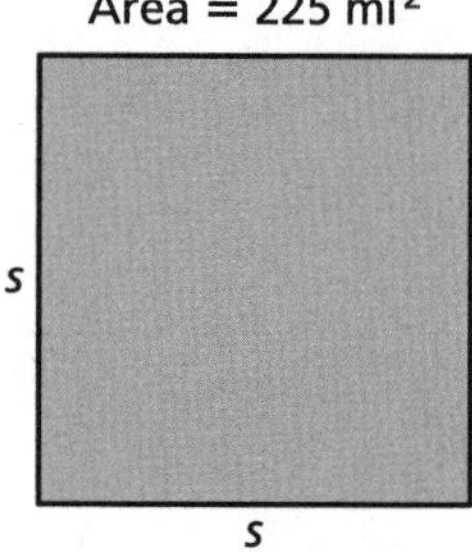

f. Area = 2.89 in.^{2}

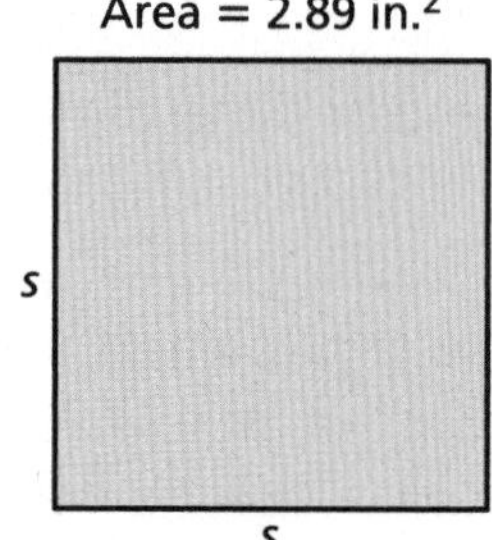

g. Area = $\frac{4}{9}$ ft^2

2 ACTIVITY: Using Square Roots

Work with a partner. Find the radius of each circle.

a.

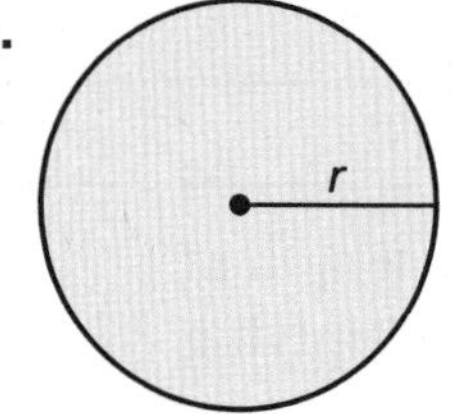

Area = 36π in.^{2}

b.

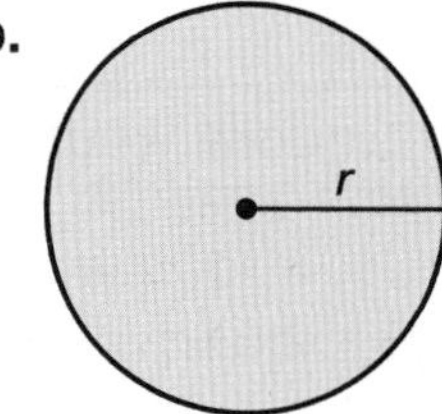

Area = π yd^2

c.

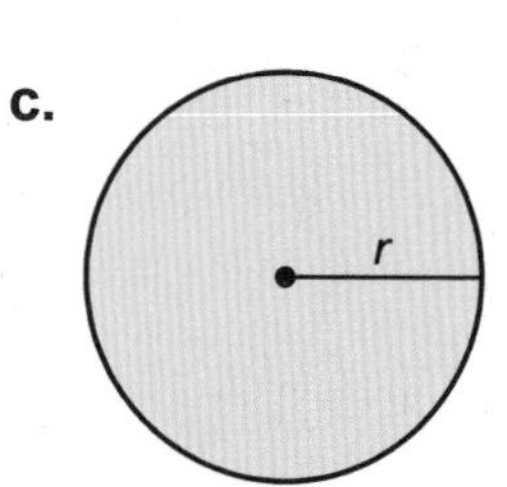

Area = 0.25π ft^2

d.

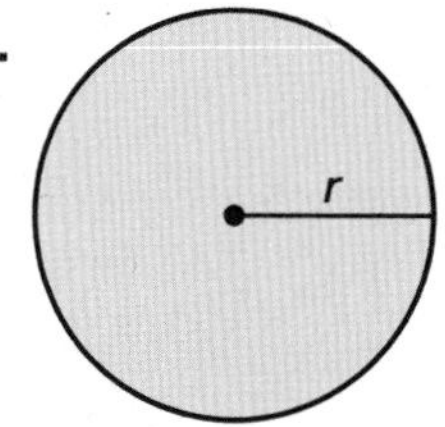

Area = $\frac{9}{16}\pi$ m^2

3 ACTIVITY: The Period of a Pendulum

Work with a partner.

The period of a pendulum is the time (in seconds) it takes the pendulum to swing back *and* forth.

The period T is represented by $T = 1.1\sqrt{L}$, where L is the length of the pendulum (in feet).

Complete the table. Then graph the function on the next page. Is the function linear?

L	1.00	1.96	3.24	4.00	4.84	6.25	7.29	7.84	9.00
T									

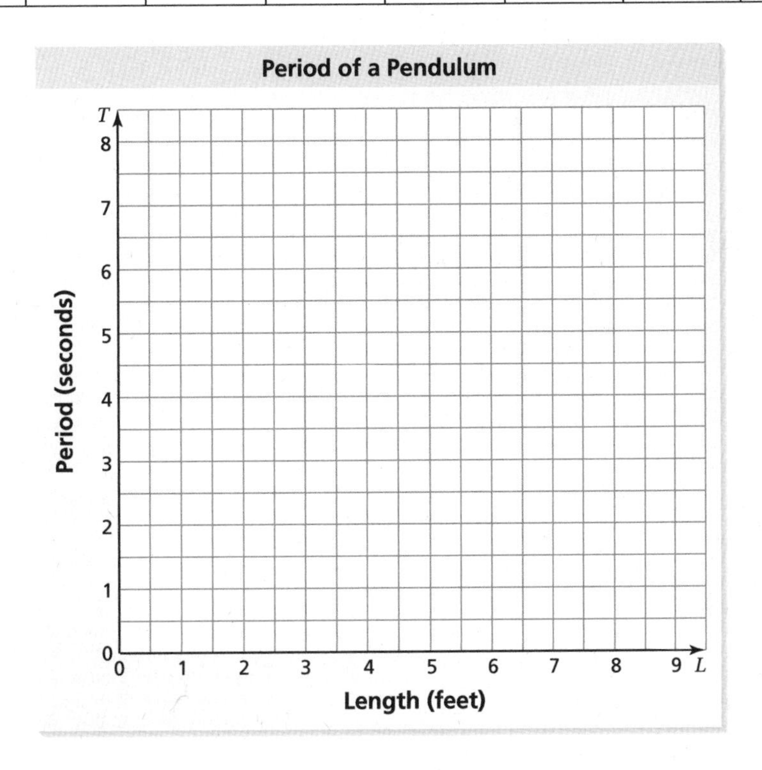

What Is Your Answer?

4. **IN YOUR OWN WORDS** How can you find the dimensions of a square or circle when you are given its area? Give an example of each. How can you check your answers?

Name ______________________________ Date __________

7.1 Practice
For use after Lesson 7.1

Find the two square roots of the number.

1. 16

2. 100

3. 196

Find the square root(s).

4. $\sqrt{169}$

5. $\sqrt{\frac{4}{225}}$

6. $-\sqrt{12.25}$

Evaluate the expression.

7. $2\sqrt{36} + 9$

8. $8 - 11\sqrt{\frac{25}{121}}$

9. $3\left(\sqrt{\frac{125}{5}} - 8\right)$

10. A trampoline has an area of 49π square feet. What is the diameter of the trampoline?

Name______________________________ Date__________

7.2 Finding Cube Roots

For use with Activity 7.2

Essential Question How is the cube root of a number different from the square root of a number?

When you multiply a number by itself twice, you cube the number.

Symbol for cubing is the exponent 3. → $4^3 = 4 \bullet 4 \bullet 4$

$= 64$

4 cubed is 64.

To "undo" this, take the *cube root* of the number.

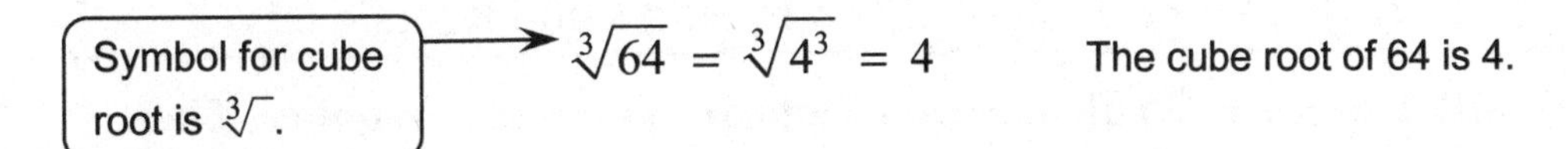

Symbol for cube root is $\sqrt[3]{}$. → $\sqrt[3]{64} = \sqrt[3]{4^3} = 4$ The cube root of 64 is 4.

1 ACTIVITY: Finding Cube Roots

Work with a partner. Use a cube root symbol to write the edge length of the cube. Then find the cube root. Check your answer by multiplying.

a. Sample: $s = \sqrt[3]{343} = \sqrt[3]{7^3} = 7$ inches

Volume = 343 in.³

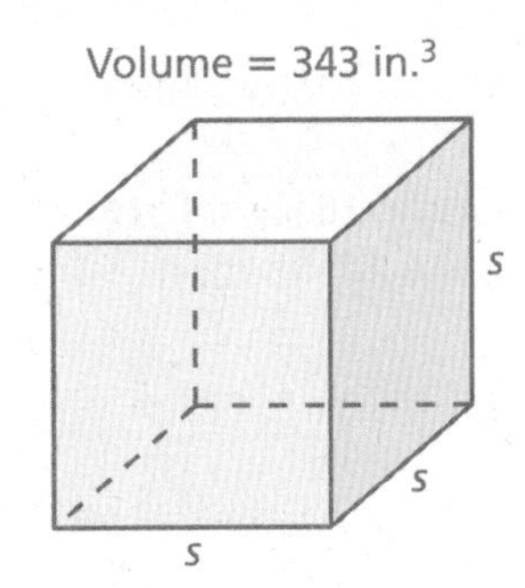

Check

$7 \bullet 7 \bullet 7 = 49 \bullet 7$

$= 343$ ✓

The edge length of the cube is 7 inches.

b. Volume = 27 ft³

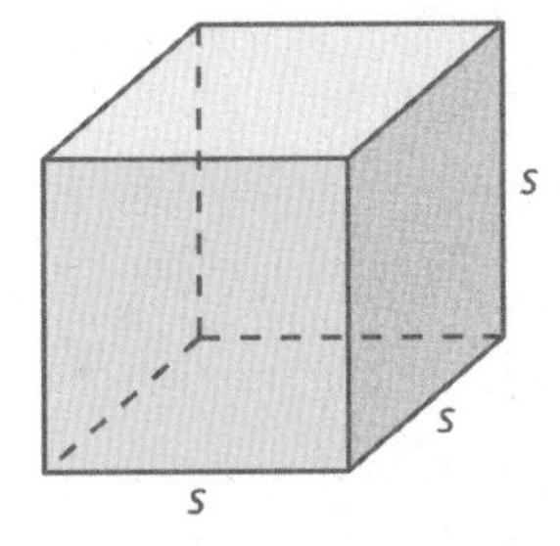

c. Volume = 125 m³

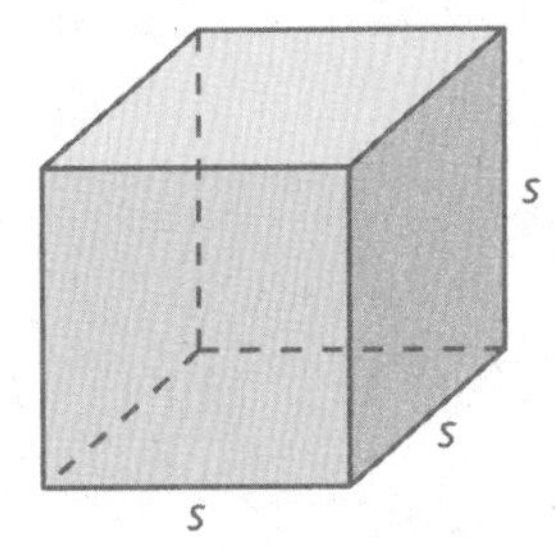

d. Volume = 0.001 cm^3

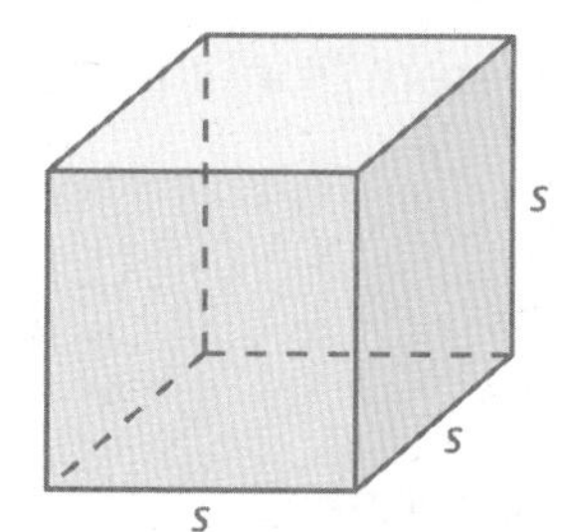

e. Volume = $\frac{1}{8}$ yd^3

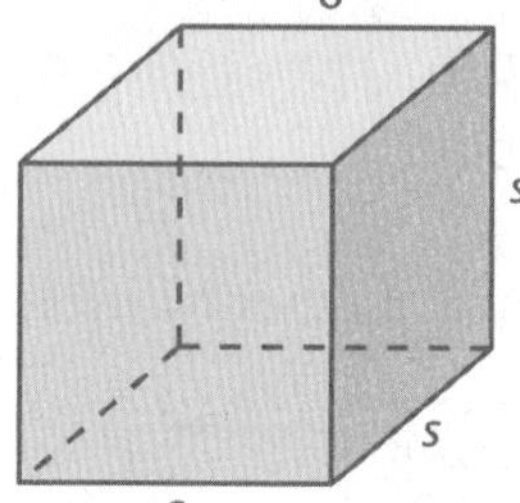

2 ACTIVITY: Use Prime Factorizations to Find Cube Roots

Work with a partner. Write the prime factorization of each number. Then use the prime factorization to find the cube root of the number.

a. 216

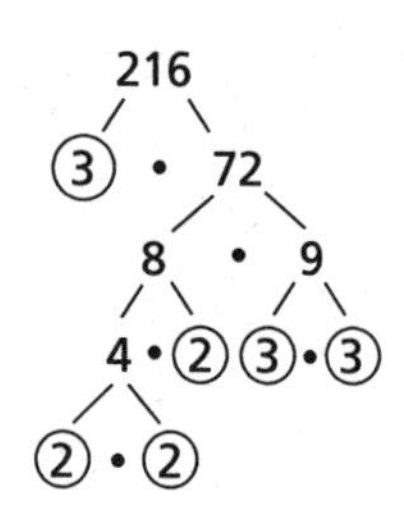

$216 = 3 \bullet 2 \bullet 3 \bullet 3 \bullet 2 \bullet 2$ Prime factorization

$= (3 \bullet \square) \bullet (3 \bullet \square) \bullet (3 \bullet \square)$ Commutative Property of Multiplication

$= \square \bullet \square \bullet \square$ Simplify.

The cube root of 216 is __________.

b. 1000

c. 3375

d. STRUCTURE Does this procedure work for every number? Explain why or why not.

What Is Your Answer?

3. Complete each statement using *positive* or *negative*.

a. A positive number times a positive number is a __________ number.

b. A negative number times a negative number is a __________ number.

c. A positive number multiplied by itself twice is a __________ number.

d. A negative number multiplied by itself twice is a __________ number.

4. REASONING Can a negative number have a cube root? Give an example to support your explanation.

5. IN YOUR OWN WORDS How is the cube root of a number different from the square root of a number?

6. Give an example of a number whose square root and cube root are equal.

7. A cube has a volume of 13,824 cubic meters. Use a calculator to find the edge length.

Name ______________________________ Date __________

7.2 Practice

For use after Lesson 7.2

Find the cube root.

1. $\sqrt[3]{27}$

2. $\sqrt[3]{8}$

3. $\sqrt[3]{-64}$

4. $\sqrt[3]{-\frac{125}{216}}$

Evaluate the expression.

5. $10 - \left(\sqrt[3]{12}\right)^3$

6. $2\sqrt[3]{512} + 10$

7. The volume of a cube is 1000 cubic inches. What is the edge length of the cube?

Name____ Date__________

7.3 The Pythagorean Theorem

For use with Activity 7.3

Essential Question How are the lengths of the sides of a right triangle related?

Pythagoras was a Greek mathematician and philosopher who discovered one of the most famous rules in mathematics. In mathematics, a rule is called a **theorem**. So, the rule that Pythagoras discovered is called the Pythagorean Theorem.

Pythagoras
(c. 570–c. 490 B.C.)

1 ACTIVITY: Discovering the Pythagorean Theorem

Work with a partner.

a. On grid paper, draw any right triangle. Label the lengths of the two shorter sides a and b.

b. Label the length of the longest side c.

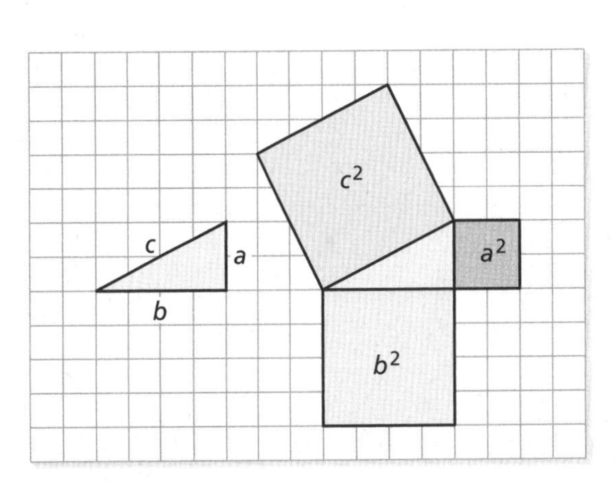

c. Draw squares along each of the three sides. Label the areas of the three squares a^2, b^2, and c^2.

d. Cut out the three squares. Make eight copies of the right triangle and cut them out. Arrange the figures to form two identical larger squares.

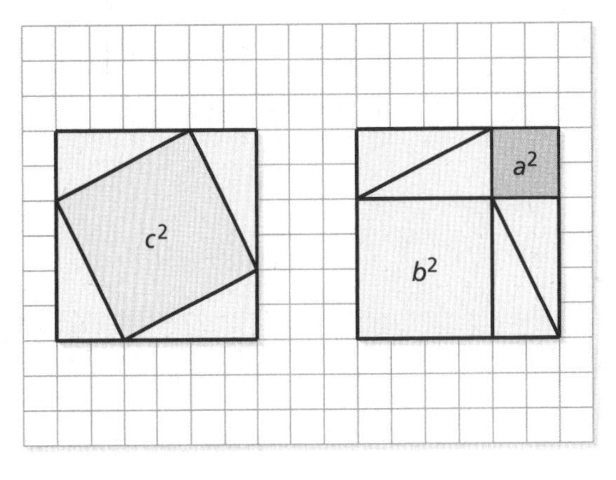

e. MODELING The Pythagorean Theorem describes the relationship among a^2, b^2, and c^2. Use your result from part (d) to write an equation that describes this relationship.

Name ______________________________ Date __________

2 ACTIVITY: Using the Pythagorean Theorem in Two Dimensions

Work with a partner. Use a ruler to measure the longest side of each right triangle. Verify the result of Activity 1 for each right triangle.

a.

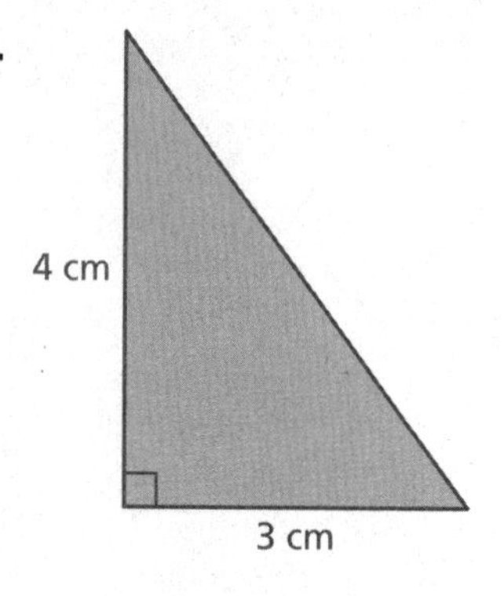

b.

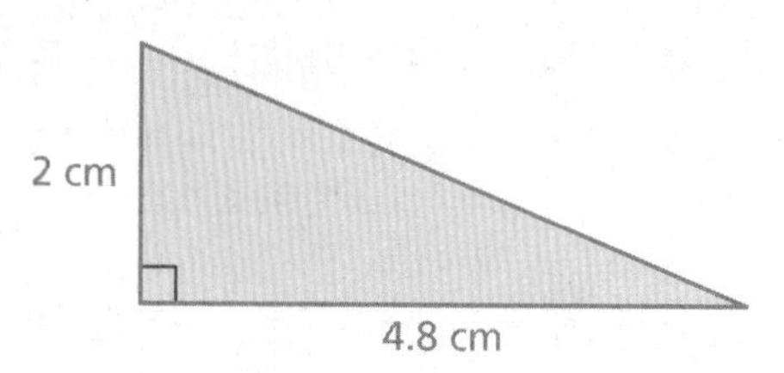

c.

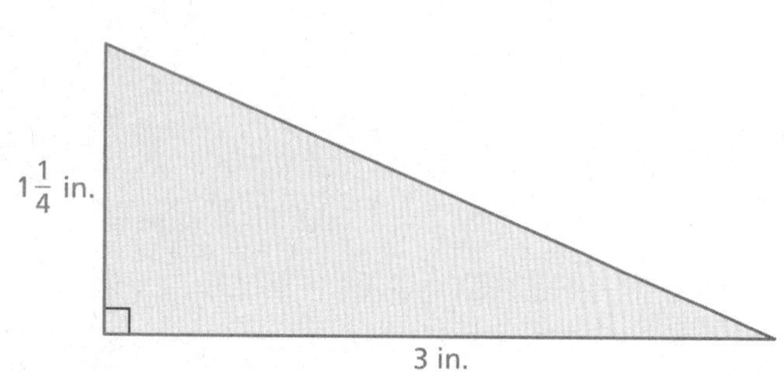

d.

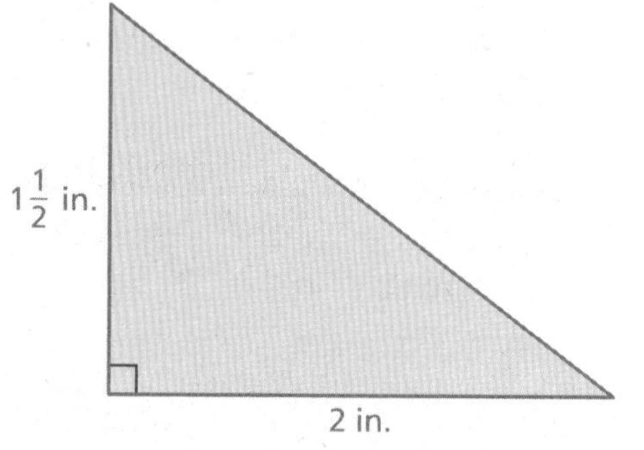

3 ACTIVITY: Using the Pythagorean Theorem in Three Dimensions

Work with a partner. A guy wire attached 24 feet above ground level on a telephone pole provides support for the pole.

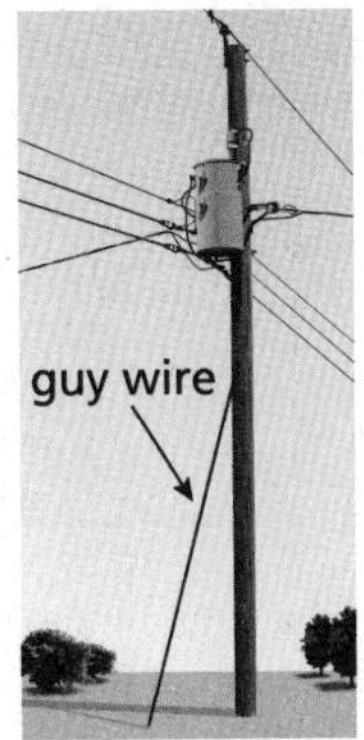

a. **PROBLEM SOLVING** Describe a procedure that you could use to find the length of the guy wire without directly measuring the wire.

b. Find the length of the wire when it meets the ground 10 feet from the base of the pole.

What Is Your Answer?

4. **IN YOUR OWN WORDS** How are the lengths of the sides of a right triangle related? Give an example using whole numbers.

Name ______________________ Date ________

7.3 Practice

For use after Lesson 7.3

Find the missing length of the triangle.

1.

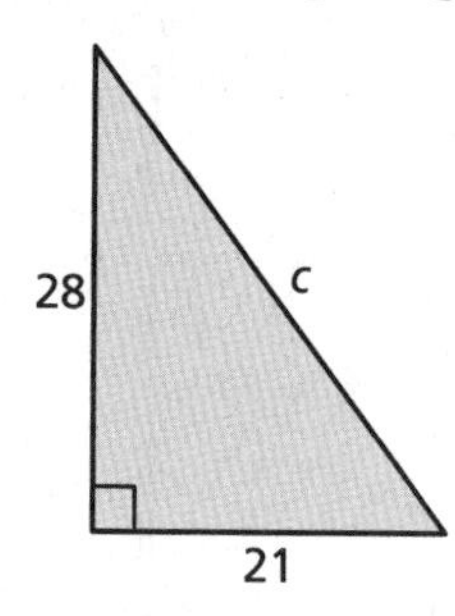

2.

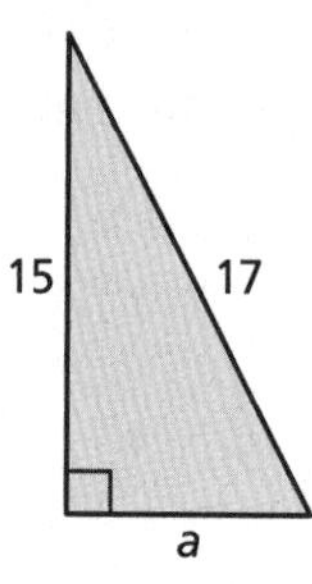

3.

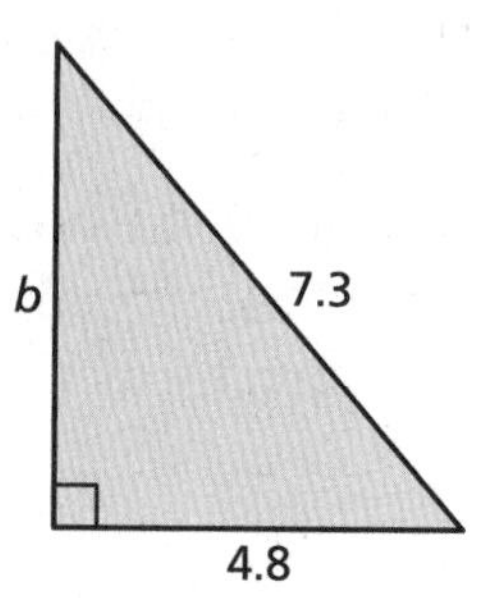

Find the missing length of the figure.

4.

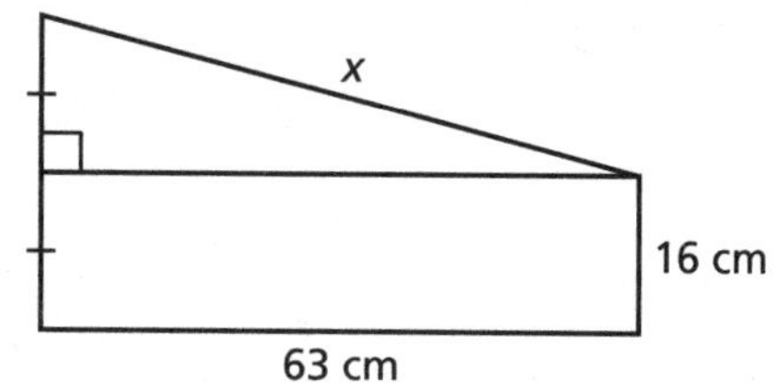

5.

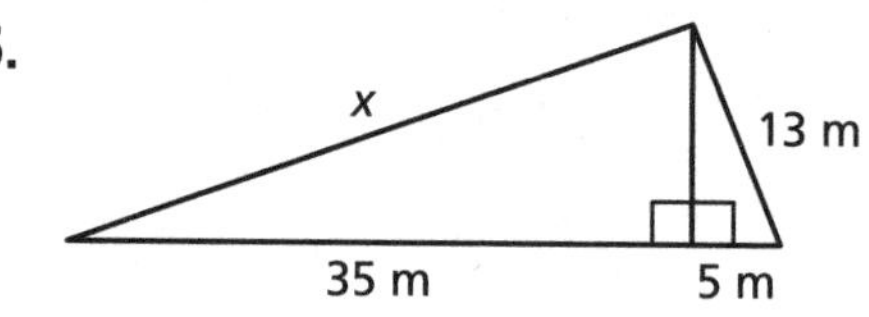

6. In wood shop, you make a bookend that is in the shape of a right triangle. What is the base b of the bookend?

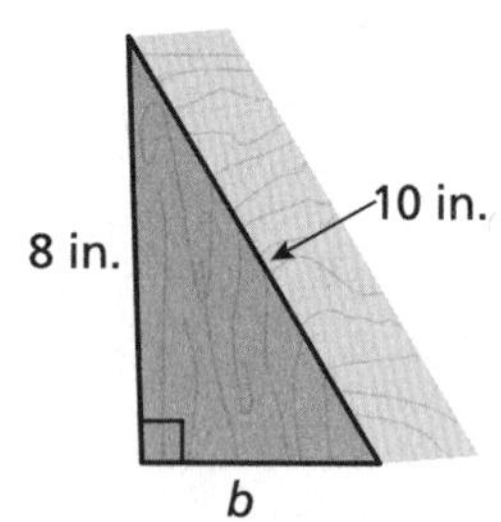

Name____ Date__________

7.4 Approximating Square Roots

For use with Activity 7.4

Essential Question How can you find decimal approximations of square roots that are not rational?

1 ACTIVITY: Approximating Square Roots

Work with a partner. Archimedes was a Greek mathematician, physicist, engineer, inventor, and astronomer. He tried to find a rational number whose square is 3. Two that he tried were $\frac{265}{153}$ and $\frac{1351}{780}$.

a. Are either of these numbers equal to $\sqrt{3}$? Explain.

b. Use a calculator to approximate $\sqrt{3}$. Write the number on a piece of paper. Enter it into the calculator and square it. Then subtract 3. Do you get 0? What does this mean?

c. The value of $\sqrt{3}$ is between which two integers?

d. Tell whether the value of $\sqrt{3}$ is between the given numbers. Explain your reasoning.

1.7 and 1.8	1.72 and 1.73	1.731 and 1.732

2 ACTIVITY: Approximating Square Roots Geometrically

Work with a partner. Refer to the square on the number line below.

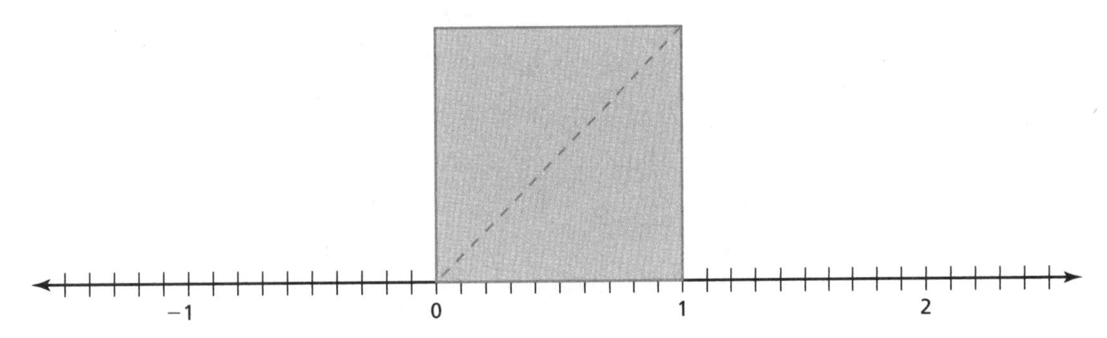

a. What is the length of the diagonal of the square?

b. Copy the square and its diagonal onto a piece of transparent paper. Rotate it about zero on the number line so that the diagonal aligns with the number line. Use the number line to estimate the length of the diagonal.

c. **STRUCTURE** How do you think your answers in parts (a) and (b) are related?

3 ACTIVITY: Approximating Square Roots Geometrically

Work with a partner.

a. Use grid paper and the given scale to draw a horizontal line segment 1 unit in length. Draw your segment near the bottom of the grid. Label this segment AC.

b. Draw a vertical line segment 2 units in length. Draw your segment near the left edge of the grid. Label this segment DC.

c. Set the point of a compass on A. Set the compass to 2 units. Swing the compass to intersect segment DC. Label this intersection as B.

d. Use the Pythagorean Theorem to find the length of segment BC.

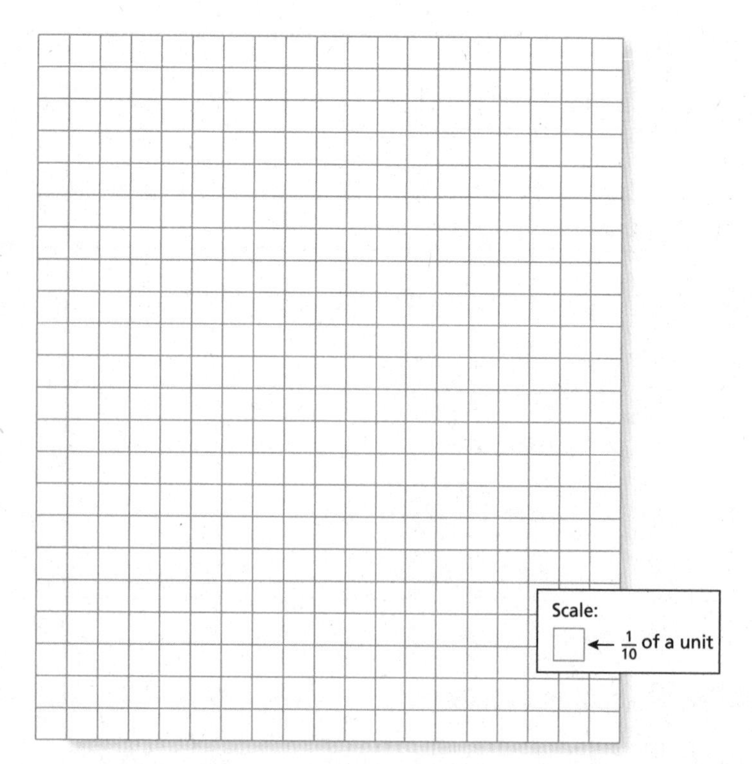

e. Use the grid paper to approximate $\sqrt{3}$ to the nearest tenth.

Name____ Date__________

4. Compare your approximation in Activity 3 with your results from Activity 1.

What Is Your Answer?

5. Repeat Activity 3 for a triangle in which segment AC is 2 units and segment BA is 3 units. Use the Pythagorean Theorem to find the length of segment BC. Use the grid paper to approximate $\sqrt{5}$ to the nearest tenth.

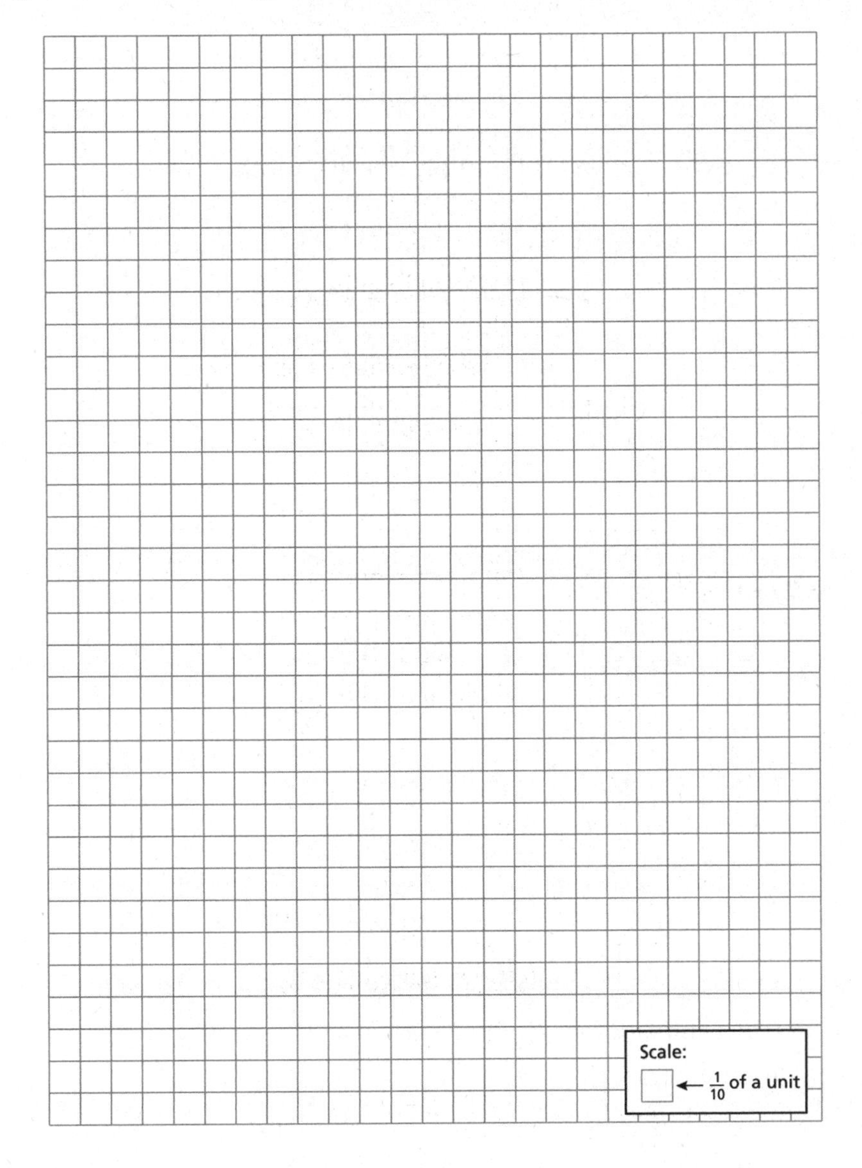

6. **IN YOUR OWN WORDS** How can you find decimal approximations of square roots that are not rational?

Name ________________________________ Date __________

7.4 Practice

For use after Lesson 7.4

Classify the real number.

1. $\sqrt{14}$

2. $-\frac{3}{7}$

3. $\frac{153}{3}$

Estimate the square root to the nearest (a) integer and (b) tenth.

4. $\sqrt{8}$

5. $\sqrt{60}$

6. $-\sqrt{\frac{172}{25}}$

Which number is greater? Explain.

7. $\sqrt{88}$, 12

8. $-\sqrt{18}$, -6

9. 14.5, $\sqrt{220}$

10. The velocity in meters per second of a ball that is dropped from a window at a height of 10.5 meters is represented by the equation $v = \sqrt{2(9.8)(10.5)}$. Estimate the velocity of the ball. Round your answer to the nearest tenth.

Name____ Date__________

Practice

For use after Extension 7.4

Write the decimal as a fraction or a mixed number.

1. $0.\overline{3}$

2. $-0.\overline{2}$

3. $1.\overline{7}$

4. $-2.\overline{6}$

5. $0.4\overline{6}$

6. $-1.8\overline{3}$

7. $-0.7\overline{3}$

8. $0.\overline{18}$

9. $-3.\overline{24}$

10. $1.\overline{09}$

11. The length of a pencil is $1.5\overline{6}$ inches. Represent the length of the pencil as a mixed number.

Name ______________________________ Date __________

7.5 Using the Pythagorean Theorem

For use with Activity 7.5

Essential Question In what other ways can you use the Pythagorean Theorem?

The *converse* of a statement switches the hypothesis and the conclusion.

Statement:	Converse of the statement:
If p, then q.	If q, then p.

1 ACTIVITY: Analyzing Converses of Statements

Work with a partner. Write the converse of the true statement. Determine whether the converse is *true* or *false*. If it is true, justify your reasoning. If it is false, give a counterexample.

a. If $a = b$, then $a^2 = b^2$.

Converse: ______________________________

b. If $a = b$, then $a^3 = b^3$.

Converse: ______________________________

c. If one figure is a translation of another figure, then the figures are congruent.

Converse: ______________________________

d. If two triangles are similar, then the triangles have the same angle measures.

Converse: ______________________________

Is the converse of a true statement always true? always false? Explain.

Name ______________________________ Date __________

2 ACTIVITY: The Converse of the Pythagorean Theorem

Work with a partner. The converse of the Pythagorean Theorem states: "If the equation $a^2 + b^2 = c^2$ is true for the side lengths of a triangle, then the triangle is a right triangle."

a. Do you think the converse of the Pythagorean Theorem is *true* or *false*? How could you use deductive reasoning to support your answer?

b. Consider $\triangle DEF$ with side lengths a, b, and c, such that $a^2 + b^2 = c^2$. Also consider $\triangle JKL$ with leg lengths a and b, where $\angle K = 90°$.

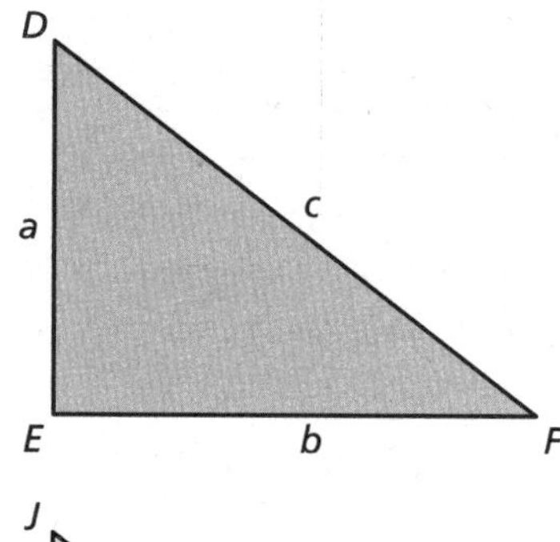

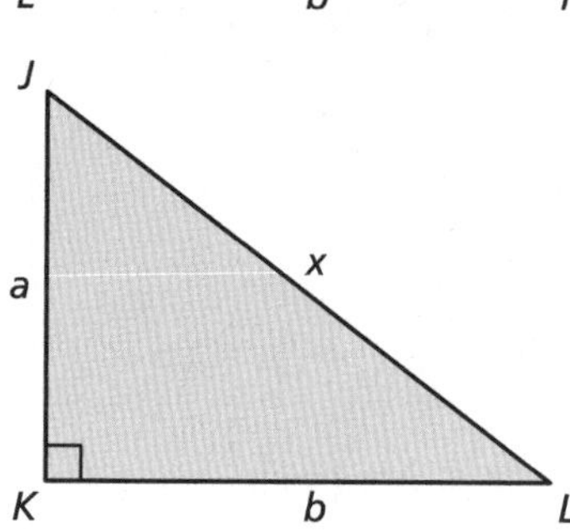

- What does the Pythagorean Theorem tell you about $\triangle JKL$?

- What does this tell you about c and x?

- What does this tell you about $\triangle DEF$ and $\triangle JKL$?

- What does this tell you about $\angle E$?

- What can you conclude?

Name____ Date__________

3 ACTIVITY: Developing the Distance Formula

Work with a partner. Follow the steps below to write a formula that you can use to find the distance between and two points in a coordinate plane.

Step 1: Choose two points in the coordinate plane that do not lie on the same horizontal or vertical line. Label the points (x_1, y_1) and (x_2, y_2).

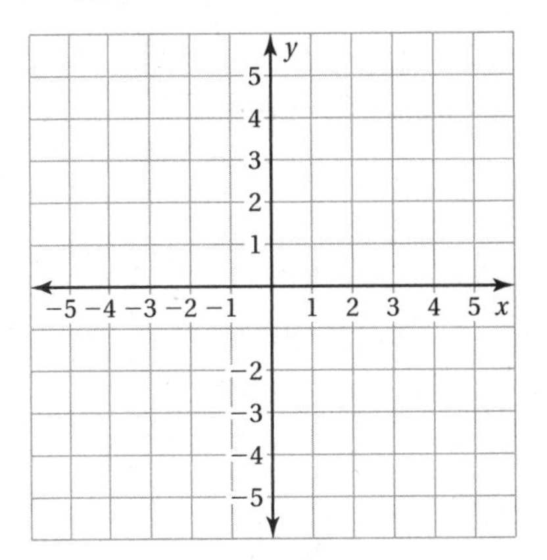

Step 2: Draw a line segment connecting the points. This will be the hypotenuse of a right triangle.

Step 3: Draw horizontal and vertical line segments from the points to form the legs of the right triangle.

Step 4: Use the x-coordinates to write an expression for the length of the horizontal leg.

Step 5: Use the y-coordinates to write an expression for the length of the vertical leg.

Step 6: Substitute the expressions for the lengths of the legs into the Pythagorean Theorem.

Step 7: Solve the equation in Step 6 for the hypotenuse c.

What does the length of the hypotenuse tell you about the two points?

What Is Your Answer?

4. **IN YOUR OWN WORDS** In what other ways can you use the Pythagorean Theorem?

5. What kind of real-life problems do you think the converse of the Pythagorean Theorem can help you solve?

Name ______________________________ Date __________

7.5 Practice

For use after Lesson 7.5

Tell whether the triangle with the given side lengths is a right triangle.

1.

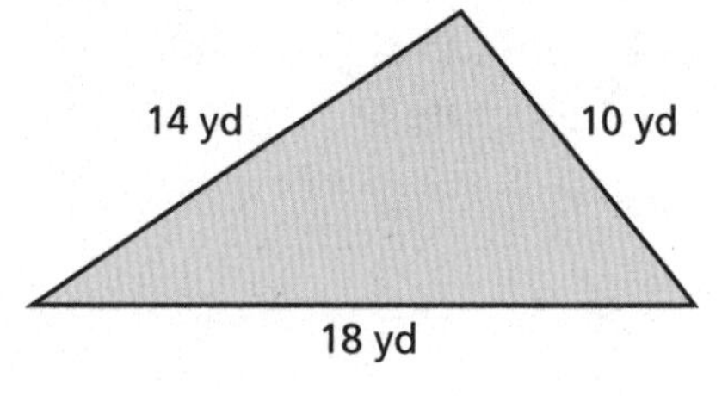

2.

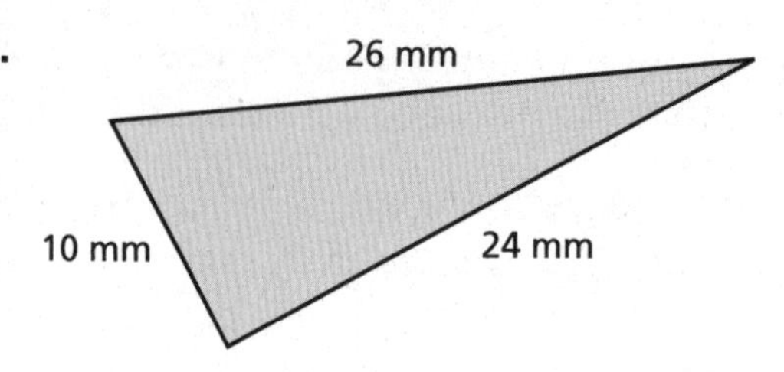

3. 4 m, 4.2 m, 5.8 m

4. 31 in., 35 in., 16 in.

Find the distance between the two points.

5. $(2, 1), (-3, 6)$

6. $(-6, -4), (2, 2)$

7. $(1, -7), (4, -5)$

8. $(-9, 3), (-5, -8)$

9. The cross-section of a wheelchair ramp is shown. Does the ramp form a right triangle?

25 in.

313 in.

312 in.

Name______________________________ Date__________

Chapter 8 Fair Game Review

Find the area of the figure.

1.

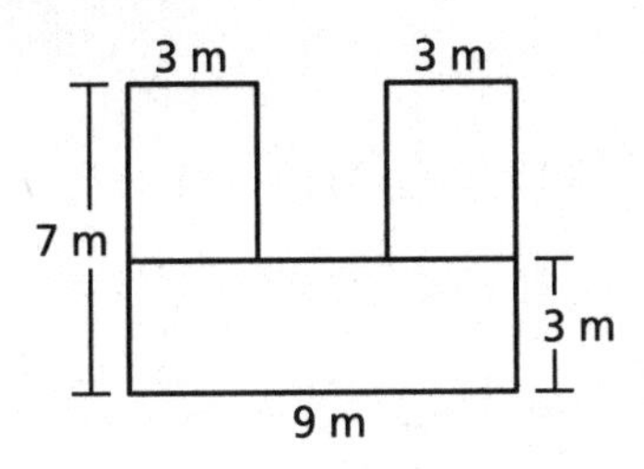

2.

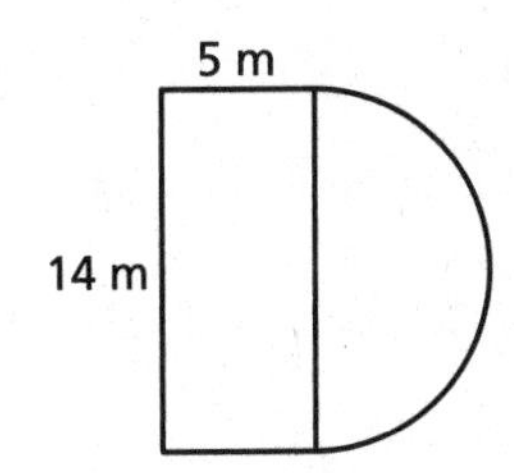

3.

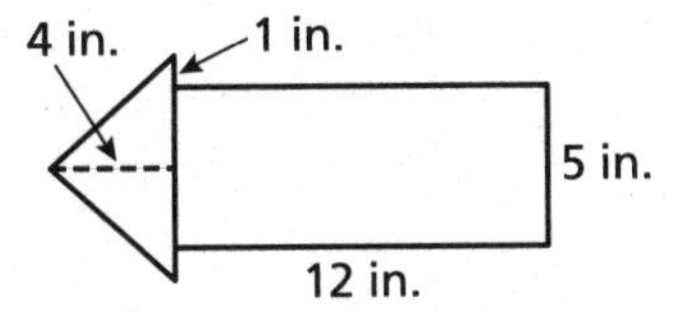

4.

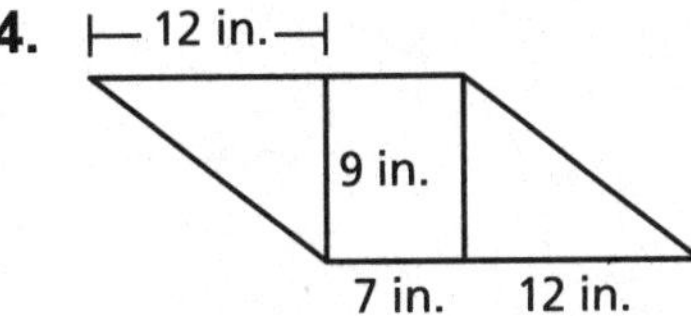

5.

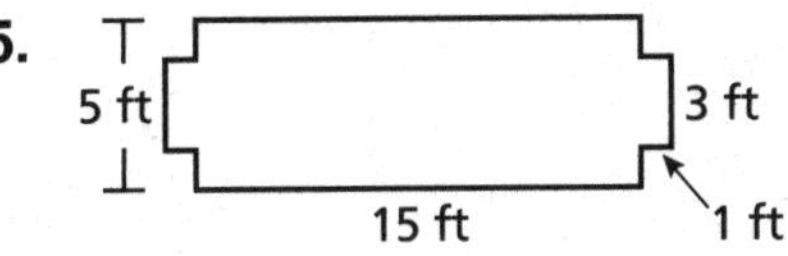

6.

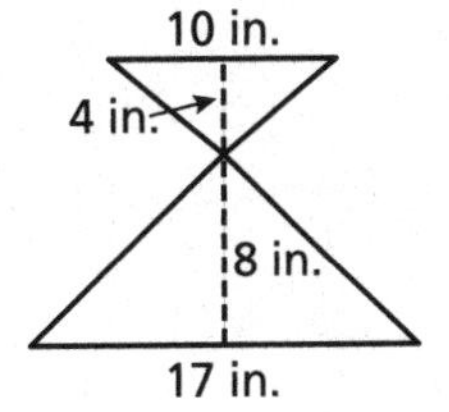

7. You are carpeting 2 rooms of your house. The carpet costs $1.48 per square foot. How much does it cost to carpet the rooms?

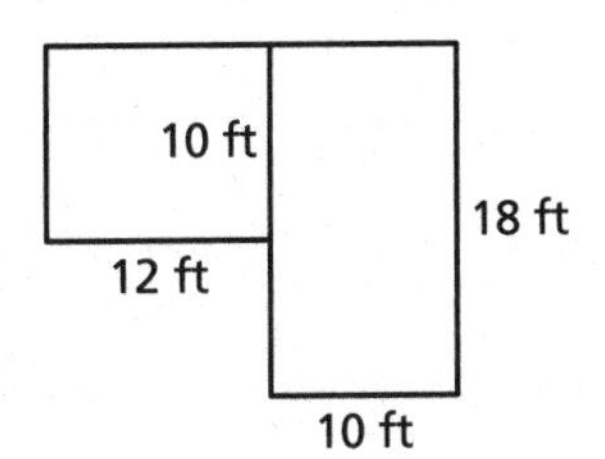

Name ______________________________ Date __________

Chapter 8 Fair Game Review (continued)

Find the area of the circle.

8.

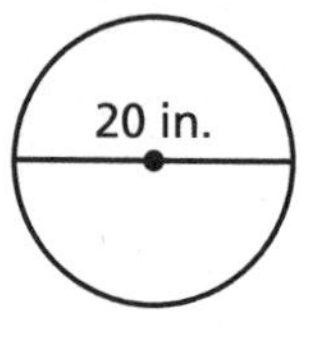

9.

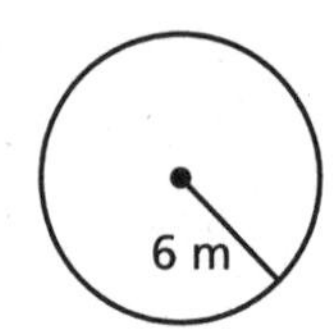

10.

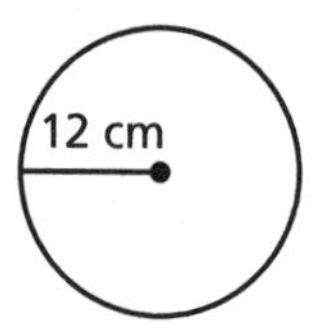

11.

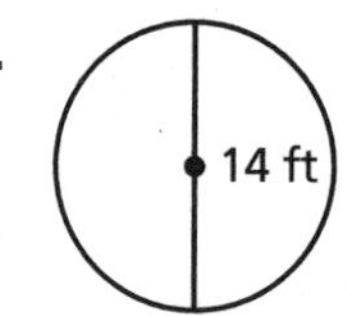

12.

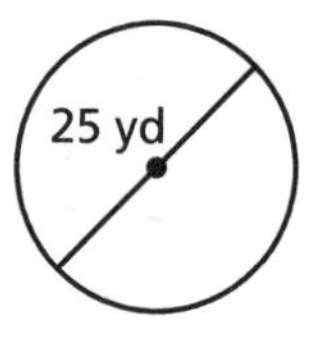

13.

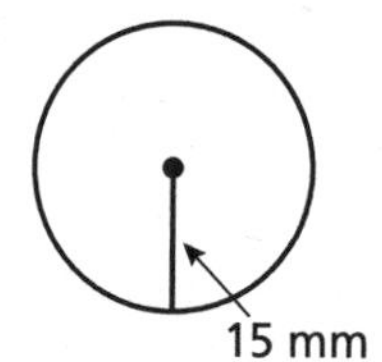

14. Find the area of the shaded region.

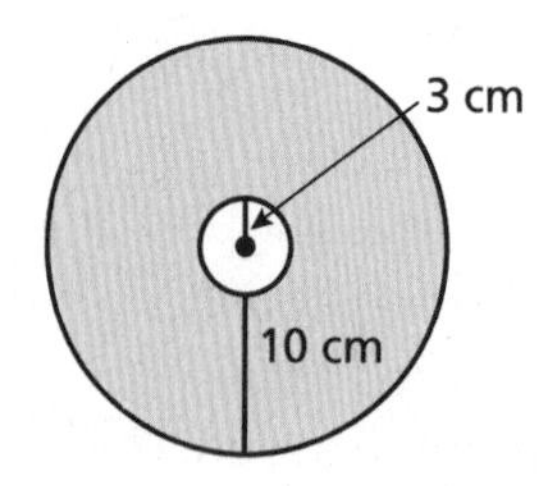

Name R Roselynn Date

8.1 Volumes of Cylinders

For use with Activity 8.1

Essential Question How can you find the volume of a cylinder?

1 ACTIVITY: Finding a Formula Experimentally

Work with a partner.

a. Find the area of the face of a coin.

b. Find the volume of a stack of a dozen coins.

c. Write a formula for the volume of a cylinder.

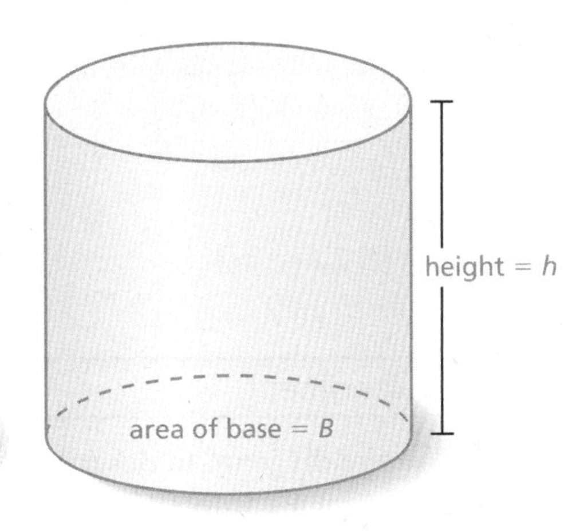

Roselynn Sim

Name ____ Date __________

8.1 Volumes of Cylinders (continued)

2 ACTIVITY: Making a Business Plan

Work with a partner. You are planning to make and sell three different sizes of cylindrical candles. You buy 1 cubic foot of candle wax for $20 to make 8 candles of each size.

a. Design the candles. What are the dimensions of each size of candle?

b. You want to make a profit of $100. Decide on a price for each size of candle.

c. Did you set the prices so that they are proportional to the volume of each size of candle? Why or why not?

3 ACTIVITY: Science Experiment

Work with a partner. Use the diagram to describe how you can find the volume of a small object.

Name____ Date__________

4 ACTIVITY: Comparing Cylinders

Work with a partner.

a. Just by looking at the two cylinders, which one do you think has the greater volume? Explain your reasoning.

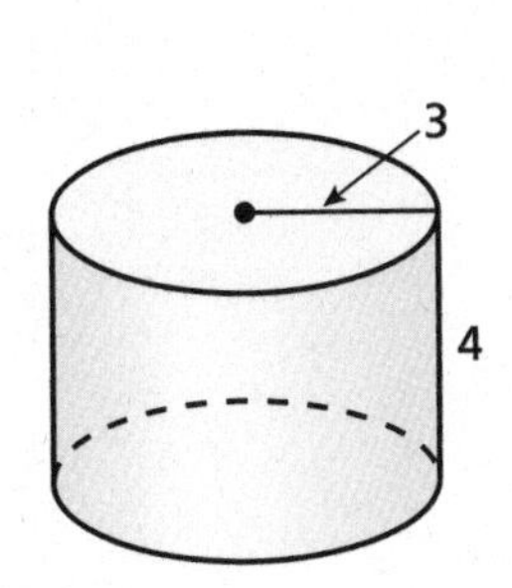

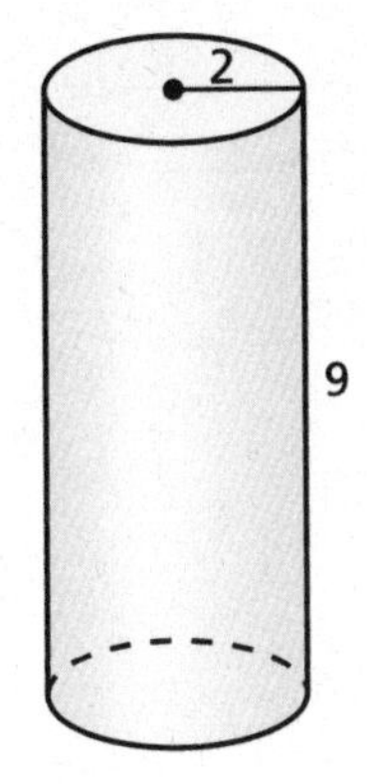

b. Find the volume of each cylinder. Was your prediction in part (a) correct? Explain your reasoning.

What Is Your Answer?

5. IN YOUR OWN WORDS How can you find the volume of a cylinder?

6. Compare your formula for the volume of a cylinder with the formula for the volume of a prism. How are they the same?

Name ______________________ Date ________

8.1 Practice

For use after Lesson 8.1

Find the volume of the cylinder. Round your answer to the nearest tenth.

1.

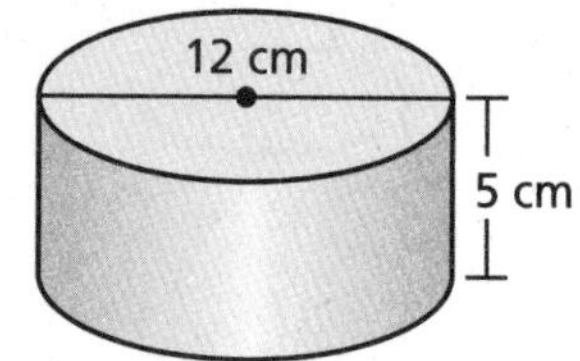

B=(3.14)(6)(6)
(36)(3.14)
(113.04)(5) = 565.2 cm

2.

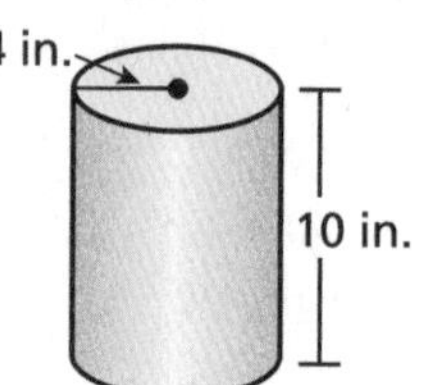

B=(3.14)(4)(4) = 50.24
(50.24)(10) = 502.4 in

Find the missing dimension of the cylinder. Round your answer to the nearest whole number.

3. Volume = 84 in.^{3}

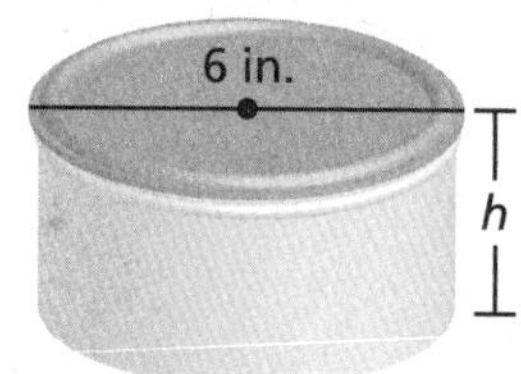

84 = BH   A=(3.14)(3)(3)
84 = 28.26 H   A = 28.26
28.26   28.26
H = 3 in

4. Volume = 650 cm^3

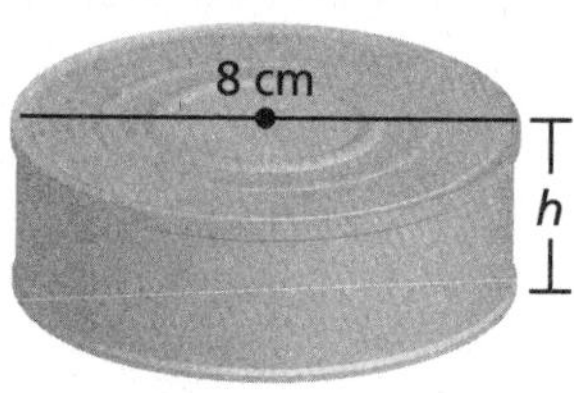

650 = BH   A = (3.14)(4)(4)
650 = 50.24H   A = 50.24
50.24   50.24
H = 13 cm

5. To make orange juice, the directions call for a can of orange juice concentrate to be mixed with three cans of water. What is the volume of orange juice that you make?

A=(3.14)(3)(3)
A = 28.26
28.26 • 5 = 141.3

V = 141.3 in

Name Roselynn Sim

Date 06/01/21

8.2 Volumes of Cones

For use with Activity 8.2

Essential Question How can you find the volume of a cone?

You already know how the volume of a pyramid relates to the volume of a prism. In this activity, you will discover how the volume of a cone relates to the volume of a cylinder.

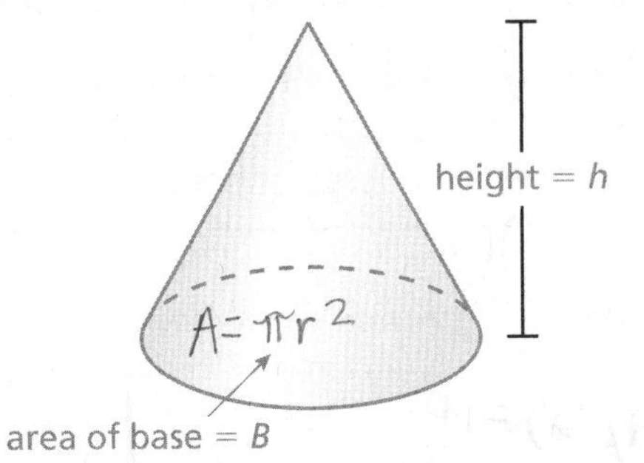

1 ACTIVITY: Finding a Formula Experimentally

Work with a partner. Use a paper cup that is shaped like a cone.

- Estimate the height of the cup.
- Trace the top of the cup on a piece of paper. Find the diameter of the circle.
- Use these measurements to draw a net for a cylinder with the same base and height as the paper cup.
- Cut out the net. Then fold and tape it to form an open cylinder.
- Fill the paper cup with rice. Then pour the rice into the cylinder. Repeat this until the cylinder is full. How many cones does it take to fill the cylinder?
- Use your result to write a formula for the volume of a cone.

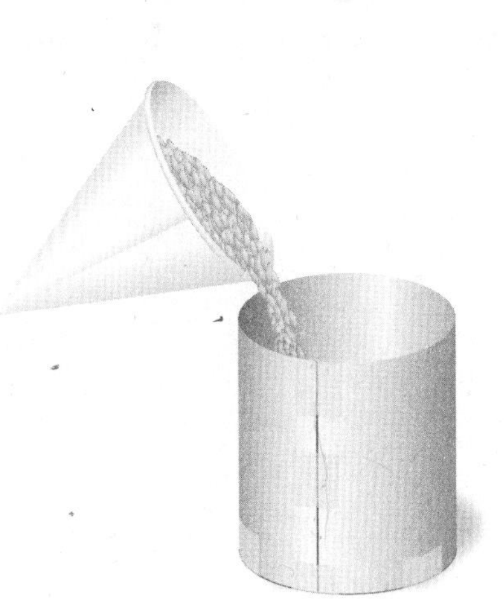

Name ______________________________ Date __________

2 ACTIVITY: Summarizing Volume Formulas

Work with a partner. You can remember the volume formulas for prisms, cylinders, pyramids, and cones with just two concepts.

Volumes of Prisms and Cylinders

Volume = | Area of base | × | ______ |

Volumes of Pyramids and Cones

Volume = | ______ | Volume of prism or cylinder with same base and height |

Make a list of all the formulas you need to remember to find the area of a base. Talk about strategies for remembering these formulas.

3 ACTIVITY: Volumes of Oblique Solids

Work with a partner. Think of a stack of paper. When you adjust the stack so that the sides are oblique (slanted), do you change the volume of the stack? If the volume of the stack does not change, then the formulas for volumes of right solids also apply to oblique solids.

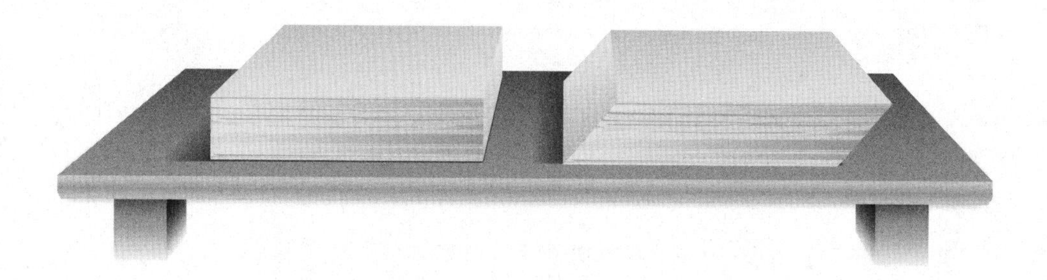

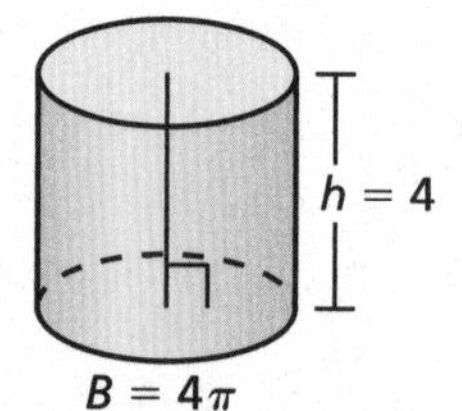

Right Cylinder

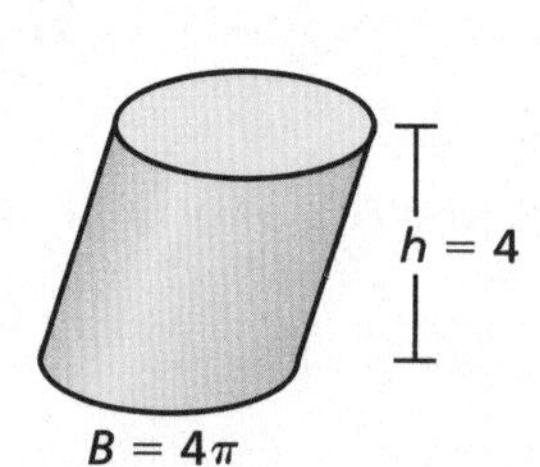

Oblique Cylinder

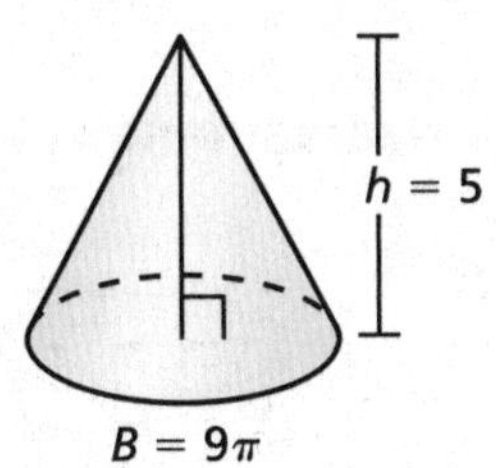

Right Cone

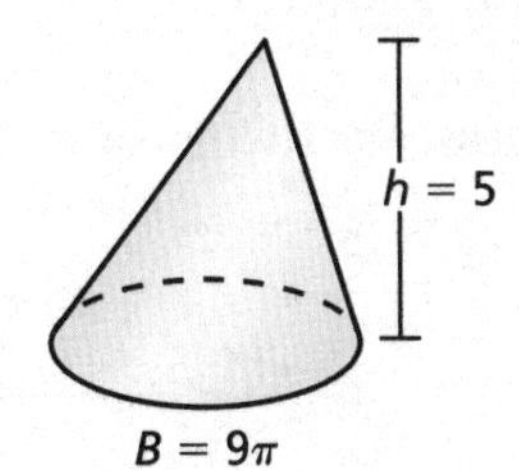

Oblique Cone

What Is Your Answer?

4. **IN YOUR OWN WORDS** How can you find the volume of a cone?

5. Describe the intersection of the plane and the cone. Then explain how to find the volume of each section of the solid.

a.

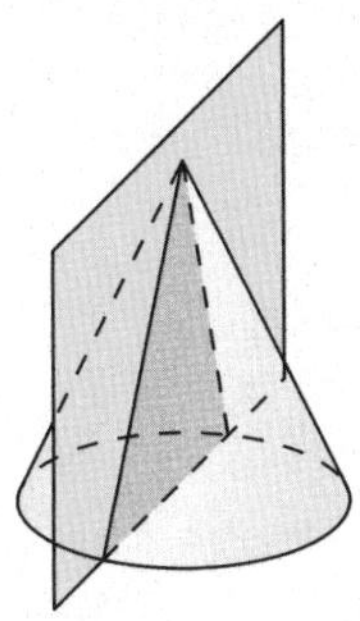

b.

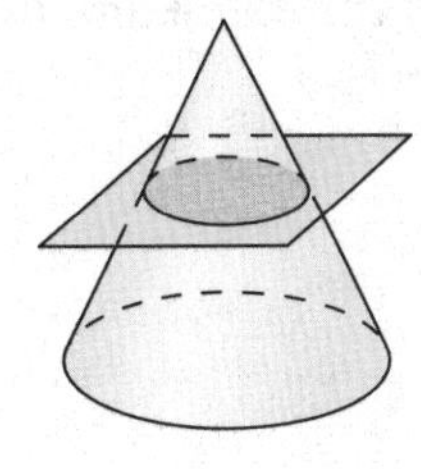

Name Roselynn Sim P.2 Date 06/01/21

8.2 Practice

For use after Lesson 8.2

Find the volume of the cone. Round your answer to the nearest tenth.

1.

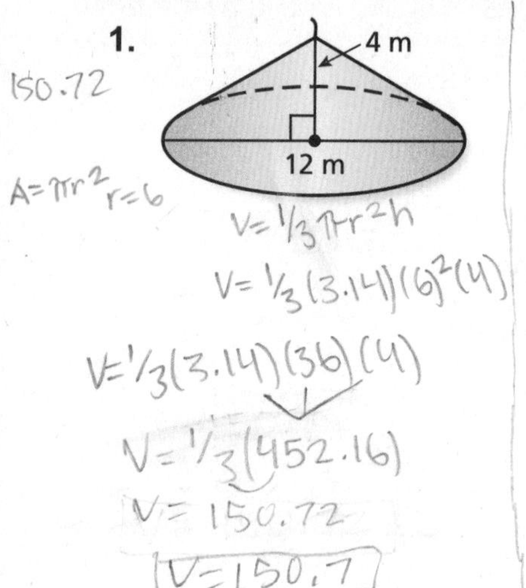

150.72

$A = \pi r^2$ $r = 6$

$V = \frac{1}{3}\pi r^2 h$

$V = \frac{1}{3}(3.14)(6)^2(4)$

$V = \frac{1}{3}(3.14)(36)(4)$

$V = \frac{1}{3}(452.16)$

$V = 150.72$

$V = 150.7$

2.

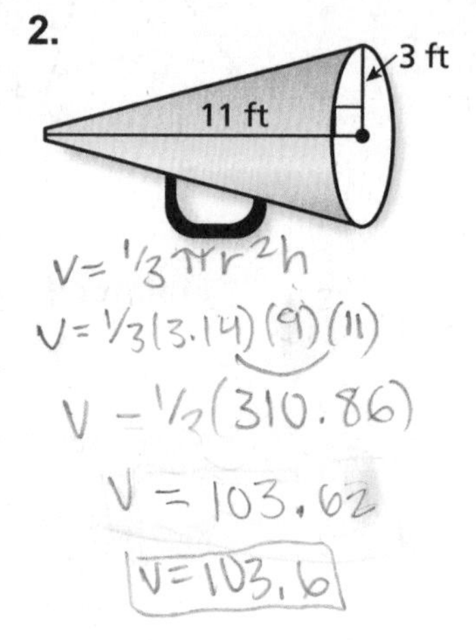

$V = \frac{1}{3}\pi r^2 h$

$V = \frac{1}{3}(3.14)(9)(11)$

$V = \frac{1}{3}(310.86)$

$V = 103.62$

$V = 103.6$

3.

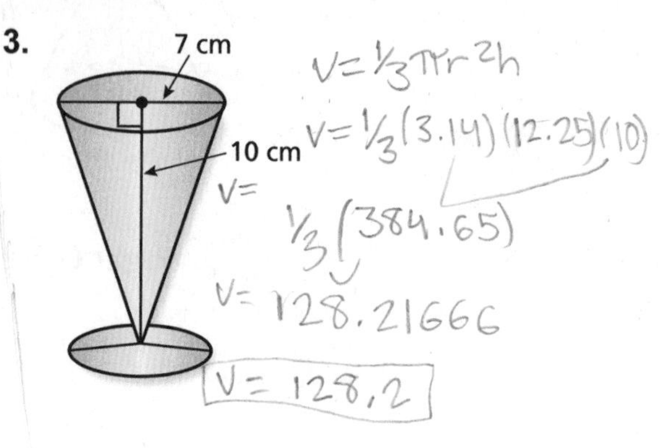

$V = \frac{1}{3}\pi r^2 h$

$V = \frac{1}{3}(3.14)(12.25)(10)$

$V =$

$\frac{1}{3}(384.65)$

$V = 128.21666$

$V = 128.2$

Find the missing dimension of the cone. Round your answer to the nearest tenth.

4. Volume $= 300\pi$ mm^3

300(3.14)

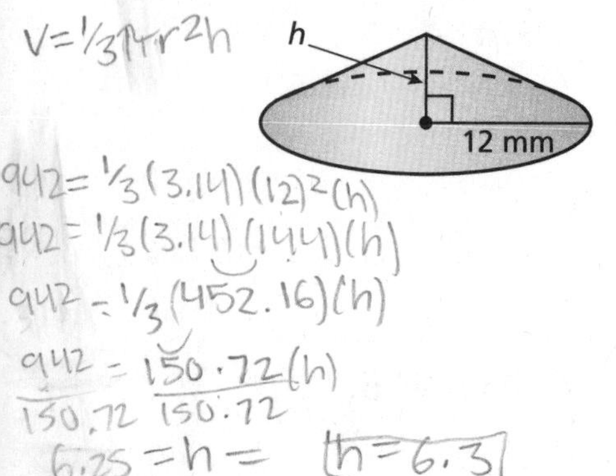

$V = \frac{1}{3}\pi r^2 h$

$942 = \frac{1}{3}(3.14)(12)^2(h)$

$942 = \frac{1}{3}(3.14)(144)(h)$

$942 = \frac{1}{3}(452.16)(h)$

$942 = 150.72(h)$

$\frac{942}{150.72} \quad \frac{150.72}{150.72}$

$6.25 = h = \quad h = 6.3$

5. Volume $= 78.5$ cm^3

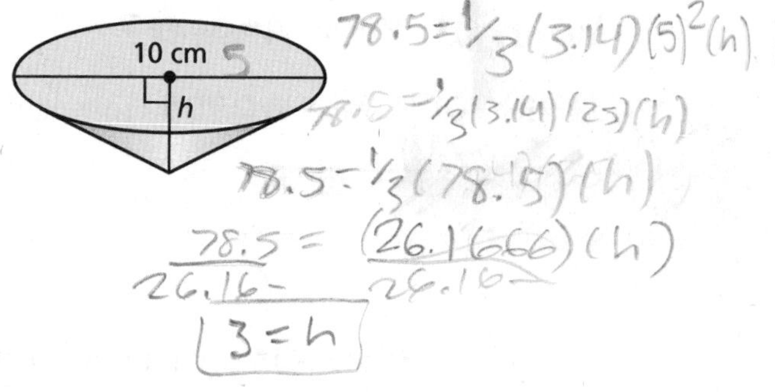

$78.5 = \frac{1}{3}(3.14)(5)^2(h)$

$78.5 = \frac{1}{3}(3.14)(25)(h)$

$78.5 = \frac{1}{3}(78.5)(h)$

$78.5 = (26.1666)(h)$

$\frac{78.5}{26.16} \quad \frac{}{26.16}$

$3 = h$

6. What is the volume of the catch and click cone?

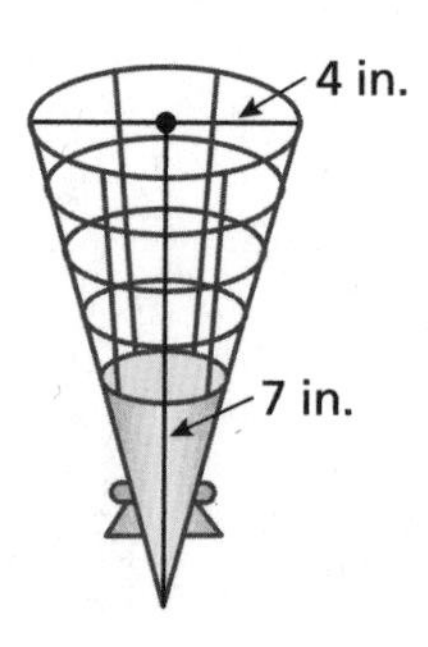

Name Roselynn R Roselynn Date

8.3 Volumes of Spheres

For use with Activity 8.3

Essential Question How can you find the volume of a sphere?

A **sphere** is the set of all points in space that are the same distance from a point called the *center*. The *radius* r is the distance from the center to any point on the sphere.

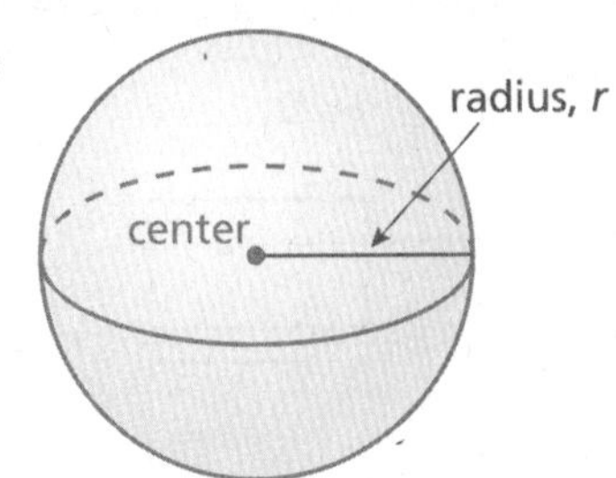

A sphere is different from the other solids you have studied so far because it does not have a base. To discover the volume of a sphere, you can use an activity similar to the one in the previous section.

1 ACTIVITY: Exploring the Volume of a Sphere

Work with a partner. Use a plastic ball similar to the one shown.

- Estimate the diameter and the radius of the ball.

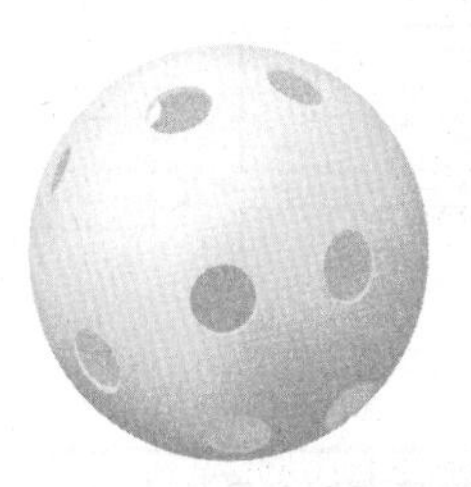

- Use these measurements to draw a net for a cylinder with a diameter and a height equal to the diameter of the ball. How is the height h of the cylinder related to the radius r of the ball? Explain.

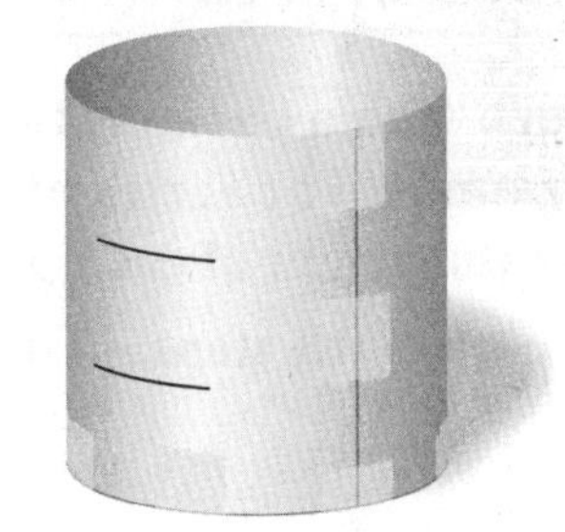

- Cut out the net. Then fold and tape it to form an open cylinder. Make two marks on the cylinder that divide it into thirds, as shown.

- Cover the ball with aluminum foil or tape. Leave one hole open. Fill the ball with rice. Then pour the rice into the cylinder. What fraction of the cylinder is filled with rice?

Name ______________________ Date ________

8.3 Volumes of Spheres (continued)

2 ACTIVITY: Deriving the Formula for the Volume of a Sphere

Work with a partner. Use the results from Activity 1 and the formula for the volume of a cylinder to complete the steps.

$V = \pi r^2 h$ Write formula for volume of a cylinder.

$= \frac{\square}{\square} \pi r^2 h$ Multiply by $\frac{\square}{\square}$ because the volume of a sphere is $\frac{\square}{\square}$ of the volume of the cylinder.

$= \frac{\square}{\square} \pi r^2 \square$ Substitute $\square$ for *h*.

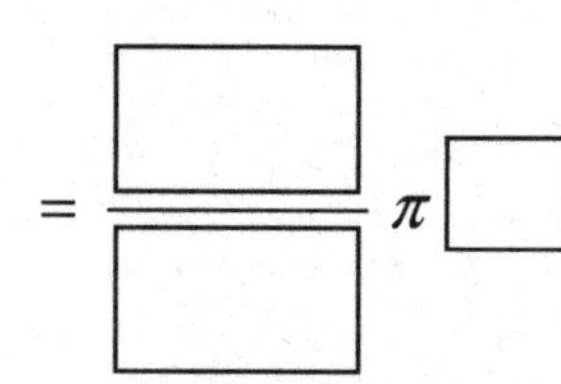

$= \frac{\square}{\square} \pi \square$ Simplify.

3 ACTIVITY: Deriving the Formula for the Volume of a Sphere

Work with a partner. Imagine filling the inside of a sphere with *n* small pyramids. The vertex of each pyramid is at the center of the sphere and the height of each pyramid is approximately equal to *r*, as shown. Complete the steps. (The surface area of a sphere is equal to $4\pi r^2$.)

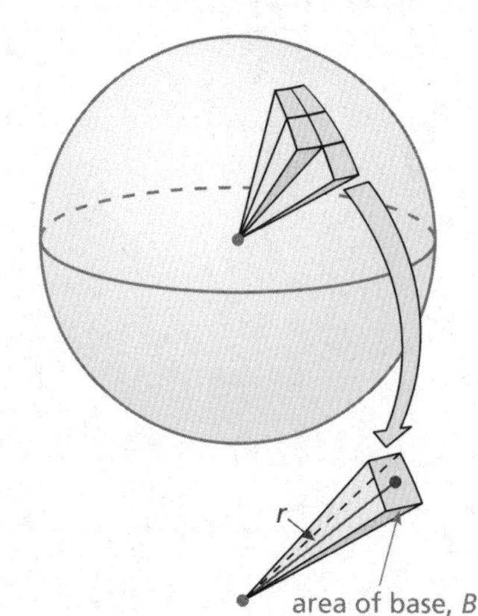

$V = \frac{1}{3}Bh$ Write formula for volume of a pyramid.

$= n\frac{1}{3}B\square$ Multiply by the number of small pyramids *n* and substitute $\square$ for *h*.

$= \frac{1}{3}(4\pi r^2)\square$ $4\pi r^2 \approx n \bullet \square$.

Show how this result is equal to the result in Activity 2.

What Is Your Answer?

4. **IN YOUR OWN WORDS** How can you find the volume of a sphere?

5. Describe the intersection of the plane and the sphere. Then explain how to find the volume of each section of the solid.

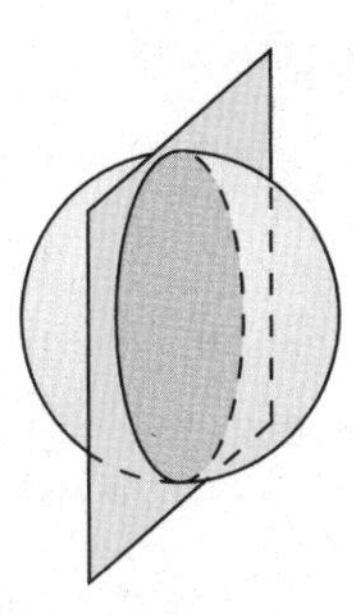

Name ______________________________ Date __________

8.3 Practice
For use after Lesson 8.3

Find the volume of the sphere. Round your answer to the nearest tenth.

1.

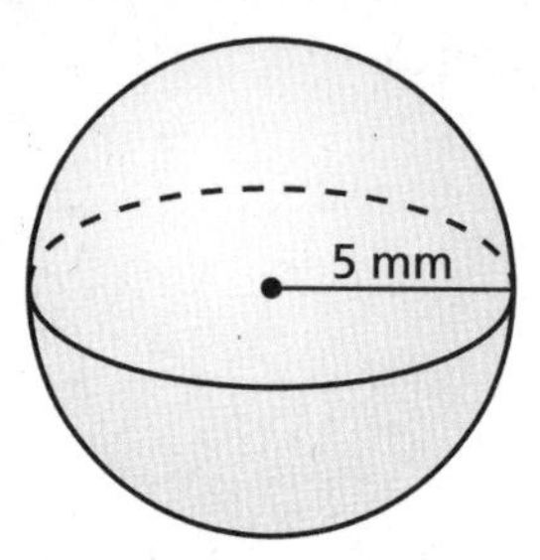

2.

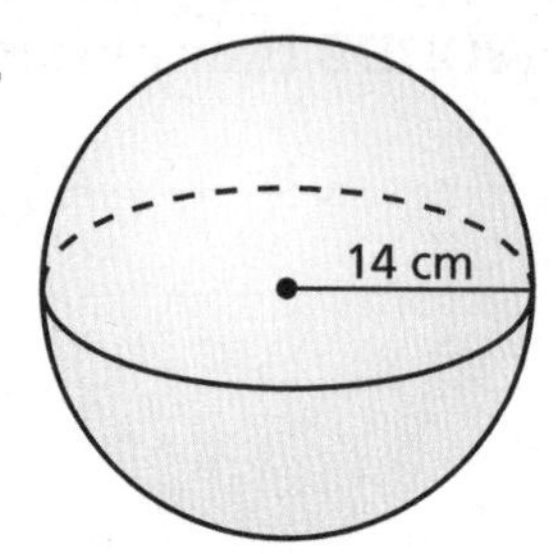

3.

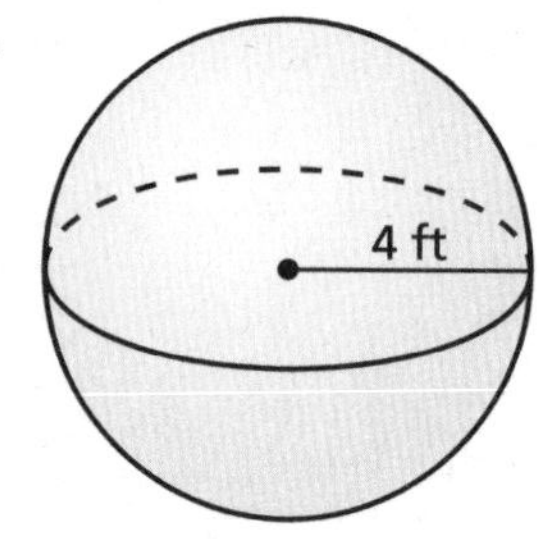

4.

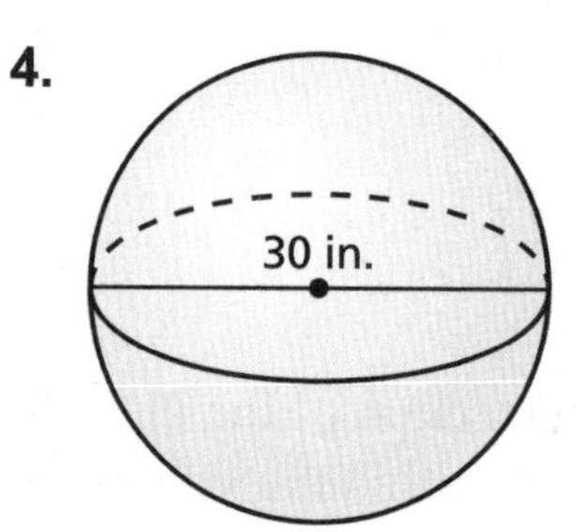

5. Find the volume of the exercise ball. Round your answer to the nearest tenth.

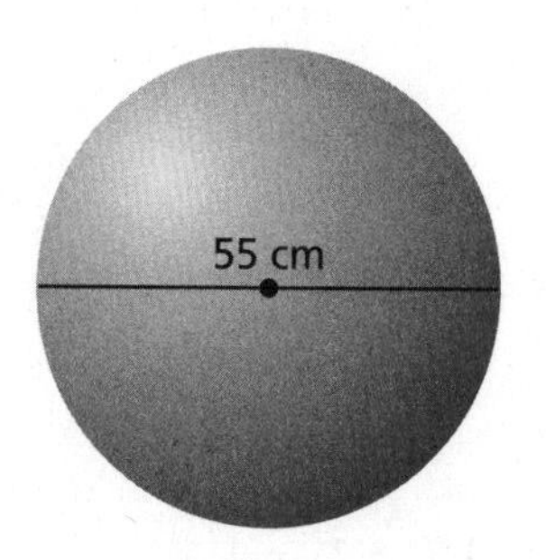

Name____ Date__________

8.4 Surface Areas and Volumes of Similar Solids

For use with Activity 8.4

Essential Question When the dimensions of a solid increase by a factor of k, how does the surface area change? How does the volume change?

1 ACTIVITY: Comparing Surface Areas and Volumes

Work with a partner. Complete the table. Describe the pattern. Are the dimensions proportional? Explain your reasoning.

a.

Radius	1	1	1	1	1
Height	1	2	3	4	5
Surface Area					
Volume					

Name ______________________________ Date __________

b.

Radius	1	2	3	4	5
Height	1	2	3	4	5
Surface Area					
Volume					

2 ACTIVITY: Comparing Surface Areas and Volumes

Work with a partner. Complete the table. Describe the pattern. Are the dimensions proportional? Explain.

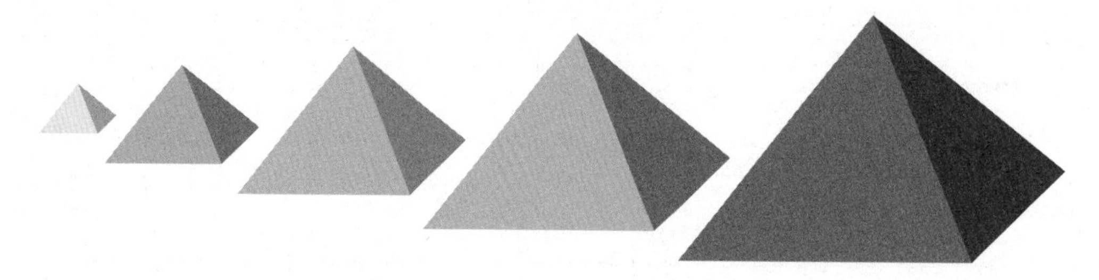

Base Side	6	12	18	24	30
Height	4	8	12	16	20
Slant Height	5	10	15	20	25
Surface Area					
Volume					

What Is Your Answer?

3. **IN YOUR OWN WORDS** When the dimensions of a solid increase by a factor of k, how does the surface area change?

4. **IN YOUR OWN WORDS** When the dimensions of a solid increase by a factor of k, how does the volume change?

5. **REPEATED REASONING** All the dimensions of a prism increase by a factor of 5.

   a. How many times greater is the surface area? Explain.

5	10	25	125

   b. How many times greater is the volume? Explain.

5	10	25	125

Name ______________________________ Date __________

8.4 Practice
For use after Lesson 8.4

Determine whether the solids are similar.

1.

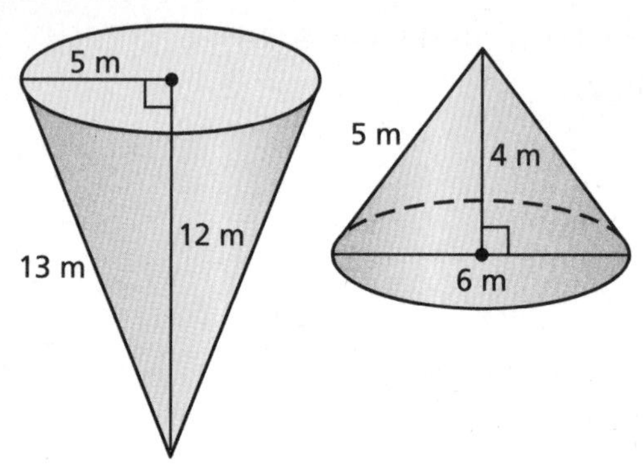

2.

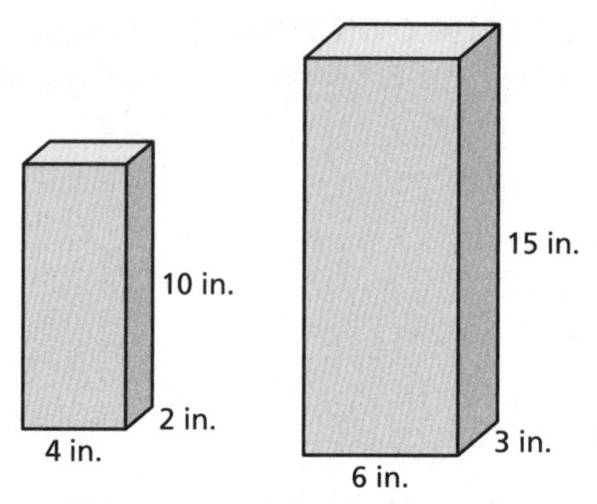

The solids are similar. Find the missing dimension(s).

3.

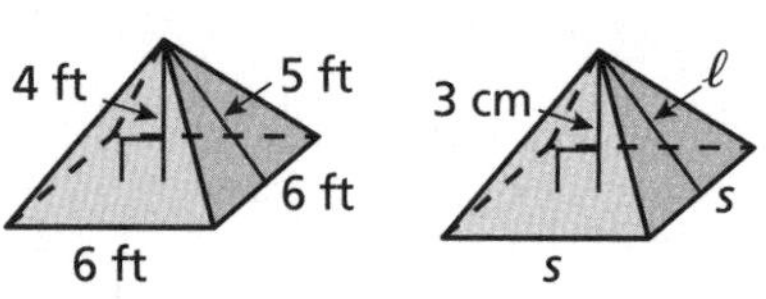

4.

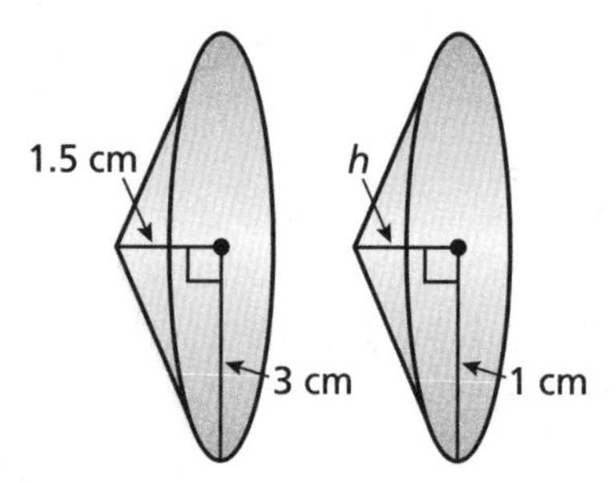

The solids are similar. Find the surface area *S* or volume *V* of the shaded solid.

5.

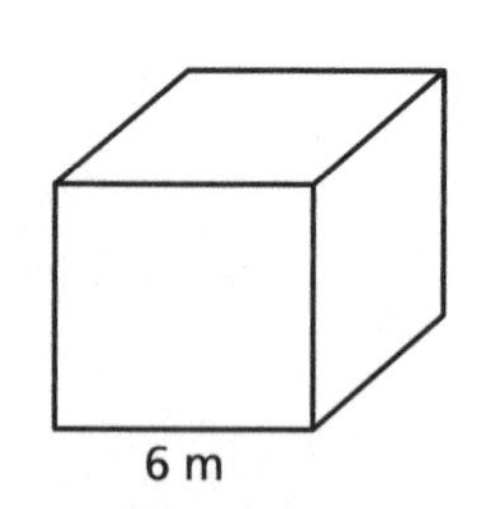

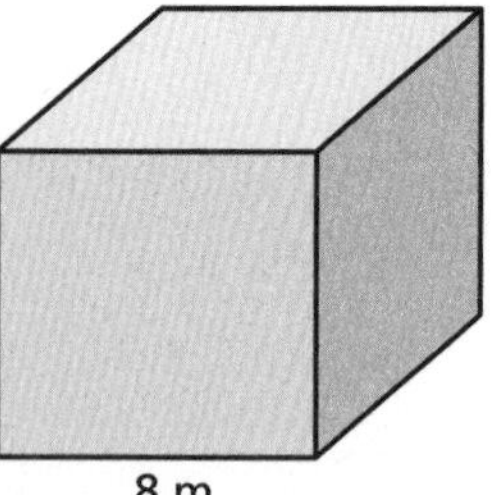

Surface Area = 198 m^2

6.

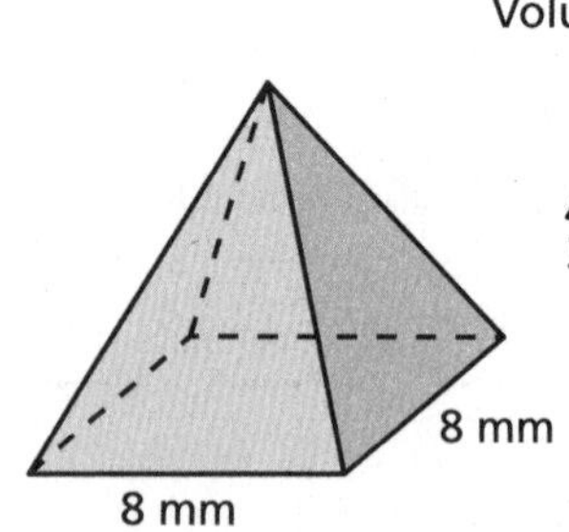

Volume = 54 mm^3

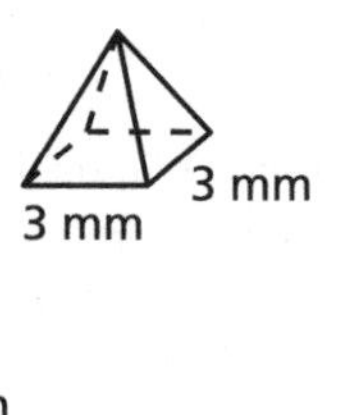

Name____ Date__________

Chapter 9 Fair Game Review

Plot the ordered pair in a coordinate plane. Describe the location of the point.

1. (2, 1)

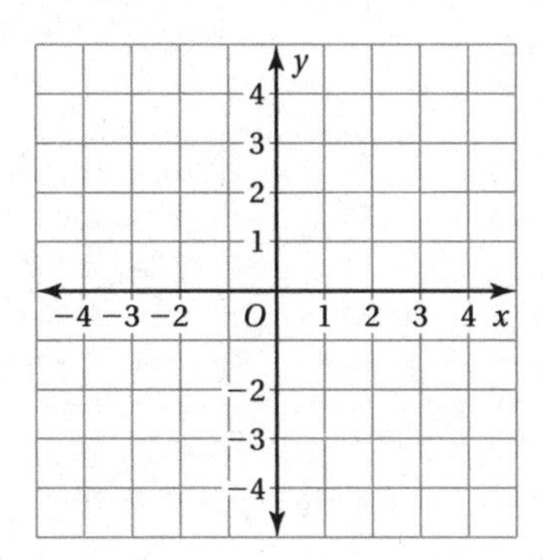

2. (−3, 3)

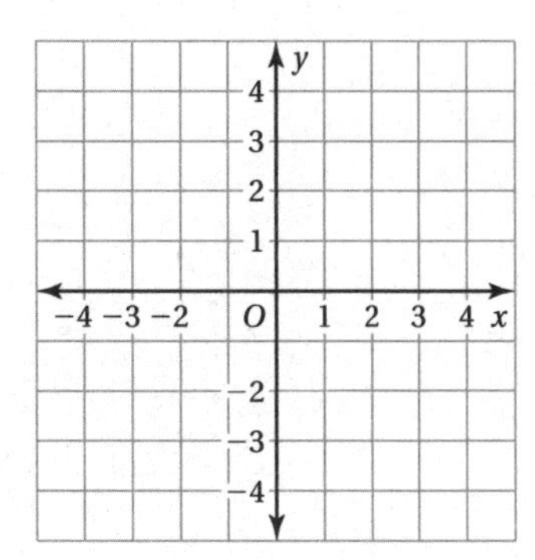

3. (4, −2)

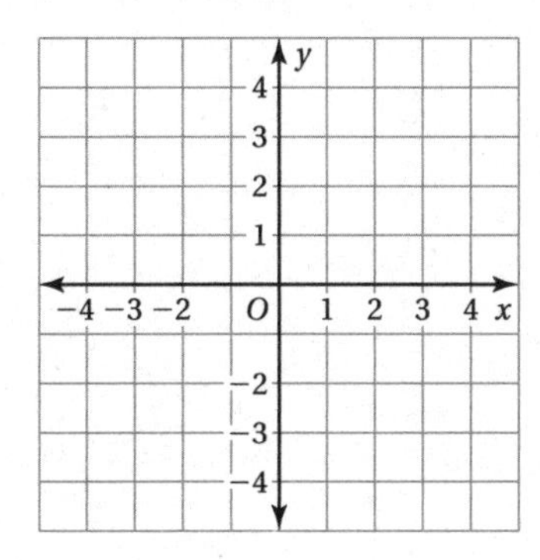

4. (−1, −1)

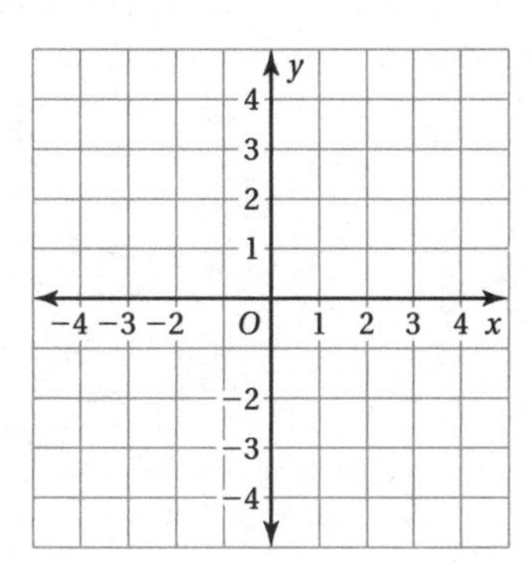

5. Describe the location of the vertices of the triangle.

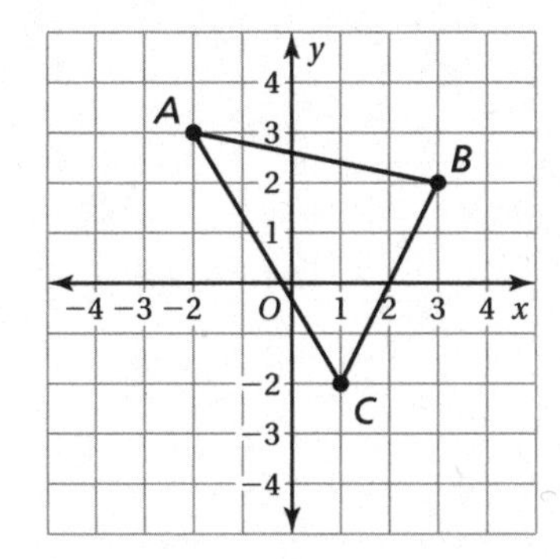

Name ________________________ Date ________

Chapter 9

Fair Game Review (continued)

Write in slope-intercept form an equation of the line that passes through the given points.

6. $(-2, -2), (1, 7)$

7. $(5, -1), (-5, 11)$

8. $(-20, -8), (5, 12)$

9. $(6, -11), (-3, 1)$

10. $(-1, -3), (2, 6)$

11. $(-3, 6), (4, -8)$

Name____ Date__________

9.1 Scatter Plots
For use with Activity 9.1

Essential Question How can you construct and interpret a scatter plot?

1 ACTIVITY: Constructing a Scatter Plot

Work with a partner. The weights x (in ounces) and circumferences C (in inches) of several sports balls are shown.

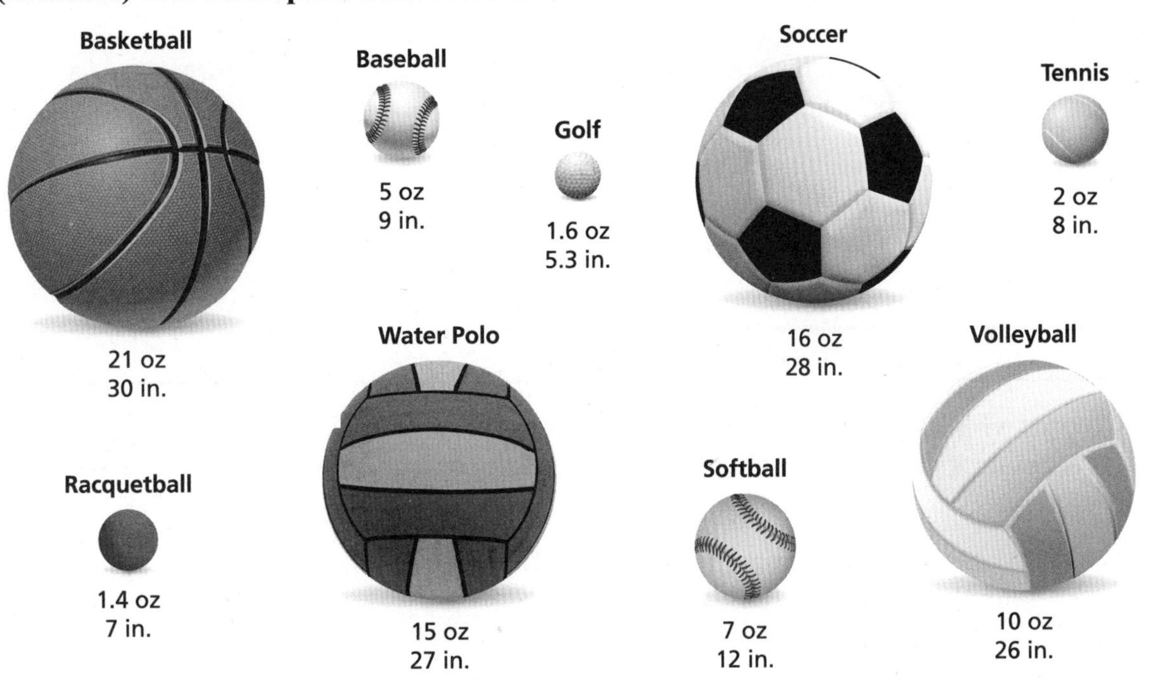

a. Choose a scale for the horizontal axis and the vertical axis of the coordinate plane shown.

b. Write the weight x and circumference C of each ball as an ordered pair. Then plot the ordered pairs in the coordinate plane.

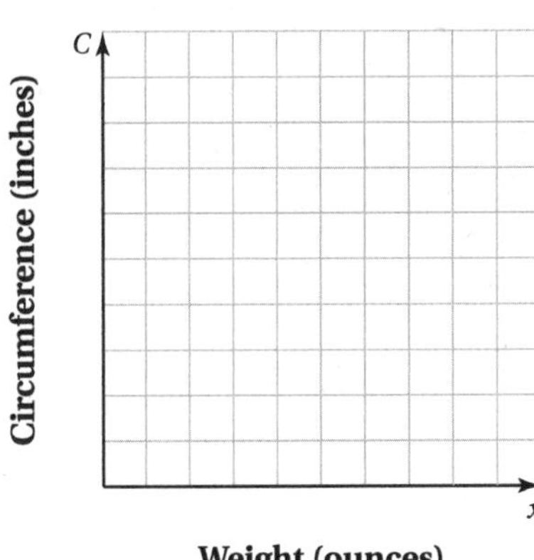

c. Describe the relationship between weight and circumference. Are any of the points close together?

d. In general, do you think you can describe this relationship as *positive* or *negative*? *linear* or *nonlinear*? Explain.

e. A bowling ball has a weight of 225 ounces and a circumference of 27 inches. Describe the location of the ordered pair that represents this data point in the coordinate plane. How does this point compare to the others? Explain your reasoning.

2 ACTIVITY: Constructing a Scatter Plot

Work with a partner. The table shows the number of absences and the final grade for each student in a sample.

Absences	Final Grade
0	95
3	88
2	90
5	83
7	79
9	70
4	85
1	94
10	65
8	75

a. Write the ordered pairs from the table. Then plot them in the coordinate plane.

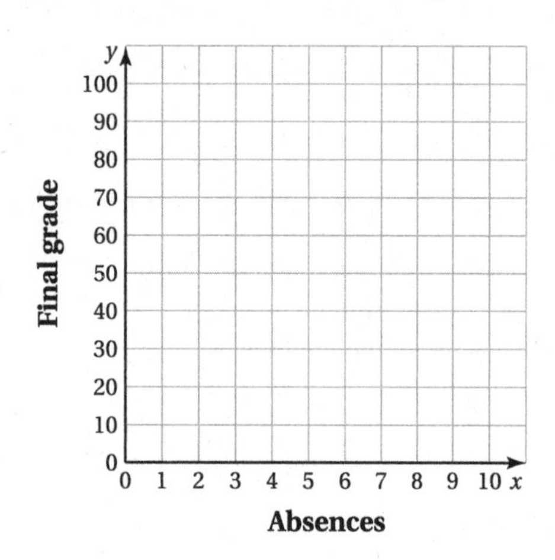

b. Describe the relationship between absences and final grade. How is this relationship similar to the relationship between weight and circumference in Activity 1? How is it different?

c. MODELING A student has been absent 6 days. Use the data to predict the student's final grade. Explain how you found your answer.

3 ACTIVITY: Identifying Scatter Plots

Work with a partner. Match the data sets with the most appropriate scatter plot. Explain your reasoning.

a. month of birth and birth weight for infants at a day care

b. quiz score and test score of each student in a class

c. age and value of laptop computers

i. y, x

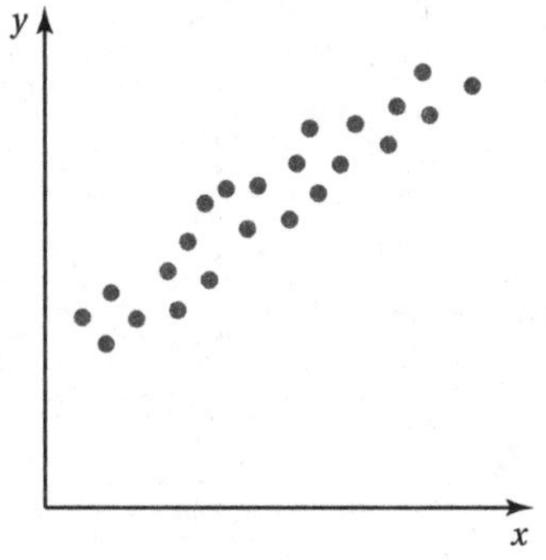

ii. y, x

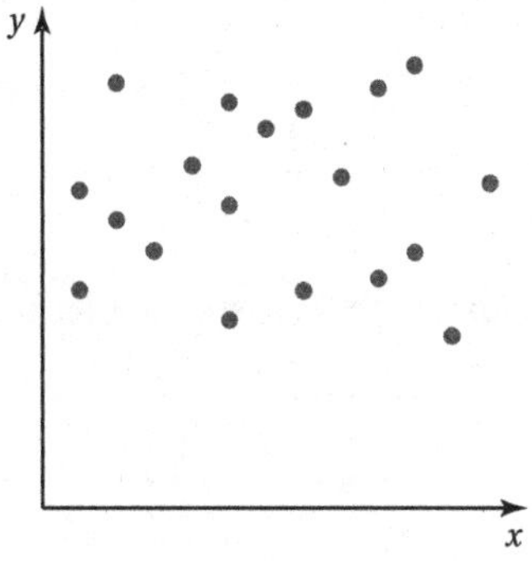

iii. y, x

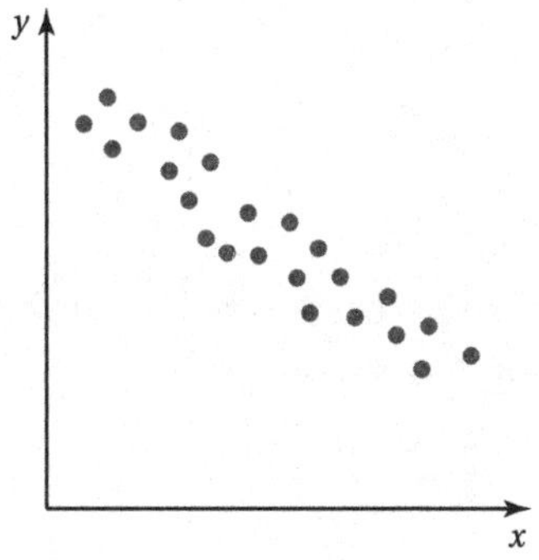

What Is Your Answer?

4. How would you define the term *scatter plot*?

5. IN YOUR OWN WORDS How can you construct and interpret a scatter plot?

Name ______________________________ Date ________

9.1 Practice
For use after Lesson 9.1

1. The scatter plot shows the participation in a bowling league over eight years.

a. About how many people were in the league in 2008?

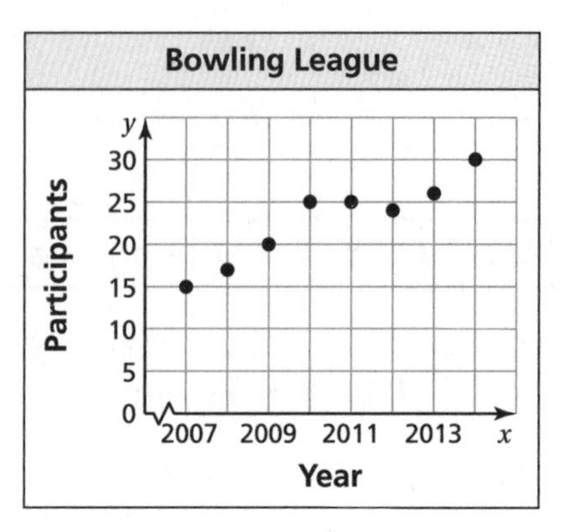

b. Describe the relationship shown by the data.

Describe the relationship between the data. Identify any outliers, gaps, or clusters.

2.

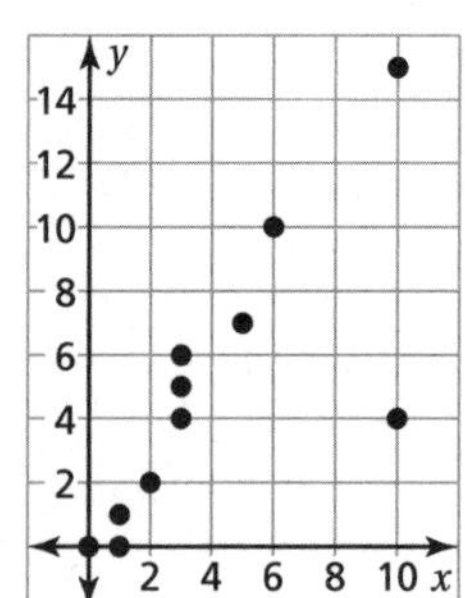

3.

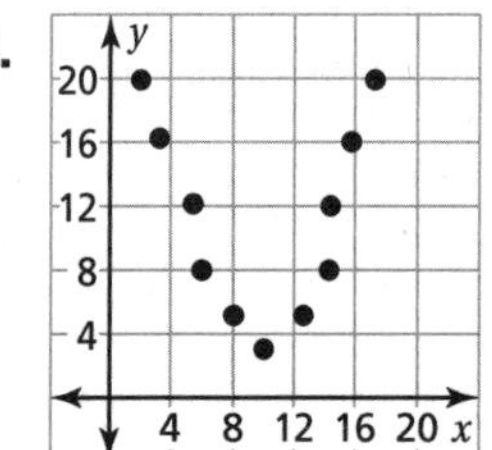

4.

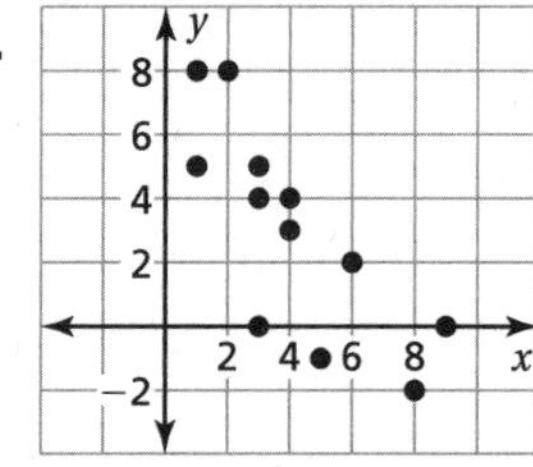

Name____ Date__________

9.2 Lines of Fit

For use with Activity 9.2

Essential Question How can you use data to predict an event?

1 ACTIVITY: Representing Data by a Linear Equation

Work with a partner. You have been working on a science project for 8 months. Each month, you measured the length of a baby alligator.

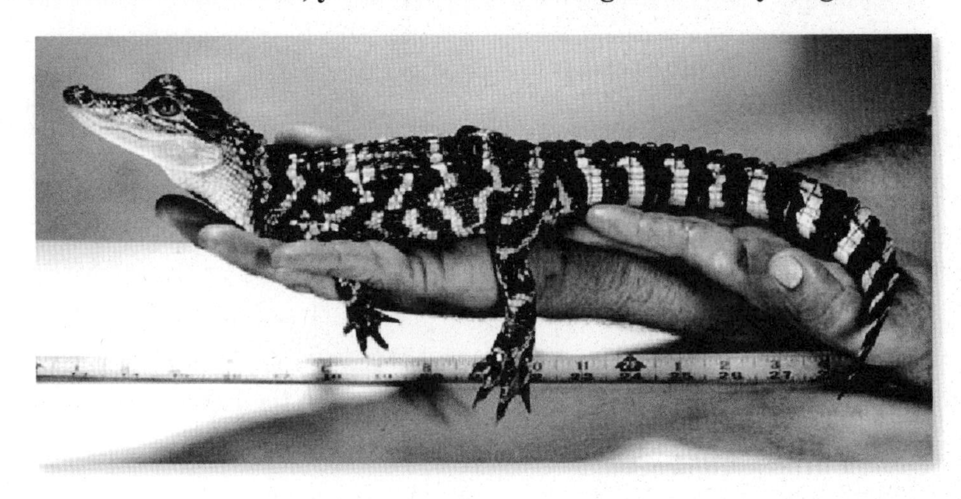

The table shows your measurements.

September → Month 0; April → Month 7

Month, x	0	1	2	3	4	5	6	7
Length (in.), y	22.0	22.5	23.5	25.0	26.0	27.5	28.5	29.5

Use the following steps to predict the baby alligator's length next September.

a. Graph the data in the table.

b. Draw a line that you think best approximates the points.

c. Write an equation for your line.

d. **MODELING** Use the equation to predict the baby alligator's length next September.

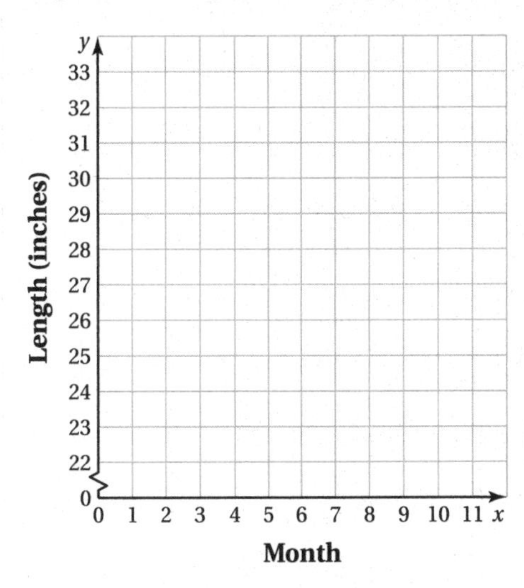

Name ______________________ Date __________

2 ACTIVITY: Representing Data by a Linear Equation

Work with a partner. You are a biologist and study bat populations.

You are asked to predict the number of bats that will be living in an abandoned mine in 3 years.

To start, you find the number of bats that have been living in the mine during the past 8 years.

The table shows the results of your research.

7 years ago → 0; this year → 7

Year, x	0	1	2	3	4	5	6	7
Bats (thousands), y	327	306	299	270	254	232	215	197

Use the following steps to predict the number of bats that will be living in the mine after 3 years.

a. Graph the data in the table.

b. Draw a line that you think best approximates the points.

c. Write an equation for your line.

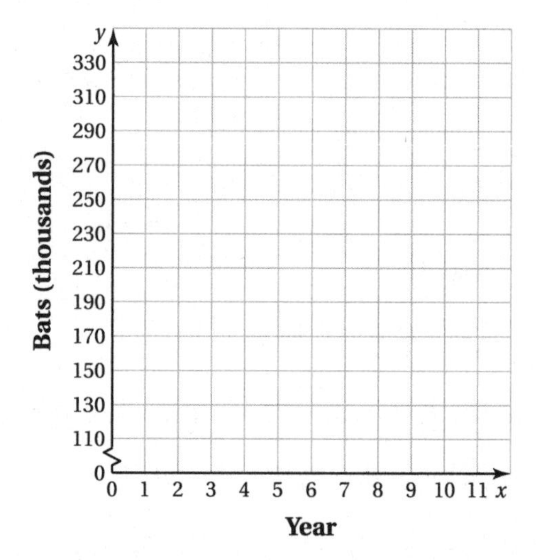

d. **MODELING** Use the equation to predict the number of bats in 3 years.

What Is Your Answer?

3. **IN YOUR OWN WORDS** How can you use data to predict an event?

4. **MODELING** Use the Internet or some other reference to find data that appear to have a linear pattern. List the data in a table and graph the data. Use an equation that is based on the data to predict a future event.

Name ______________________________ Date __________

9.2 Practice

For use after Lesson 9.2

1. The table shows the money you owe to pay off a credit card bill over five months.

a. Make a scatter plot of the data and draw a line of fit.

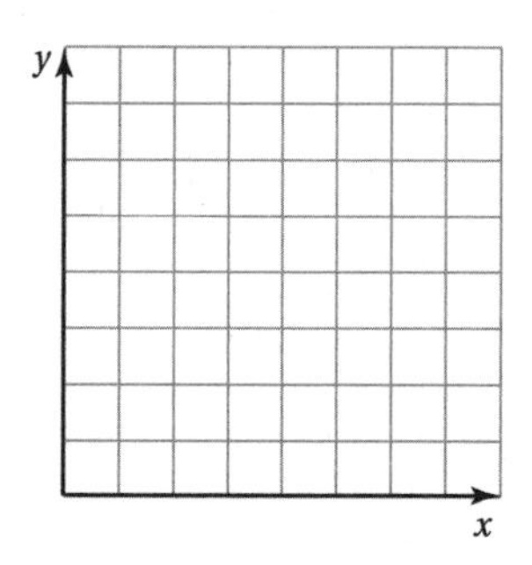

Months, x	Money owed (dollars), y
1	1200
2	1000
3	850
4	600
5	410

b. Write an equation of the line of fit.

c. Interpret the slope and y-intercept of the line of fit.

d. Predict the amount of money you will owe in six months.

Use a graphing calculator to find an equation of the line of best fit. Identify and interpret the correlation coefficient.

2.

x	−8	−6	−4	−2	0	2	4	6	8
y	10	7	1	0	−3	−5	−4	−14	−11

3.

x	1	2	3	4	5	6	7	8
y	8	6	4	2	0	2	4	6

Name______________________________ Date__________

9.3 Two-Way Tables

For use with Activity 9.3

Essential Question How can you read and make a two-way table?

Two categories of data can be displayed in a *two-way table.*

1 ACTIVITY: Reading a Two-Way Table

Work with a partner. You are the manager of a sports shop. The two-way table shows the numbers of soccer T-shirts that your shop has left in stock at the end of the season.

		T-Shirt Size					
		S	M	L	XL	XXL	Total
Color	Blue/White	5	4	1	0	2	
	Blue/Gold	3	6	5	2	0	
	Red/White	4	2	4	1	3	
	Black/White	3	4	1	2	1	
	Black/Gold	5	2	3	0	2	
	Total						65

a. Complete the totals for the rows and columns.

b. Are there any black-and-gold XL T-shirts in stock? Justify your answer.

c. The numbers of T-shirts you ordered at the beginning of the season are shown below. Complete the two-way table.

		T-Shirt Size					
		S	M	L	XL	XXL	Total
Color	Blue/White	5	6	7	6	5	
	Blue/Gold	5	6	7	6	5	
	Red/White	5	6	7	6	5	
	Black/White	5	6	7	6	5	
	Black/Gold	5	6	7	6	5	
	Total						

d. **REASONING** How would you alter the numbers of T-shirts you order for next season? Explain your reasoning.

2 ACTIVITY: Analyzing Data

Work with a partner. The three-dimensional two-way table shows information about the numbers of hours students at a high school work at part-time jobs during the school year.

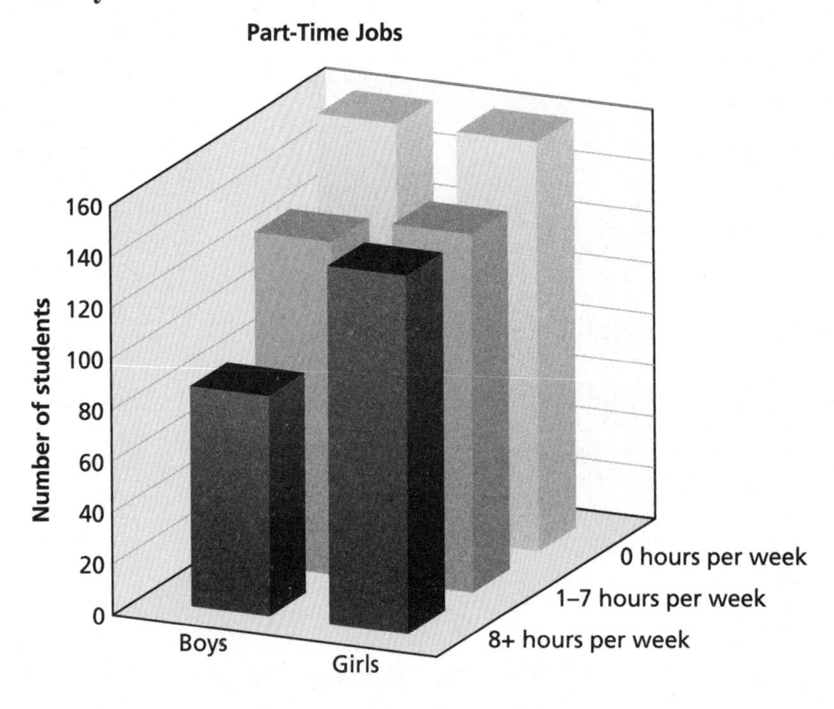

a. Make a two-way table showing the data. Use estimation to find the entries in your table.

b. Write two observations you can make that summarize the data in your table.

c. **REASONING** A newspaper article claims that more boys than girls drop out of high school to work full-time. Do the data support this claim? Explain your reasoning.

What Is Your Answer?

3. **IN YOUR OWN WORDS** How can you read and make a two-way table?

4. Find a real-life data set that can be represented by a two-way table. Then make a two-way table for the data set.

Name ______________________________ Date __________

9.3 Practice

For use after Lesson 9.3

1. You randomly survey students in a school about whether they got the flu after receiving a flu shot. The results of the survey are shown in the two-way table.

   a. How many of the students in the survey received a flu shot and still got the flu?

   b. Find and interpret the marginal frequencies for the survey.

		Flu Shot		
		Yes	No	Total
Flu	Yes	8	13	
	No	27	32	
	Total			

2. You randomly survey students in a school about whether they eat breakfast at home or at school.

   Grade 6 Students: 28 eat breakfast at home, 12 eat breakfast at school

   Grade 7 Students: 15 eat breakfast at home, 15 eat breakfast at school

   Grade 8 Students: 9 eat breakfast at home, 21 eat breakfast at school

   a. Make a two-way table that includes the marginal frequencies.

   b. For each grade level, what percent of the students in the survey eat breakfast at home? eat breakfast at school? Organize the results in a two-way table. Explain what one of the entries represents.

Name____ Date__________

Choosing a Data Display
For use with Activity 9.4

Essential Question How can you display data in a way that helps you make decisions?

1 ACTIVITY: Displaying Data

Work with a partner. Analyze and display each data set in a way that best describes the data. Explain your choice of display.

a. ROADKILL A group of schools in New England participated in a 2-month study and reported 3962 dead animals.

Birds: 307
Mammals: 2746
Amphibians: 145
Reptiles: 75
Unknown: 689

b. BLACK BEAR ROADKILL The data below show the numbers of black bears killed on a state's roads from 1993 to 2012.

1993:	30	2003:	74
1994:	37	2004:	88
1995:	46	2005:	82
1996:	33	2006:	109
1997:	43	2007:	99
1998:	35	2008:	129
1999:	43	2009:	111
2000:	47	2010:	127
2001:	49	2011:	141
2002:	61	2012:	135

Name ______________________ Date __________

c. RACCOON ROADKILL A 1-week study along a 4-mile section of road found the following weights (in pounds) of raccoons that had been killed by vehicles.

13.4	14.8	17.0	12.9	21.3	21.5	16.8	14.8
15.2	18.7	18.6	17.2	18.5	9.4	19.4	15.7
14.5	9.5	25.4	21.5	17.3	19.1	11.0	12.4
20.4	13.6	17.5	18.5	21.5	14.0	13.9	19.0

d. What do you think can be done to minimize the number of animals killed by vehicles?

2 ACTIVITY: Statistics Project

ENDANGERED SPECIES PROJECT Use the Internet or some other reference to write a report about an animal species that is (or has been) endangered. Include graphical displays of the data you have gathered.

Sample: Florida Key Deer In 1939, Florida banned the hunting of Key deer. The numbers of Key deer fell to about 100 in the 1940s.

About half of Key deer deaths are due to vehicles.

In 1947, public sentiment was stirred by 11-year-old Glenn Allen from Miami. Allen organized Boy Scouts and others in a letter-writing campaign that led to the establishment of the National Key Deer Refuge in 1957. The approximately 8600-acre refuge includes 2280 acres of designated wilderness.

One of two Key deer wildlife underpasses on Big Pine Key.

The Key Deer Refuge has increased the population of Key deer. A recent study estimated the total Key deer population to be approximately 800.

What Is Your Answer?

3. **IN YOUR OWN WORDS** How can you display data in a way that helps you make decisions? Use the Internet or some other reference to find examples of the following types of data displays.

- Bar graph
- Circle graph
- Scatter plot
- Stem-and-leaf plot
- Box-and-whisker plot

Name ______________________________ Date __________

9.4 Practice

For use after Lesson 9.4

Choose an appropriate data display for the situation. Explain your reasoning.

1. the number of people that donated blood over the last 5 years

2. percent of class participating in school clubs

Explain why the data display is misleading.

3.

Temperature in Naples

Temperature (degrees Fahrenheit)

y: 70, 75, 80, 85, 90, 95

x: 0, 1, 2, 3, 4, 5, 6, 7

Days

4.

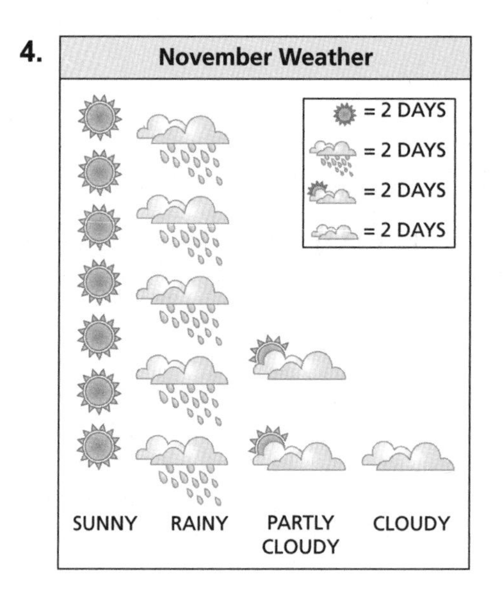

5. A team statistician wants to use a data display to show the points scored per game during the season. Choose an appropriate data display for the situation. Explain your reasoning.

Name____ Date__________

Chapter 10 Fair Game Review

Evaluate the expression.

1. $2 + 1 \bullet 4^2 - 12 \div 3$

2. $8^2 \div 16 \bullet 2 - 5$

3. $7(9 - 3) + 6^2 \bullet 10 - 8$

4. $3 \bullet 5 - 10 + 9(2 + 1)^2$

5. $8(6 + 5) - (9^2 + 3) \div 7$

6. $5[3(12 - 8)] - 6 \bullet 8 + 2^2$

7. $4 + 4 + 5 \times 2 \times 5 + (3 + 3 + 3) \times 6 \times 6 + 2 + 2$

a. Evaluate the expression.

b. Rewrite the expression using what you know about order of operations. Then evaluate.

Name ______________________________ Date ________

Fair Game Review (continued)

Find the product or quotient.

8. $3.92 \bullet 0.6$

9. $0.78 \bullet 0.13$

10. $\begin{array}{r} 5.004 \\ \times \quad 1.2 \\ \hline \end{array}$

11. $6.3 \div 0.7$

12. $2.25 \div 1.5$

13. $0.003\overline{)8.1}$

14. Grapes cost $1.98 per pound. You buy 3.5 pounds of grapes. How much do you pay for the grapes?

Name________________________________ Date__________

10.1 Exponents

For use with Activity 10.1

Essential Question How can you use exponents to write numbers?

The expression 3^5 is called a *power*. The *base* is 3. The *exponent* is 5.

base → 3^5 ← exponent

1 ACTIVITY: Using Exponent Notation

Work with a partner.

a. Complete the table.

Power	Repeated Multiplication Form	Value
$(-3)^1$		
$(-3)^2$		
$(-3)^3$		
$(-3)^4$		
$(-3)^5$		
$(-3)^6$		
$(-3)^7$		

b. REPEATED REASONING Describe what is meant by the expression $(-3)^n$.

How can you find the value of $(-3)^n$?

Name ____________________ Date ________

10.1 Exponents (continued)

2 ACTIVITY: Using Exponent Notation

Work with a partner.

a. The cube at the right has $3 in each of its small cubes. Write a power that represents the total amount of money in the large cube.

b. Evaluate the power to find the total amount of money in the large cube.

3 ACTIVITY: Writing Powers as Whole Numbers

Work with a partner. Write each distance as a whole number. Which numbers do you know how to write in words? For instance, in words, 10^3 is equal to *one thousand*.

a. 10^{26} meters: diameter of observable universe

b. 10^{21} meters: diameter of Milky Way galaxy

c. 10^{16} meters: diameter of solar system

d. 10^{7} meters: diameter of Earth

e. 10^{6} meters: length of Lake Erie shoreline

f. 10^{5} meters: width of Lake Erie

4 ACTIVITY: Writing a Power

Work with a partner. Write the number of kits, cats, sacks, and wives as a power.

As I was going to St. Ives
I met a man with seven wives
Each wife had seven sacks
Each sack had seven cats
Each cat had seven kits
Kits, cats, sacks, wives
How many were going to St. Ives?

Nursery Rhyme, 1730

What Is Your Answer?

5. **IN YOUR OWN WORDS** How can you use exponents to write numbers? Give some examples of how exponents are used in real life.

Name ______________________________ Date __________

10.1 Practice

For use after Lesson 10.1

Write the product using exponents.

1. $4 \bullet 4 \bullet 4 \bullet 4 \bullet 4$

2. $\left(-\frac{1}{8}\right) \bullet \left(-\frac{1}{8}\right) \bullet \left(-\frac{1}{8}\right)$

3. $5 \bullet 5 \bullet (-x) \bullet (-x) \bullet (-x) \bullet (-x)$

4. $9 \bullet 9 \bullet y \bullet y \bullet y \bullet y \bullet y \bullet y$

Evaluate the expression.

5. 10^3

6. $(-7)^4$

7. $-\left(\frac{1}{6}\right)^5$

8. $3 + 6 \bullet (-5)^2$

9. $\left|-\frac{1}{3}\left(1^{10} + 9 - 2^3\right)\right|$

10. A foam toy is 2 inches wide. It doubles in size for every minute it is in water. Write an expression for the width of the toy after 5 minutes. What is the width after 5 minutes?

Name______________________________ Date__________

10.2 Product of Powers Property

For use with Activity 10.2

Essential Question How can you use inductive reasoning to observe patterns and write general rules involving properties of exponents?

1 ACTIVITY: Finding Products of Powers

Work with a partner.

a. Complete the table.

Product	Repeated Multiplication Form	Power
$2^2 \bullet 2^4$		
$(-3)^2 \bullet (-3)^4$		
$7^3 \bullet 7^2$		
$5.1^1 \bullet 5.1^6$		
$(-4)^2 \bullet (-4)^2$		
$10^3 \bullet 10^5$		
$\left(\frac{1}{2}\right)^5 \bullet \left(\frac{1}{2}\right)^5$		

b. INDUCTIVE REASONING Describe the pattern in the table. Then write a *general rule* for multiplying two powers that have the same base.

$$a^m \bullet a^n = a^{\_\_\_}$$

c. Use your rule to simplify the products in the first column of the table above. Does your rule give the results in the third column?

d. Most calculators have *exponent* keys that are used to evaluate powers. Use a calculator with an exponent key to evaluate the products in part (a).

2 ACTIVITY: Writing a Rule for Powers of Powers

Work with a partner. Write the expression as a single power. Then write a *general rule* for finding a power of a power.

a. $\left(3^2\right)^3 =$

b. $\left(2^2\right)^4 =$

c. $\left(7^3\right)^2 =$

d. $\left(y^3\right)^3 =$

e. $\left(x^4\right)^2 =$

3 ACTIVITY: Writing a Rule for Powers of Products

Work with a partner. Write the expression as the product of two powers. Then write a *general rule* for finding a power of a product.

a. $(2 \bullet 3)^3 =$

b. $(2 \bullet 5)^2 =$

c. $(5 \bullet 4)^3 =$

d. $(6a)^4 =$

e. $(3x)^2 =$

Name____ Date__________

4 ACTIVITY: The Penny Puzzle

Work with a partner.

- The rows y and columns x of a chess board are numbered as shown.
- Each position on the chess board has a stack of pennies. (Only the first row is shown.)
- The number of pennies in each stack is $2^x \bullet 2^y$.

a. How many pennies are in the stack in location $(3, 5)$?

b. Which locations have 32 pennies in their stacks?

c. How much money (in dollars) is in the location with the tallest stack?

d. A penny is about 0.06 inch thick. About how tall (in inches) is the tallest stack?

What Is Your Answer?

5. **IN YOUR OWN WORDS** How can you use inductive reasoning to observe patterns and write general rules involving properties of exponents?

Name ______________________________ Date ________

10.2 Practice

For use after Lesson 10.2

Simplify the expression. Write your answer as a power.

1. $(-6)^5 \bullet (-6)^4$

2. $x^1 \bullet x^9$

3. $\left(\frac{4}{5}\right)^3 \bullet \left(\frac{4}{5}\right)^{12}$

4. $(-1.5)^{11} \bullet (-1.5)^{11}$

5. $\left(y^{10}\right)^{20}$

6. $\left(\left(-\frac{2}{9}\right)^8\right)^7$

Simplify the expression.

7. $(2a)^6$

8. $(-4b)^4$

9. $\left(-\frac{9}{10}p\right)^2$

10. $(xy)^{15}$

11. $10^5 \bullet 10^3 - \left(10^1\right)^8$

12. $7^2\left(7^4 \bullet 7^4\right)$

13. The surface area of the Sun is about $4 \times 3.141 \times \left(7 \times 10^5\right)^2$ square kilometers. Simplify the expression.

Name______________________________ Date__________

10.3 Quotient of Powers Property

For use with Activity 10.3

Essential Question How can you divide two powers that have the same base?

1 ACTIVITY: Finding Quotients of Powers

Work with a partner.

a. Complete the table.

Quotient	Repeated Multiplication Form	Power
$\frac{2^4}{2^2}$		
$\frac{(-4)^5}{(-4)^2}$		
$\frac{7^7}{7^3}$		
$\frac{8.5^9}{8.5^6}$		
$\frac{10^8}{10^5}$		
$\frac{3^{12}}{3^4}$		
$\frac{(-5)^7}{(-5)^5}$		
$\frac{11^4}{11^1}$		

b. INDUCTIVE REASONING Describe the pattern in the table. Then write a rule for dividing two powers that have the same base.

$$\frac{a^m}{a^n} = a^{\_\_\_}$$

Name ______________________________ Date __________

c. Use your rule to simplify the quotients in the first column of the table on the previous page. Does your rule give the results in the third column?

2 ACTIVITY: Comparing Volumes

Work with a partner.

How many of the smaller cubes will fit inside the larger cube? Record your results in the table on the next page. Describe the pattern in the table.

a.

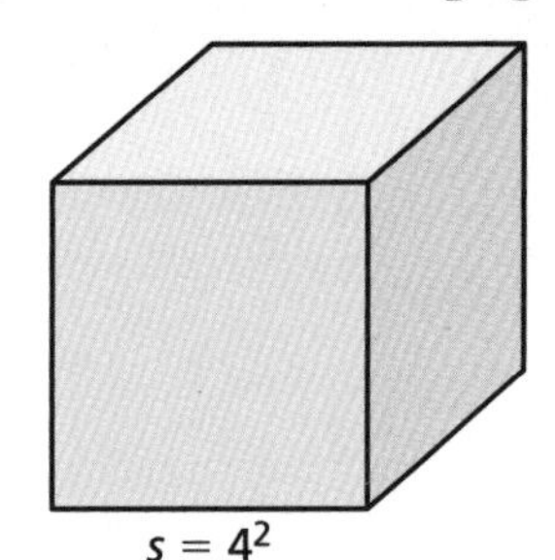

b.

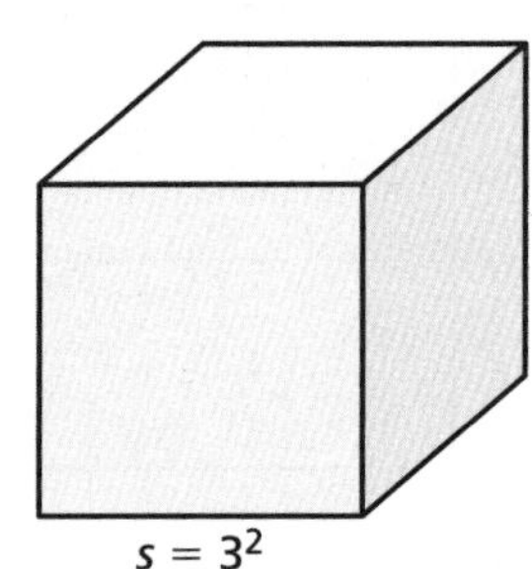

c.

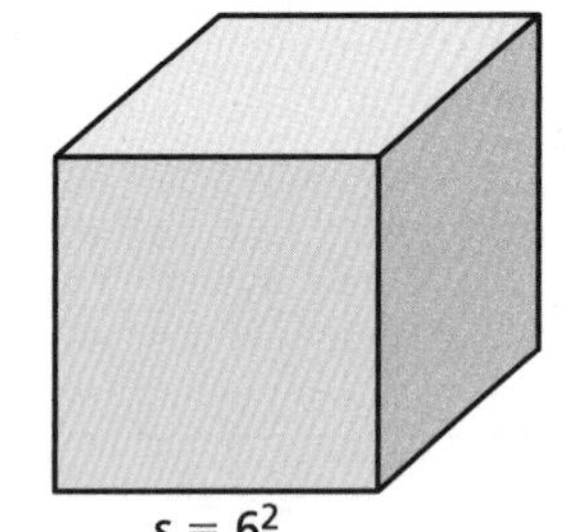

d.

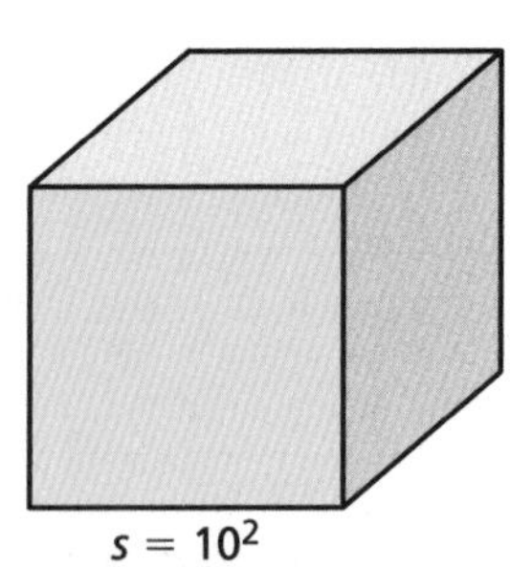

Name____ Date__________

10.3 Quotient of Powers Property (continued)

	Volume of Smaller Cube	Volume of Larger Cube	$\frac{\text{Larger Volume}}{\text{Smaller Volume}}$	Answer
a.				
b.				
c.				
d.				

What Is Your Answer?

3. **IN YOUR OWN WORDS** How can you divide two powers that have the same base? Give two examples of your rule.

Name ______________________________ Date __________

10.3 Practice

For use after Lesson 10.3

Simplify the expression. Write your answer as a power.

1. $\dfrac{7^6}{7^5}$

2. $\dfrac{(-21)^{15}}{(-21)^9}$

3. $\dfrac{(3.9)^{20}}{(3.9)^{10}}$

4. $\dfrac{t^7}{t^3}$

5. $\dfrac{8^7 \bullet 8^4}{8^9}$

6. $\dfrac{(-1.1)^{13} \bullet (-1.1)^{12}}{(-1.1)^{10} \bullet (-1.1)^1}$

Simplify the expression.

7. $\dfrac{k \bullet 3^9}{3^5}$

8. $\dfrac{x^4 \bullet y^{10} \bullet 2^{11}}{y^8 \bullet 2^7}$

9. The radius of a basketball is about 3.6 times greater than the radius of a tennis ball. How many times greater is the volume of a basketball than the volume of a tennis ball? $\left(\text{Note: The volume of a sphere is } V = \dfrac{4}{3}\pi r^3.\right)$

Name____ Date__________

10.4 Zero and Negative Exponents

For use with Activity 10.4

Essential Question How can you evaluate a nonzero number with an exponent of zero? How can you evaluate a nonzero number with a negative integer exponent?

1 ACTIVITY: Using the Quotient of Powers Property

Work with a partner.

a. Complete the table.

Quotient	Quotient of Powers Property	Power
$\frac{5^3}{5^3}$		
$\frac{6^2}{6^2}$		
$\frac{(-3)^4}{(-3)^4}$		
$\frac{(-4)^5}{(-4)^5}$		

b. REPEATED REASONING Evaluate each expression in the first column of the table. What do you notice?

c. How can you use these results to define a^0 where $a \neq 0$?

Name ______________________________ Date __________

10.4 Zero and Negative Exponents (continued)

2 ACTIVITY: Using the Product of Powers Property

Work with a partner.

a. Complete the table.

Product	Product of Powers Property	Power
$3^0 \bullet 3^4$		
$8^2 \bullet 8^0$		
$(-2)^3 \bullet (-2)^0$		
$\left(-\frac{1}{3}\right)^0 \bullet \left(-\frac{1}{3}\right)^5$		

b. Do these results support your definition in Activity 1(c)?

3 ACTIVITY: Using the Product of Powers Property

Work with a partner.

a. Complete the table.

Product	Product of Powers Property	Power
$5^{-3} \bullet 5^3$		
$6^2 \bullet 6^{-2}$		
$(-3)^4 \bullet (-3)^{-4}$		
$(-4)^{-5} \bullet (-4)^5$		

b. According to your results from Activities 1 and 2, the products in the first column are equal to what value?

Name______________________________ Date__________

c. REASONING How does the Multiplicative Inverse Property help you to rewrite the numbers with negative exponents?

d. STRUCTURE Use these results to define a^{-n} where $a \neq 0$ and n is an integer.

4 ACTIVITY: Using a Place Value Chart

Work with a partner. Use the place value chart that shows the number 3452.867.

Place Value Chart

thousands	hundreds	tens	ones	and	tenths	hundredths	thousandths
10^3	10^2	10^1	$10^{\square}$		$10^{\square}$	$10^{\square}$	$10^{\square}$
3	4	5	2	.	8	6	7

a. REPEATED REASONING What pattern do you see in the exponents? Continue the pattern to find the other exponents.

b. STRUCTURE Show how to write the expanded form of 3452.867.

What Is Your Answer?

5. IN YOUR OWN WORDS How can you evaluate a nonzero number with an exponent of zero? How can you evaluate a nonzero number with a negative integer exponent?

Name ______________________________ Date __________

10.4 Practice

For use after Lesson 10.4

Evaluate the expression.

1. 29^0

2. 12^{-1}

3. $10^{-4} \bullet 10^{-6}$

4. $\frac{1}{3^{-3}} \bullet \frac{1}{3^5}$

Simplify. Write the expression using only positive exponents.

5. $19x^{-6}$

6. $\frac{14a^{-5}}{a^{-8}}$

7. $3t^6 \bullet 8t^{-6}$

8. $\frac{12s^{-1} \bullet 4^{-2} \bullet r^3}{s^2 \bullet r^5}$

9. The density of a proton is about $\frac{1.64 \times 10^{-24}}{3.7 \times 10^{-38}}$ grams per cubic centimeter. Simplify the expression.

Name____ Date__________

10.5 Reading Scientific Notation
For use with Activity 10.5

Essential Question How can you read numbers that are written in scientific notation?

1 ACTIVITY: Very Large Numbers

Work with a partner.

- Use a calculator. Experiment with multiplying large numbers until your calculator displays an answer that is *not* in standard form.

- When the calculator at the right was used to multiply 2 billion by 3 billion, it listed the result as

  6.0E+18.

- Multiply 2 billion by 3 billion by hand. Use the result to explain what 6.0E+18 means.

- Check your explanation using products of other large numbers.
- Why didn't the calculator show the answer in standard form?

- Experiment to find the maximum number of digits your calculator displays. For instance, if you multiply 1000 by 1000 and your calculator shows 1,000,000, then it can display 7 digits.

Name ______________________________ Date ________

2 ACTIVITY: Very Small Numbers

Work with a partner.

- Use a calculator. Experiment with multiplying very small numbers until your calculator displays an answer that is *not* in standard form.

- When the calculator at the right was used to multiply 2 billionths by 3 billionths, it listed the result as

  6.0E−18.

- Multiply 2 billionths by 3 billionths by hand. Use the result to explain what 6.0E−18 means.

- Check your explanation by calculating the products of other very small numbers.

3 ACTIVITY: Powers of 10 Matching Game

Work with a partner. Match each picture with its power of 10. Explain your reasoning.

10^5 m	10^2 m	10^0 m	10^{-1} m	10^{-2} m	10^{-5} m

A. 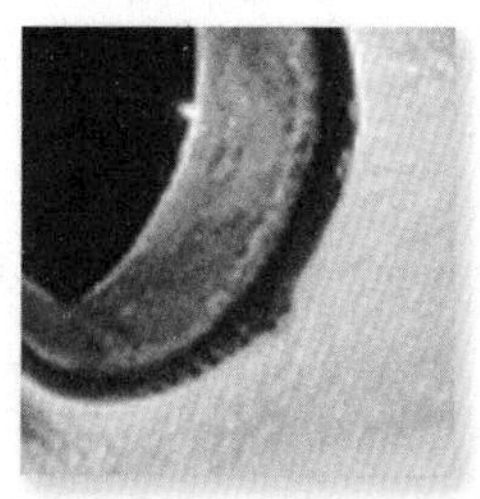

B. 

C. 

D. 

E. 

F. 

Name____ Date__________

10.5 Reading Scientific Notation (continued)

4 ACTIVITY: Choosing Appropriate Units

Work with a partner. Match each unit with its most appropriate measurement.

inches	centimeters	feet	millimeters	meters

a. Height of a door:

2×10^0

b. Height of a volcano

1.6×10^4

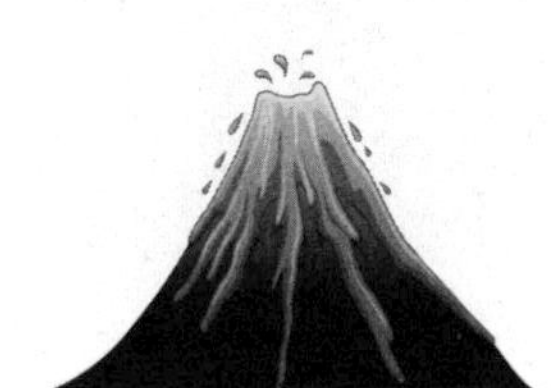

c. Length of a pen:

1.4×10^2

d. Diameter of a steel ball bearing:

6.3×10^{-1}

e. Circumference of a beach ball:

7.5×10^1

What Is Your Answer?

5. **IN YOUR OWN WORDS** How can you read numbers that are written in scientific notation? Why do you think this type of notation is called "scientific notation"? Why is scientific notation important?

Name ______________________ Date ________

10.5 Practice
For use after Lesson 10.5

Tell whether the number is written in scientific notation. Explain.

1. 14×10^8

2. 2.6×10^{12}

3. 4.79×10^{-8}

4. 3.99×10^{16}

5. 0.15×10^{22}

6. 6×10^3

Write the number in standard form.

7. 4×10^9

8. 2×10^{-5}

9. 3.7×10^6

10. 4.12×10^{-3}

11. 7.62×10^{10}

12. 9.908×10^{-12}

13. Light travels at 3×10^8 meters per second.

a. Write the speed of light in standard form.

b. How far has light traveled after 5 seconds?

Name________________________________ Date__________

10.6 Writing Scientific Notation

For use with Activity 10.6

Essential Question How can you write a number in scientific notation?

1 ACTIVITY: Finding pH Levels

Work with a partner. In chemistry, pH is a measure of the activity of dissolved hydrogen ions (H^+). Liquids with low pH values are called *acids*. Liquids with high pH values are called *bases*.

Find the pH of each liquid. Is the liquid a base, neutral, or an acid?

a. Lime juice: $[H^+] = 0.01$

b. Egg: $[H^+] = 0.00000001$

c. Distilled water: $[H^+] = 0.0000001$

d. Ammonia water: $[H^+] = 0.00000000001$

e. Tomato juice: $[H^+] = 0.0001$

f. Hydrochloric acid: $[H^+] = 1$

pH	$[H^+]$	
14	1×10^{-14}	Bases
13	1×10^{-13}	
12	1×10^{-12}	
11	1×10^{-11}	
10	1×10^{-10}	
9	1×10^{-9}	
8	1×10^{-8}	
7	1×10^{-7}	Neutral
6	1×10^{-6}	Acids
5	1×10^{-5}	
4	1×10^{-4}	
3	1×10^{-3}	
2	1×10^{-2}	
1	1×10^{-1}	
0	1×10^{0}	

Name ______________________________ Date __________

2 ACTIVITY: Writing Scientific Notation

Work with a partner. Match each planet with its distance from the Sun. Then write each distance in scientific notation. Do you think it is easier to match the distances when they are written in standard form or in scientific notation? Explain.

a. 1,800,000,000 miles

b. 67,000,000 miles

c. 890,000,000 miles

d. 93,000,000 miles

e. 140,000,000 miles

f. 2,800,000,000 miles

g. 480,000,000 miles

h. 36,000,000 miles

Name____ Date__________

3 ACTIVITY: Making a Scale Drawing

Work with a partner. The illustration in Activity 2 is not drawn to scale. Use the instructions below to make a scale drawing of the distances in our solar system.

- **Cut a sheet of paper into three strips of equal width. Tape the strips together.**
- **Draw a long number line. Label the number line in hundreds of millions of miles.**
- **Locate each planet's position on the number line.**

What Is Your Answer?

4. **IN YOUR OWN WORDS** How can you write a number in scientific notation?

Name ____________________ Date ________

10.6 Practice

For use after Lesson 10.6

Write the number in scientific notation.

1. 4,200,000

2. 0.038

3. 600,000

4. 0.0000808

5. 0.0007

6. 29,010,000,000

Order the numbers from least to greatest.

7. $6.4 \times 10^8, 5.3 \times 10^9, 2.3 \times 10^8$

8. $9.1 \times 10^{-3}, 9.6 \times 10^{-3}, 9.02 \times 10^{-3}$

9. $7.3 \times 10^7, 5.6 \times 10^{10}, 3.7 \times 10^9$

10. $1.4 \times 10^{-5}, 2.01 \times 10^{-15}, 6.3 \times 10^{-2}$

11. A patient has 0.0000075 gram of iron in 1 liter of blood. The normal level is between 6×10^{-7} gram and 1.6×10^{-5} gram. Is the patient's iron level normal? Write the patient's amount of iron in scientific notation.

Name____ Date__________

10.7 Operations in Scientific Notation

For use with Activity 10.7

Essential Question How can you perform operations with numbers written in scientific notation?

1 ACTIVITY: Adding Numbers in Scientific Notation

Work with a partner. Consider the numbers 2.4×10^3 and 7.1×10^3.

a. Explain how to use order of operations to find the sum of these numbers. Then find the sum.

$$2.4 \times 10^3 + 7.1 \times 10^3$$

b. The factor ____ is common to both numbers. How can you use the Distributive Property to rewrite the sum $(2.4 \times 10^3) + (7.1 \times 10^3)$?

$(2.4 \times 10^3) + (7.1 \times 10^3) =$ ________________ Distributive Property

c. Use order of operations to evaluate the expression you wrote in part (b). Compare the result with your answer in part (a).

d. **STRUCTURE** Write a rule you can use to add numbers written in scientific notation where the powers of 10 are the same. Then test your rule using the sums below.

- $(4.9 \times 10^5) + (1.8 \times 10^5) =$ ____________________
- $(3.85 \times 10^4) + (5.72 \times 10^4) =$ ____________________

2 ACTIVITY: Adding Numbers in Scientific Notation

Work with a partner. Consider the numbers 2.4×10^3 and 7.1×10^4.

a. Explain how to use order of operations to find the sum of these numbers. Then find the sum.

$$2.4 \times 10^3 + 7.1 \times 10^4$$

Name ______________________ Date ________

b. How is this pair of numbers different from the pair of numbers in Activity 1?

c. Explain why you cannot immediately use the rule you wrote in Activity 1(d) to find this sum.

d. **STRUCTURE** How can you rewrite one of the numbers so that you can use the rule you wrote in Activity 1(d)? Rewrite one of the numbers. Then find the sum using your rule and compare the result with your answer in part (a).

e. **REASONING** Does this procedure work when subtracting numbers written in scientific notation? Justify your answer by evaluating the differences below.

- $(8.2 \times 10^5) - (4.6 \times 10^5) =$ ______________
- $(5.88 \times 10^5) - (1.5 \times 10^4) =$ ______________

3 ACTIVITY: Multiplying Numbers in Scientific Notation

Work with a partner. Match each step with the correct description.

Step	Description
$(2.4 \times 10^3) \times (7.1 \times 10^3)$	**Original expression**
1. $= 2.4 \times 7.1 \times 10^3 \times 10^3$	**A.** Write in standard form.
2. $= (2.4 \times 7.1) \times (10^3 \times 10^3)$	**B.** Product of Powers Property
3. $= 17.04 \times 10^6$	**C.** Write in scientific notation.
4. $= 1.704 \times 10^1 \times 10^6$	**D.** Commutative Property of Multiplication
5. $= 1.704 \times 10^7$	**E.** Simplify.
6. $= 17{,}040{,}000$	**F.** Associative Property of Multiplication

Does this procedure work when the numbers have different powers of 10? Justify your answer by using this procedure to evaluate the products below.

- $(1.9 \times 10^2) \times (2.3 \times 10^5) =$

- $(8.4 \times 10^6) \times (5.7 \times 10^{-4}) =$

4 ACTIVITY: Using Scientific Notation to Estimate

Work with a partner. A person normally breathes about 6 liters of air per minute. The life expectancy of a person in the United States at birth is about 80 years. Use scientific notation to estimate the total amount of air a person born in the United States breathes over a lifetime.

What Is Your Answer?

5. IN YOUR OWN WORDS How can you perform operations with numbers written in scientific notation?

6. Use a calculator to evaluate the expression. Write your answer in scientific notation and in standard form.

a. $(1.5 \times 10^4) + (6.3 \times 10^4)$

b. $(7.2 \times 10^5) - (2.2 \times 10^3)$

c. $(4.1 \times 10^{-3}) \times (4.3 \times 10^{-3})$

d. $(4.75 \times 10^{-6}) \times (1.34 \times 10^7)$

Name ______________________________ Date __________

10.7 Practice

For use after Lesson 10.7

Find the sum or difference. Write your answer in scientific notation.

1. $(2 \times 10^4) + (7.2 \times 10^4)$

2. $(3.2 \times 10^{-2}) + (9.4 \times 10^{-2})$

3. $(6.7 \times 10^5) - (4.3 \times 10^5)$

4. $(8.9 \times 10^{-3}) - (1.9 \times 10^{-3})$

Find the product or quotient. Write your answer in scientific notation.

5. $(6 \times 10^8) \times (4 \times 10^6)$

6. $(9 \times 10^{-3}) \times (9 \times 10^{-3})$

7. $(8 \times 10^3) \div (2 \times 10^2)$

8. $(2.34 \times 10^5) \div (7.8 \times 10^5)$

9. How many times greater is the radius of a basketball than the radius of a marble?

Radius $= 1.143 \times 10^1$ cm

Radius $= 5 \times 10^{-1}$ cm

Name______________________________ Date__________

Chapter 11 Fair Game Review

Graph the inequality.

1. $x < -3$

2. $x \geq -5$

3. $x \leq 2$

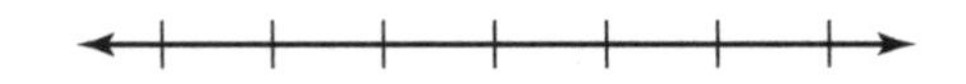

4. $x > 7$

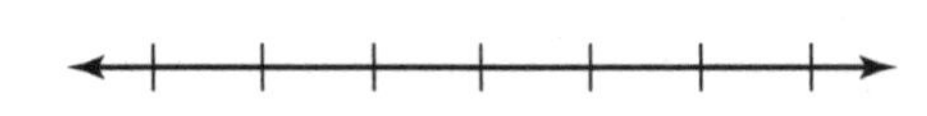

5. $x \leq -2.3$

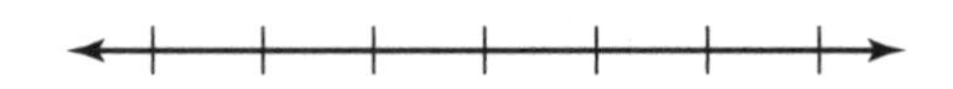

6. $x > \frac{2}{5}$

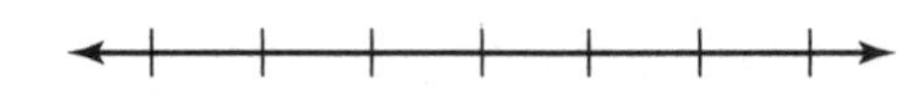

7. The deepest free dive by a human in the ocean is 417 feet. The depth humans have been in the ocean can be represented by the inequality $x \leq 417$. Graph the inequality.

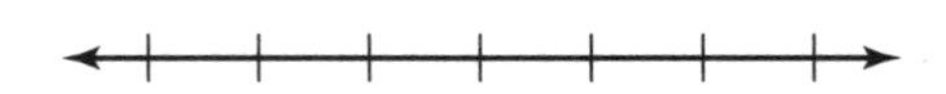

Name ______________________________ Date __________

Chapter 11 Fair Game Review (continued)

Complete the number sentence with < or >.

8. $\frac{3}{4}$ _____ 0.2

9. $\frac{7}{8}$ _____ 0.7

10. -0.6 _____ $-\frac{2}{3}$

11. -1.76 _____ 1.75

12. $\frac{17}{3}$ _____ 6

13. 1.8 _____ $\frac{31}{16}$

14. Your height is 5 feet and $1\frac{5}{8}$ inches. Your friend's height is 5.6 feet. Who is taller? Explain.

Name_____________________________ Date__________

11.1 Writing and Graphing Inequalities

For use with Activity 11.1

Essential Question How can you use a number line to represent solutions of an inequality?

1 ACTIVITY: Understanding Inequality Statements

Work with a partner. Read the statement. Circle each number that makes the statement true, and then answer the questions.

a. "You are in **at least** 5 of the photos."

–3 –2 –1 0 1 2 3 4 5 6

- What do you notice about the numbers that you circled?

- Is the number 5 included? Why or why not?

- Write four other numbers that make the statement true.

b. "The temperature is **less than** –4 degrees Fahrenheit."

–7 –6 –5 –4 –3 –2 –1 0 1 2

- What do you notice about the numbers that you circled?

- Can the temperature be exactly –4 degrees Fahrenheit? Explain.

- Write four other numbers that make the statement true.

c. "**More than** 3 students from our school are in the chess tournament."

–3 –2 –1 0 1 2 3 4 5 6

- What do you notice about the numbers that you circled?

11.1 Writing and Graphing Inequalities (continued)

- Is the number 3 included? Why or why not?

- Write four other numbers that make the statement true.

d. "The balance in a yearbook fund is **no more than** $-\$5$."

-7 -6 -5 -4 -3 -2 -1 0 1 2

- What do you notice about the numbers that you circled?

- Is the number -5 included? Why or why not?

- Write four other numbers that make the statement true.

2 ACTIVITY: Understanding Inequality Symbols

Work with a partner.

a. Consider the statement "x is a number such that $x > -1.5$."

- Can the number be exactly -1.5? Explain.

- Make a number line. Shade the part of the number line that shows the numbers that make the statement true.

- Write four other numbers that are not integers that make the statement true.

b. Consider the statement "x is a number such that $x \leq \frac{5}{2}$."

- Can the number be exactly $\frac{5}{2}$? Explain.

11.1 Writing and Graphing Inequalities (continued)

- Make a number line. Shade the part of the number line that shows the numbers that make the statement true.

- Write four other numbers that are not integers that make the statement true.

3 ACTIVITY: Writing and Graphing Inequalities

Work with a partner. Write an inequality for each graph. Then, in words, describe all the values of x that make the inequality true.

a. −8 −6 −4 −2 0 2 4 6 8

b. −8 −6 −4 −2 0 2 4 6 8

c. −8 −6 −4 −2 0 2 4 6 8

d.

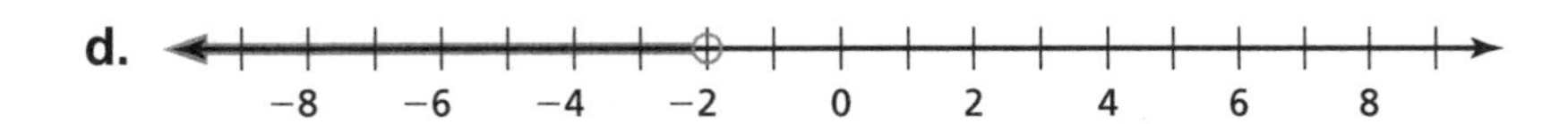

What Is Your Answer?

4. IN YOUR OWN WORDS How can you use a number line to represent solutions of an inequality?

5. STRUCTURE Is $x \geq -1.4$ the same as $-1.4 \leq x$? Explain.

Name ______________________________ Date __________

11.1 Practice

For use after Lesson 11.1

Write the word sentence as an inequality.

1. A number t is less than or equal to 5.

2. A number g subtracted from 6 is no more than $\frac{3}{4}$.

Tell whether the given value is a solution of the inequality.

3. $r - 3 \leq 9; r = 8$

4. $4h > -12; h = -5$

Graph the inequality on a number line.

5. $y > -1$

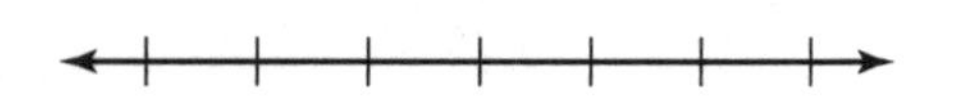

6. $d \leq 2.5$

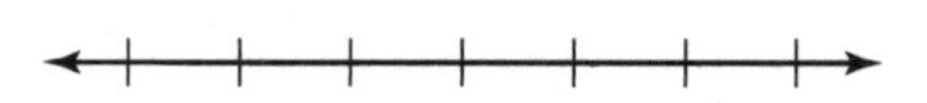

7. $s \geq 3\frac{3}{4}$

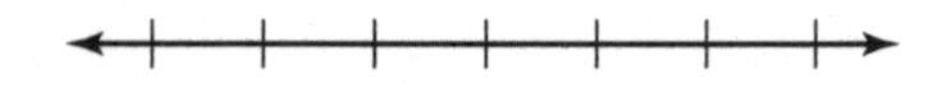

8. $p < 9$

9. You have at most 30 games on your smart phone. Write an inequality that represents this situation.

Name________________________________ Date__________

11.2 Solving Inequalities Using Addition or Subtraction

For use with Activity 11.2

Essential Question How can you use addition or subtraction to solve an inequality?

1 ACTIVITY: Writing an Inequality

Work with a partner. Members of the Boy Scouts must be less than 18 years old. In 4 years, your friend will still be eligible to be a scout.

a. Which of the following represents your friend's situation? What does x represent? Explain your reasoning.

$x + 4 > 18$ $x + 4 < 18$ $x + 4 \geq 18$ $x + 4 \leq 18$

b. Graph the possible ages of your friend on a number line. Explain how you decided what to graph.

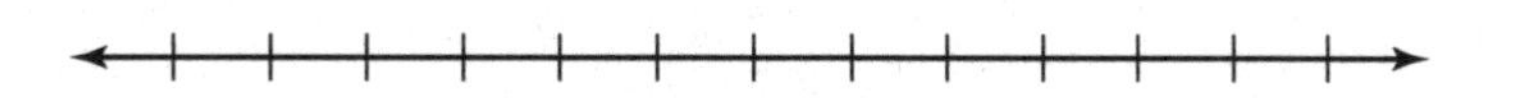

2 ACTIVITY: Writing an Inequality

Work with a partner. Supercooling is the process of lowering the temperature of a liquid or a gas below its freezing point without it becoming a solid. Water can be supercooled to 86°F below its normal freezing point (32°F) and still not freeze.

a. Let x represent the temperature of water. Which inequality represents the temperature at which water can be a liquid or a gas? Explain your reasoning.

$x - 32 > -86$ $x - 32 < -86$ $x - 32 \geq -86$ $x - 32 \leq -86$

11.2 Solving Inequalities Using Addition or Subtraction (continued)

b. On a number line, graph the possible temperatures at which water can be a liquid or a gas. Explain how you decided what to graph.

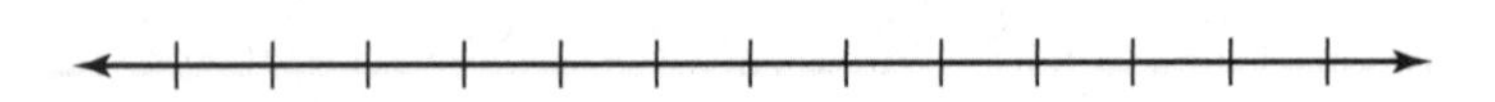

3 ACTIVITY: Solving Inequalities

Work with a partner. Complete the following steps for Activity 1. Then repeat the steps for Activity 2.

- Use your inequality from part (a). Replace the inequality symbol with an equal sign.

- Solve the equation.

- Replace the equal sign with the original inequality symbol.

- Graph this new inequality.

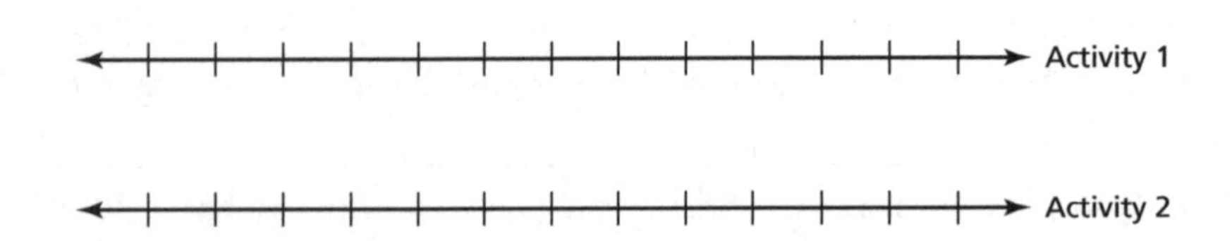

- Compare the graph with your graph in part (b). What do you notice?

Name______________________________ Date__________

4 ACTIVITY: Temperatures of Continents

Work with a partner. The table shows the lowest recorded temperature on each continent. Write an inequality that represents each statement. Then solve and graph the inequality.

Continent	Lowest Temperature
Africa	–11°F
Antarctica	–129°F
Asia	–90°F
Australia	–9.4°F
Europe	–67°F
North America	–81.4°F
South America	–27°F

a. The temperature at a weather station in Asia is more than 150°F greater than the record low in Asia.

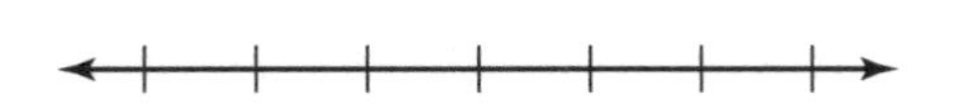

b. The temperature at a research station in Antarctica is at least 80°F greater than the record low in Antarctica.

What Is Your Answer?

5. IN YOUR OWN WORDS How can you use addition or subtraction to solve an inequality?

6. Describe a real-life situation that you can represent with an inequality. Write the inequality. Graph the solution on a number line.

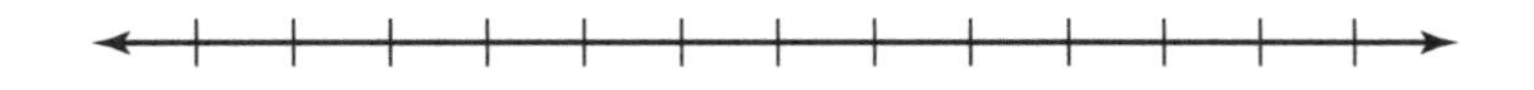

Name ______________________________ Date __________

11.2 Practice

For use after Lesson 11.2

Solve the inequality. Graph the solution.

1. $y - 3 \geq -12$

2. $-14 \leq 8 + x$

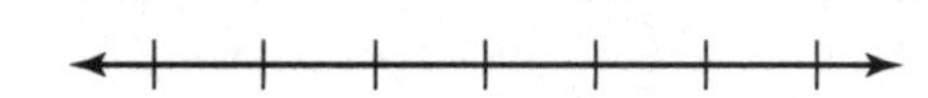

3. $t - 4 < -4$

4. $-9 \geq 2 + d$

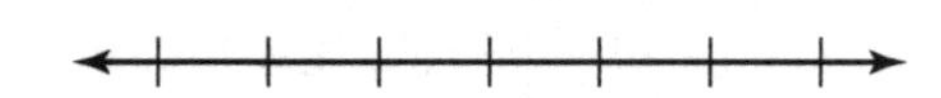

5. $-3.4 > c - 1.2$

6. $j + \frac{5}{12} < -\frac{3}{4}$

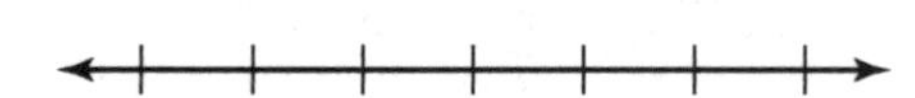

7. A bounce house can hold 15 children. Seven children go in the bounce house. Write and solve an inequality that represents the additional number of children that can go in the bounce house.

Name______________________________ Date__________

11.3 Solving Inequalities Using Multiplication or Division

For use with Activity 11.3

Essential Question How can you use multiplication or division to solve an inequality?

1 ACTIVITY: Using a Table to Solve an Inequality

Work with a partner.

- **Complete the table.**
- **Decide which graph represents the solution of the inequality.**
- **Write the solution of the inequality.**

a. $4x > 12$

x	-1	0	1	2	3	4	5
$4x$							
$4x \overset{?}{>} 12$							

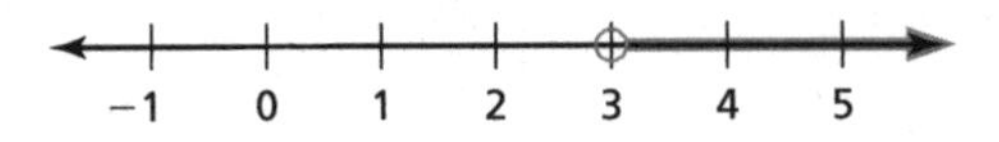

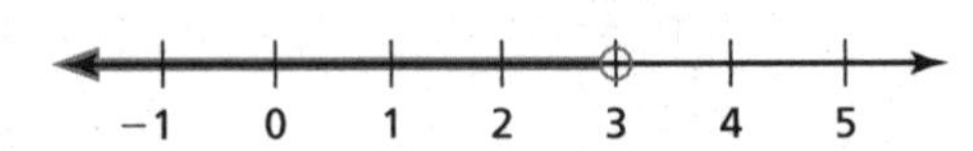

b. $-3x \le 9$

x	-5	-4	-3	-2	-1	0	1
$-3x$							
$-3x \overset{?}{\le} 9$							

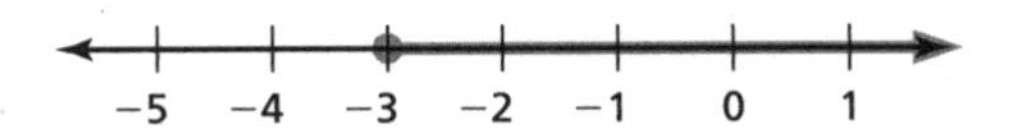

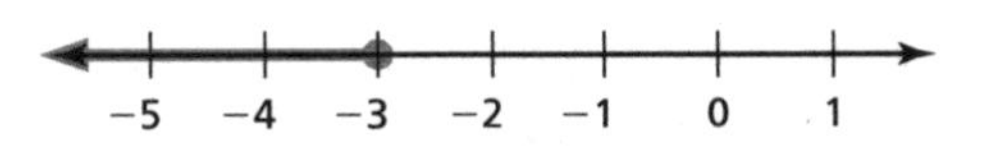

Name ______________________________ Date __________

2 ACTIVITY: Solving an Inequality

Work with a partner.

a. Solve $-3x \leq 9$ by adding $3x$ to each side of the inequality first. Then solve the resulting inequality.

b. Compare the solution in part (a) with the solution in Activity 1(b).

3 ACTIVITY: Using a Table to Solve an Inequality

Work with a partner.

- **Complete the table.**
- **Decide which graph represents the solution of the inequality.**
- **Write the solution of the inequality.**

a. $\frac{x}{3} < 1$

x	-1	0	1	2	3	4	5
$\frac{x}{3}$							
$\frac{x}{3} \overset{?}{<} 1$							

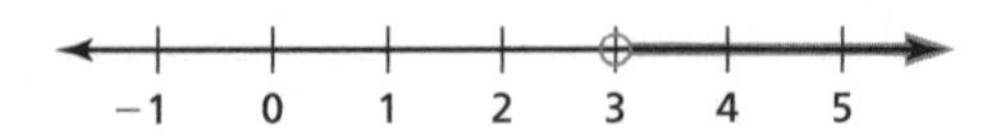

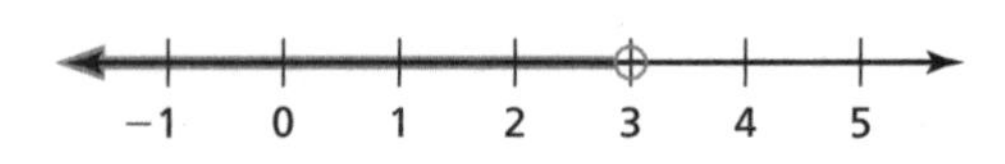

b. $\dfrac{x}{-4} \geq \dfrac{3}{4}$

x	-5	-4	-3	-2	-1	0	1
$\dfrac{x}{-4}$							
$\dfrac{x}{-4} \overset{?}{\geq} \dfrac{3}{4}$							

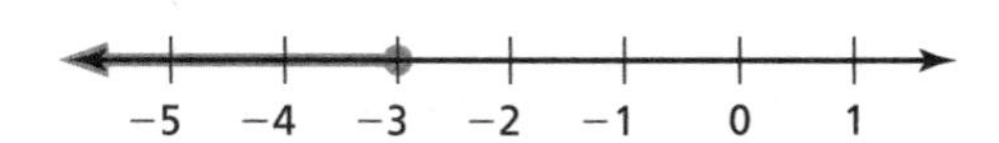

4 ACTIVITY: Writing Rules

Work with a partner. Use a table to solve each inequality.

a. $-2x \leq 10$ **b.** $-6x > 0$ **c.** $\dfrac{x}{-4} < 1$ **d.** $\dfrac{x}{-8} \geq \dfrac{1}{8}$

x											
$-2x$											
$-6x$											
$\dfrac{x}{-4}$											
$\dfrac{x}{-8}$											

Write a set of rules that describes how to solve inequalities like those in Activities 1 and 3. Then use your set of rules to solve each of the four inequalities above.

What Is Your Answer?

5. IN YOUR OWN WORDS How can you use multiplication or division to solve an inequality?

Name ______________________________ Date __________

11.3 Practice
For use after Lesson 11.3

Solve the inequality. Graph the solution.

1. $6n < 90$

2. $\frac{x}{4} \le -18$

3. $-20t > -80$

4. $-3q \ge 91.5$

5. $-4p < \frac{2}{3}$

6. $-8 \ge 1.6m$

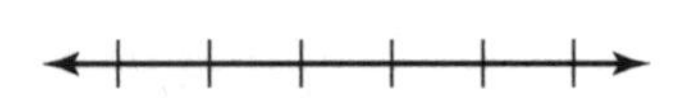

7. $-\frac{r}{4} \le -10$

8. $-\frac{t}{5} > 2.5$

9. $-2 \ge \frac{q}{-0.3}$

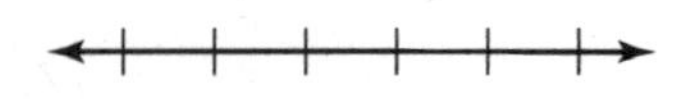

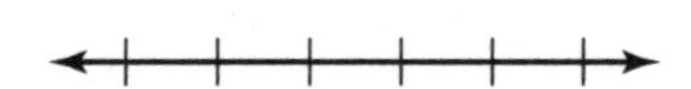

10. To win a game, you need at least 45 points. Each question is worth 3 points. Write and solve an inequality that represents the number of questions you need to answer correctly to win the game.

Name____ Date__________

11.4 Solving Two-Step Inequalities
For use with Activity 11.4

Essential Question How can you use an inequality to describe the dimensions of a figure?

ACTIVITY: Areas and Perimeters of Figures

Work with a partner.

- **Use the given condition to choose the inequality that you can use to find the possible values of the variable. Justify your answer.**
- **Write four values of the variable that satisfy the inequality you chose.**

a. You want to find the values of x so that the area of the rectangle is more than 22 square units.

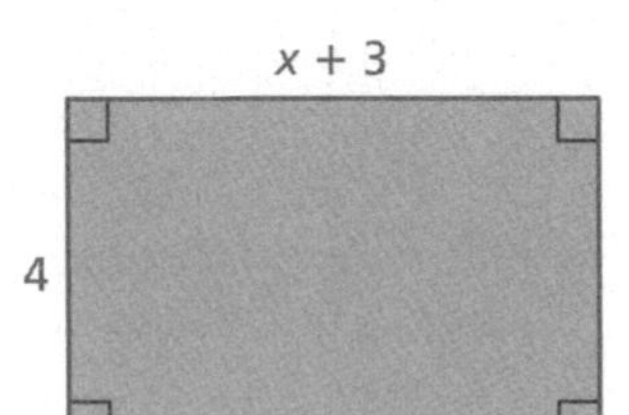

$4x + 12 > 22$	$4x + 3 > 22$
$4x + 12 \geq 22$	$2x + 14 > 22$

b. You want to find the values of x so that the perimeter of the rectangle is greater than or equal to 28 units.

$x + 7 \geq 28$	$4x + 12 \geq 28$	$2x + 14 \geq 28$	$2x + 14 \leq 28$

c. You want to find the values of y so that the area of the parallelogram is fewer than 41 square units.

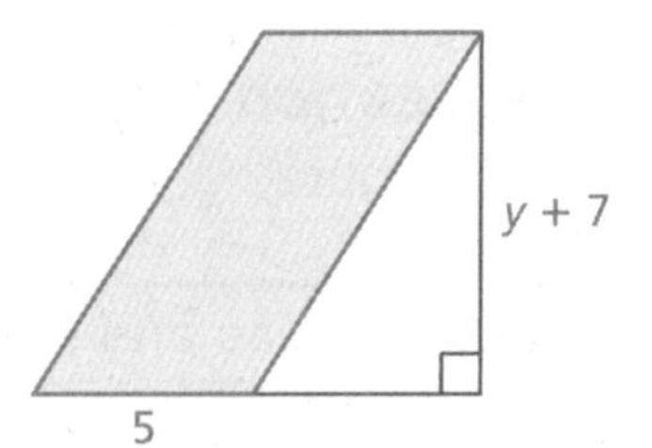

$5y + 7 < 41$	$5y + 35 < 41$
$5y + 7 \leq 41$	$5y + 35 \leq 41$

d. You want to find the values of z so that the area of the trapezoid is at most 100 square units.

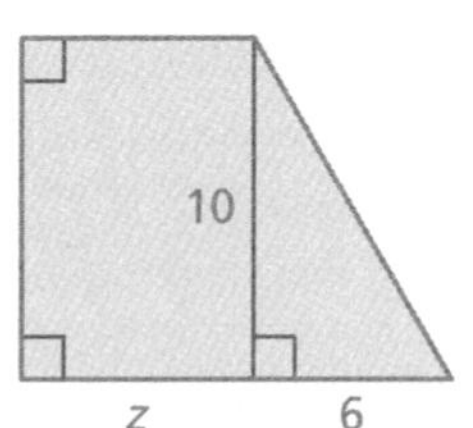

$5z + 30 \leq 100$	$10z + 30 \leq 100$
$5z + 30 < 100$	$10z + 30 < 100$

2 ACTIVITY: Volumes of Rectangular Prisms

Work with a partner.

- **Use the given condition to choose the inequality that you can use to find the possible values of the variable. Justify your answer.**
- **Write four values of the variable that satisfy the inequality you chose.**

a. You want to find the values of x so that the volume of the rectangular prism is at least 50 cubic units.

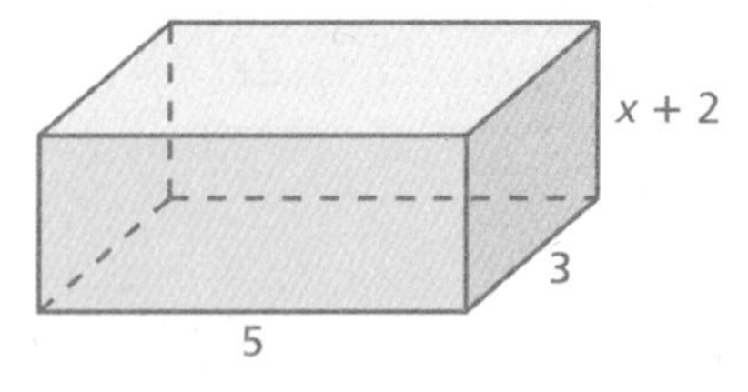

$15x + 30 > 50$	$x + 10 \geq 50$
$15x + 30 \geq 50$	$15x + 2 \geq 50$

Name____ Date__________

b. You want to find the values of x so that the volume of the rectangular prism is no more than 36 cubic units.

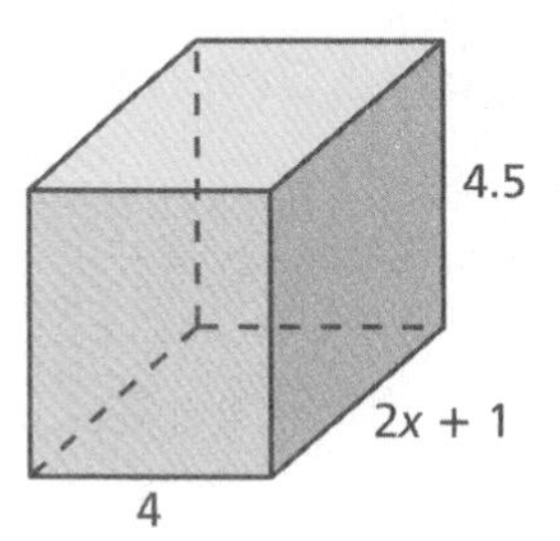

$8x + 4 < 36$	$36x + 18 < 36$
$2x + 9.5 \leq 36$	$36x + 18 \leq 36$

What Is Your Answer?

3. **IN YOUR OWN WORDS** How can you use an inequality to describe the dimensions of a figure?

4. Use what you know about solving equations and inequalities to describe how you can solve a two-step inequality. Give an example to support your explanation.

Name ______________________ Date ________

11.4 Practice

For use after Lesson 11.4

Solve the inequality. Graph the solution.

1. $5 - 3x > 8$

2. $-4x - 7 \leq 9$

3. $3 + 4.5x \geq 21$

4. $-2y - 5 > \frac{5}{2}$

5. $2(y - 4) < -18$

6. $-6 \geq -6(y - 3)$

7. You borrow \$200 from a friend to help pay for a new laptop computer. You pay your friend back \$12 per week. Write and solve an inequality to find when you will owe your friend less than \$60.

Name____ Date__________

Chapter 12 Fair Game Review

Use a protractor to find the measure of the angle. Then classify the angle as *acute*, *obtuse*, *right*, or *straight*.

1.

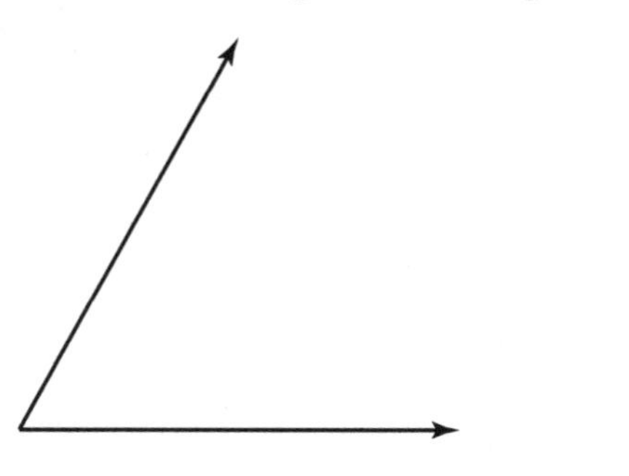

2.

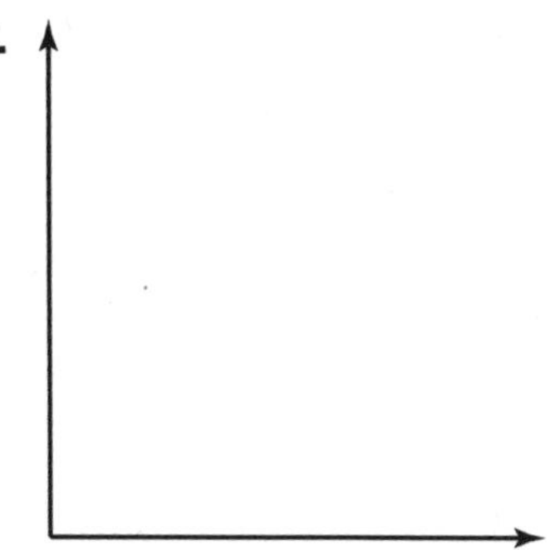

3.

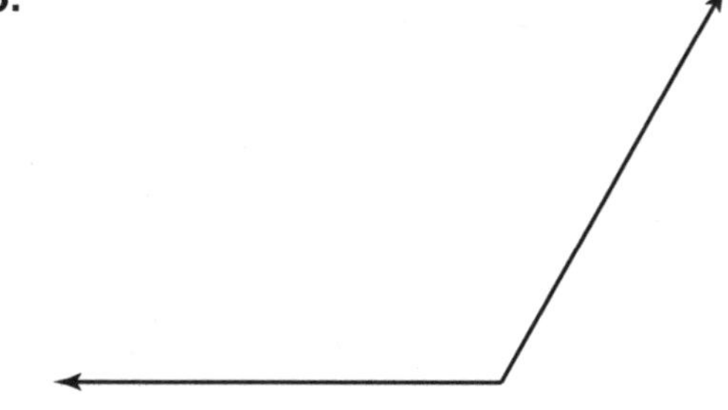

4.

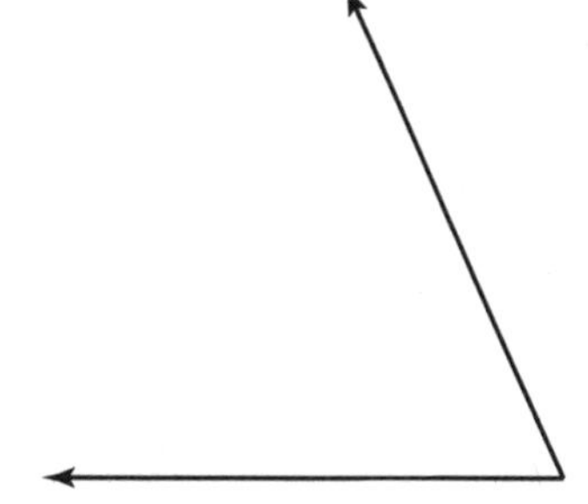

5.

6.

Name ____________________ Date ________

Fair Game Review (continued)

Use a protractor to draw an angle with the given measure.

7. 80°

8. 35°

9. 100°

10. 175°

11. 57°

12. 122°

Name____ Date__________

12.1 Adjacent and Vertical Angles
For use with Activity 12.1

Essential Question What can you conclude about the angles formed by two intersecting lines?

Classification of Angles

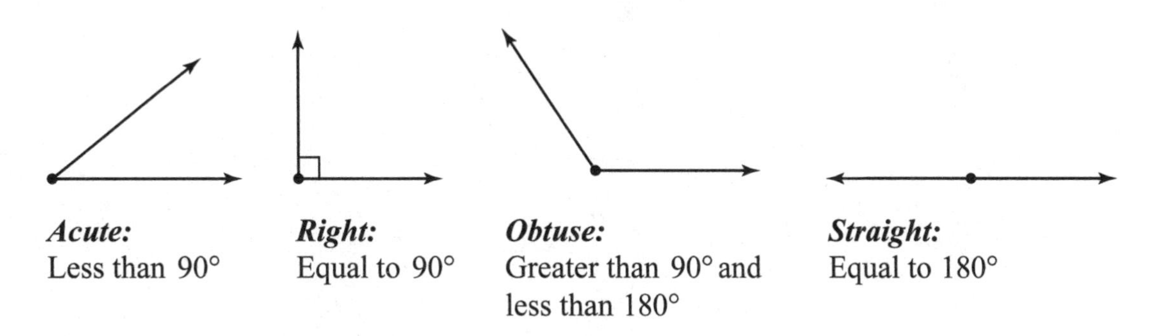

1 ACTIVITY: Drawing Angles

Work with a partner.

a. Draw the hands of the clock to represent the given type of angle.

b. What is the measure of the angle formed by the hands of the clock at the given time?

9:00 6:00 12:00

Name ______________________________ Date __________

2 ACTIVITY: Naming Angles

Work with a partner. Some angles, such as $\angle A$, can be named by a single letter. When this does not clearly identify an angle, you should use three letters, as shown.

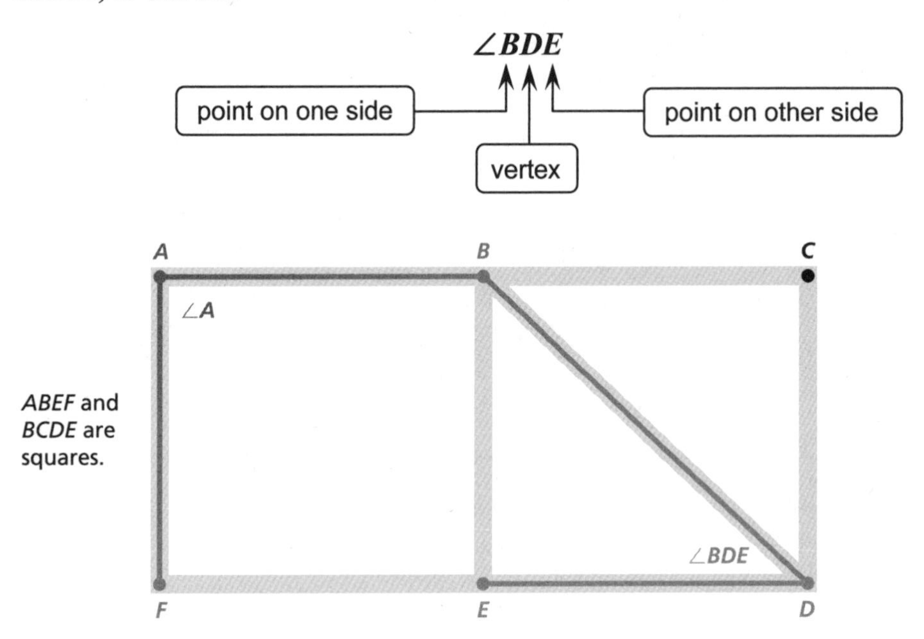

a. Name all of the right angles, acute angles, and obtuse angles.

b. Which pairs of angles do you think are *adjacent*? Explain.

Name________________________________ Date__________

3 ACTIVITY: Measuring Angles

Work with a partner.

a. How many angles are formed by the intersecting roads? Number the angles.

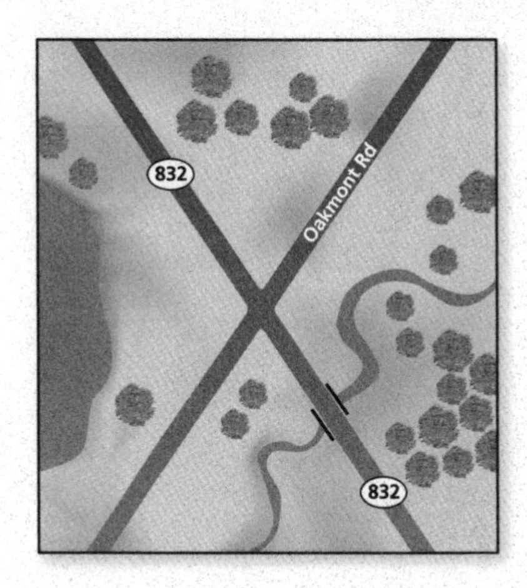

b. **CHOOSE TOOLS** Measure each angle formed by the intersecting roads. What do you notice?

What Is Your Answer?

4. **IN YOUR OWN WORDS** What can you conclude about the angles formed by two intersecting lines?

5. Draw two acute angles that are adjacent.

Name ______________________________ Date __________

12.1 Practice

For use after Lesson 12.1

Name two pairs of adjacent angles and two pairs of vertical angles in the figure.

1.

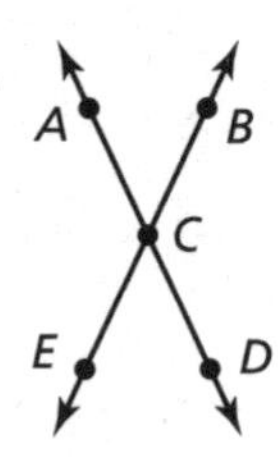

2.

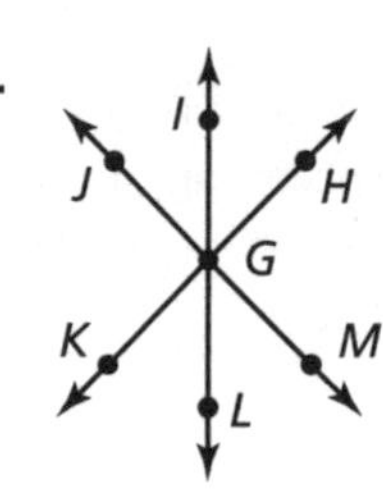

Tell whether the angles are *adjacent* or *vertical*. Then find the value of *x*.

3.

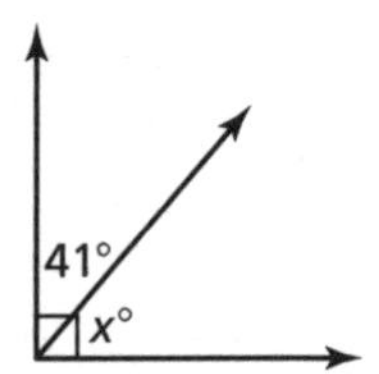

4.

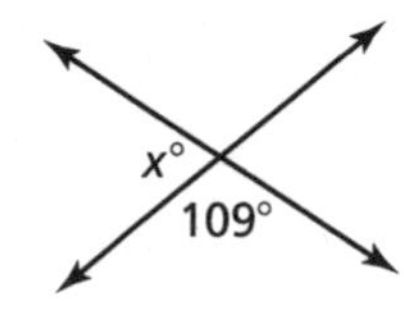

5.

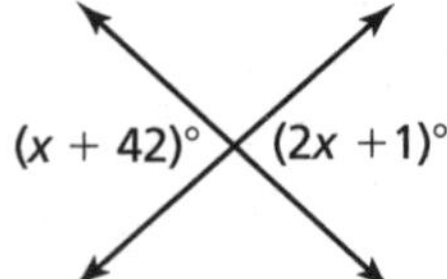

6. $(x + 96)^\circ$ $5x^\circ$

7. A tree is leaning toward the ground. How many degrees does the tree have to fall before hitting the ground?

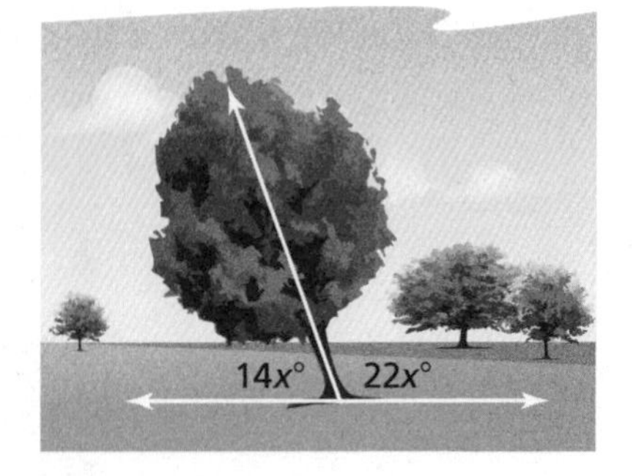

Name____ Date__________

12.2 Complementary and Supplementary Angles

For use with Activity 12.2

Essential Question How can you classify two angles as complementary or supplementary?

1 ACTIVITY: Complementary and Supplementary Angles

Work with a partner.

a. The graph represents the measures of *complementary angles*. Use the graph to complete the table.

x		20°		30°	45°		75°
y	80°		65°	60°		40°	

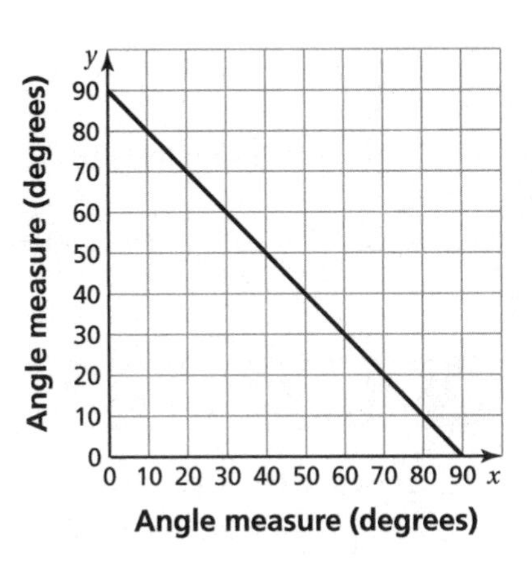

b. How do you know when two angles are complementary? Explain.

c. The graph represents the measures of *supplementary angles*. Use the graph to complete the table.

x	20°		60°	90°		140°	
y		150°		90°	50°		30°

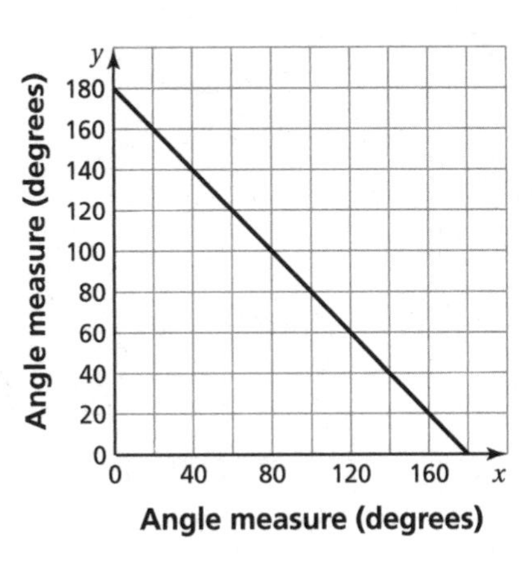

d. How do you know when two angles are supplementary? Explain.

Name ______________________________ Date __________

12.2 Complementary and Supplementary Angles (continued)

2 ACTIVITY: Exploring Rules About Angles

Work with a partner. Complete each sentence with *always*, *sometimes*, or *never*.

a. If x and y are complementary angles, then both x and y are ______________ acute.

b. If x and y are supplementary angles, then x is ______________ acute.

c. If x is a right angle, then x is ______________ acute.

d. If x and y are complementary angles, then x and y are ______________ adjacent.

e. If x and y are supplementary angles, then x and y are ______________ vertical.

3 ACTIVITY: Classifying Pairs of Angles

Work with a partner. Tell whether the two angles shown on the clocks are *complementary*, *supplementary*, or *neither*. Explain your reasoning.

a.

b.

c.

d.

Name____ Date__________

12.2 Complementary and Supplementary Angles (continued)

4 ACTIVITY: Identifying Angles

Work with a partner. Use a protractor and the figure shown.

a. Name four pairs of complementary angles and four pairs of supplementary angles.

b. Name two pairs of vertical angles.

What Is Your Answer?

5. IN YOUR OWN WORDS How can you classify two angles as complementary or supplementary? Give examples of each type.

Name ______________________________ Date __________

12.2 Practice

For use after Lesson 12.2

Tell whether the angles are *complementary*, *supplementary*, or *neither*.

1.

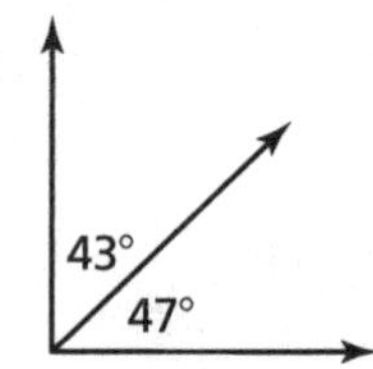

2.

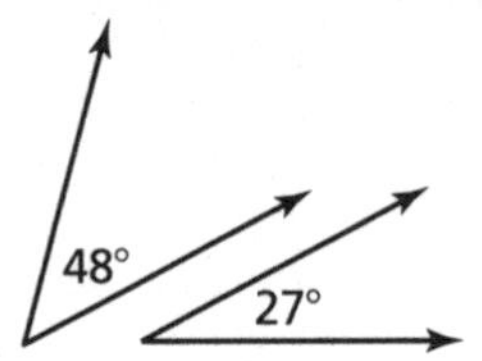

3.

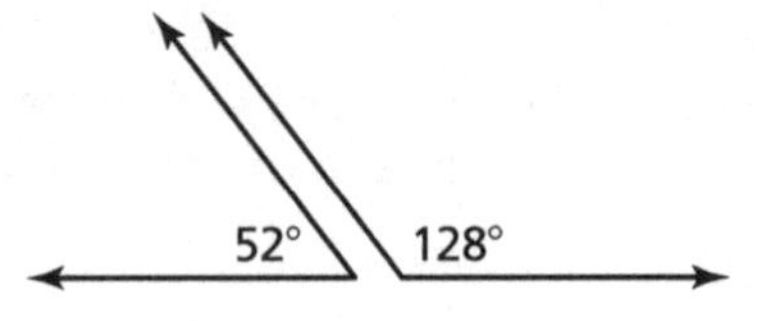

Tell whether the angles are *complementary* or *supplementary*. Then find the value of *x*.

4.

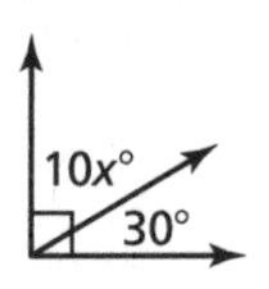

5.

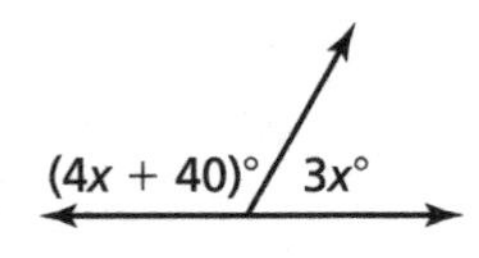

6. Find the value of x needed to hit the ball in the hole.

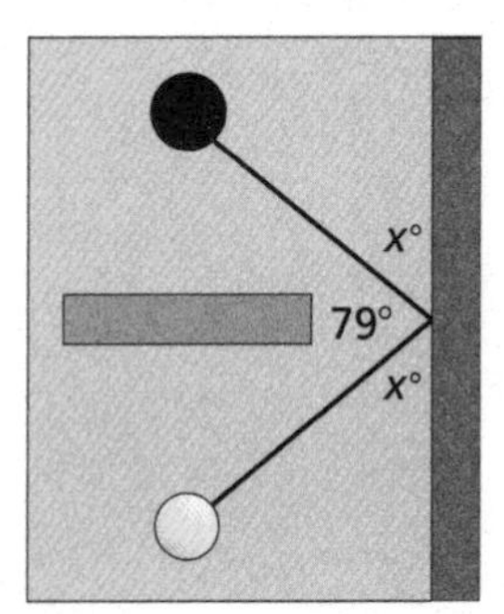

Name______________________________ Date__________

12.3 Triangles

For use with Activity 12.3

Essential Question How can you construct triangles?

1 ACTIVITY: Constructing Triangles Using Side Lengths

Work with a partner. Cut different-colored straws to the lengths shown. Then construct a triangle with the specified straws, if possible. Compare your results with those of others in your class.

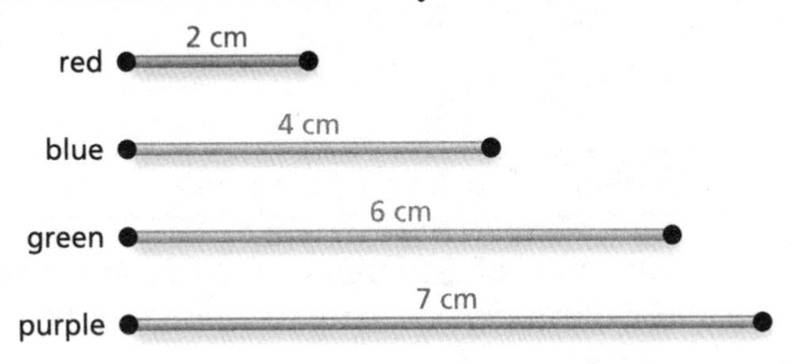

a. blue, green, purple

b. red, green, purple

c. red, blue, purple

d. red, blue, green

2 ACTIVITY: Using Technology to Draw Triangles (Side Lengths)

Work with a partner. Use geometry software to draw a triangle with the two given side lengths. What is the length of the third side of your triangle? Compare your results with those of others in your class.

a. 4 units, 7 units

Begin by drawing the side length of 4 units.

A

4

B

7

C

Then draw the side length of 7 units.

b. 3 units, 5 units **c.** 2 units, 8 units **d.** 1 unit, 1 unit

3 ACTIVITY: Constructing Triangles Using Angle Measures

Work with a partner. Two angle measures of a triangle are given. Draw the triangle. What is the measure of the third angle? Compare your results with those of others in your class.

a. 40°, 70°

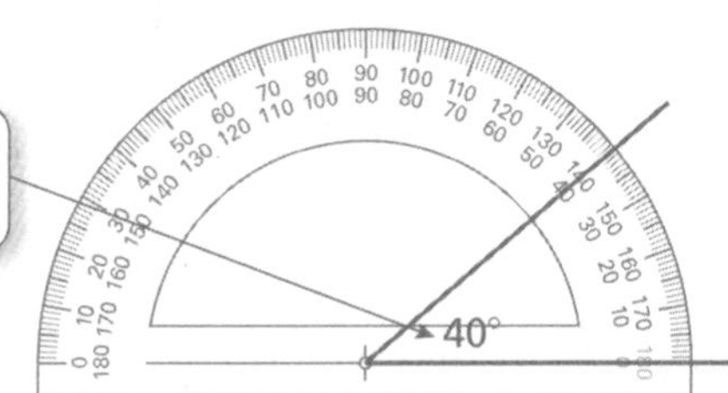

b. 60°, 75° **c.** 90°, 30° **d.** 100°, 40°

4 ACTIVITY: Using Technology to Draw Triangles (Angle Measures)

Work with a partner. Use geometry software to draw a triangle with the two given angle measures. What is the measure of the third angle? Compare your results with those of others in your class.

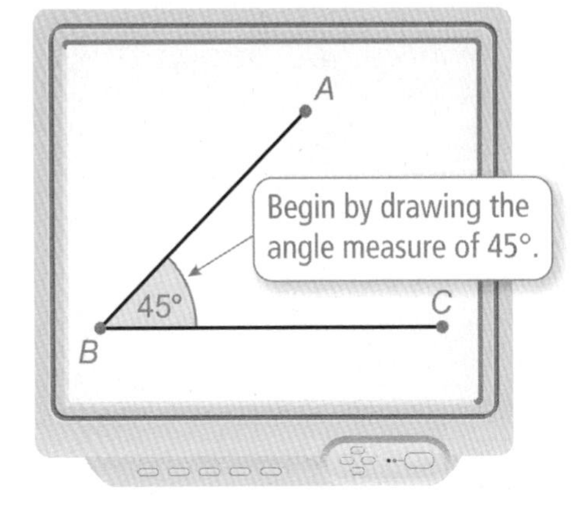

a. 45°, 55°

b. 50°, 40°

c. 110°, 35°

What Is Your Answer?

5. IN YOUR OWN WORDS How can you construct triangles?

6. REASONING Complete the table below for each set of side lengths in Activity 2. Write a rule that compares the sum of any two side lengths to the third side length.

Side Length			
Sum of Other Two Side Lengths			

7. REASONING Use a table to organize the angle measures of each triangle you formed in Activity 3. Include the sum of the angle measures. Then describe the pattern in the table and write a conclusion based on the pattern.

	$\angle 1$	$\angle 2$	$\angle 3$	$\angle 1 + \angle 2 + \angle 3$
a.				
b.				
c.				
d.				

Name ______________________________ Date __________

12.3 Practice

For use after Lesson 12.3

Classify the triangle.

1. 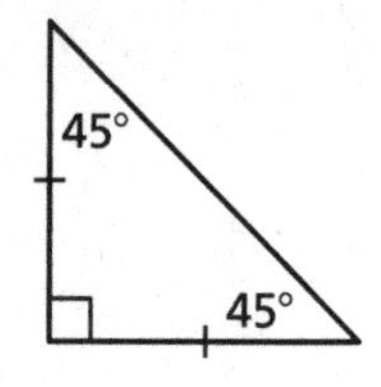

2. 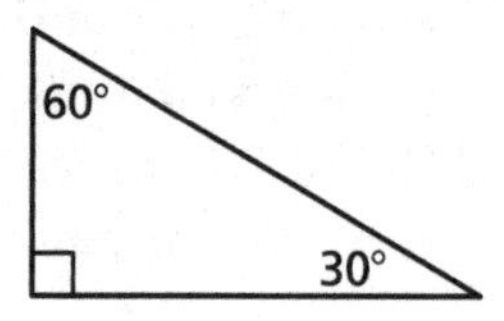

3. 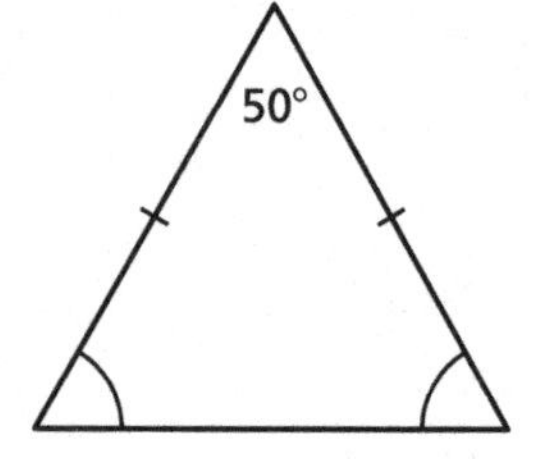

4. 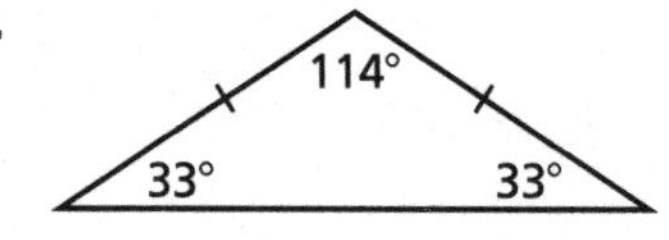

Draw a triangle with the given angle measures.

5. 28°, 42°, 110°

6. 67°, 98°, 15°

7. 31°, 59°, 90°

8. What type of triangle must the hanger be to hang clothes evenly?

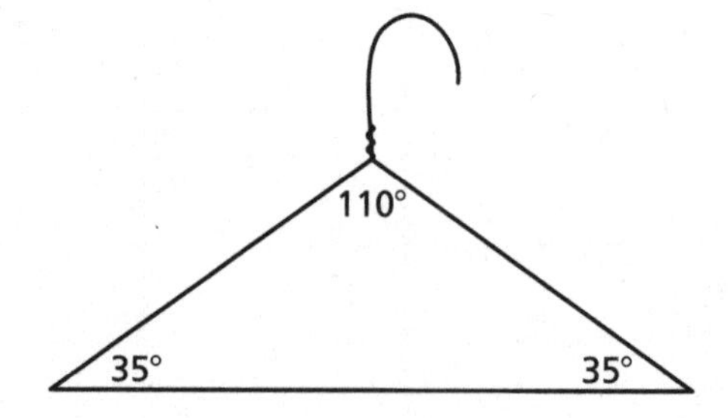

Name____ Date__________

Extension 12.3 Practice

For use after Extension 12.3

Find the value of *x*. Then classify the triangle.

1.

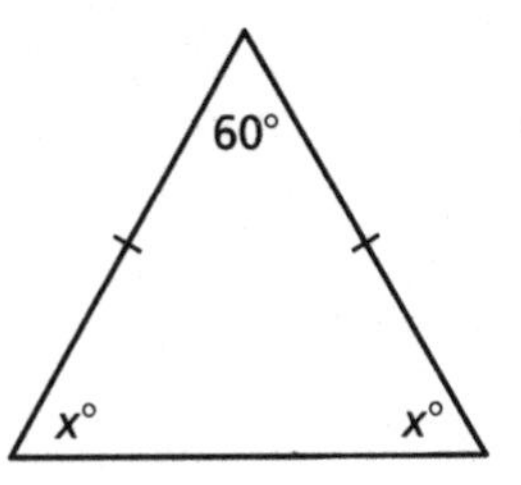

2.

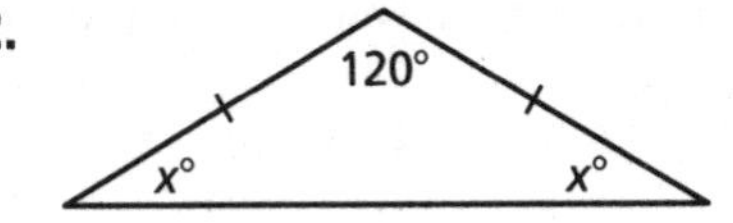

3.

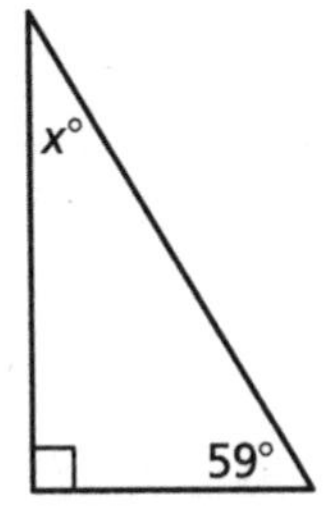

4.

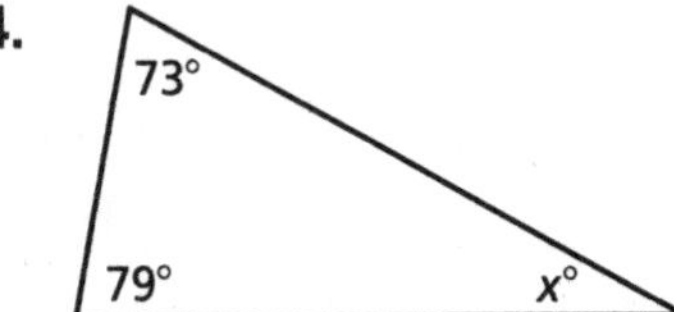

5. Find the value of x.

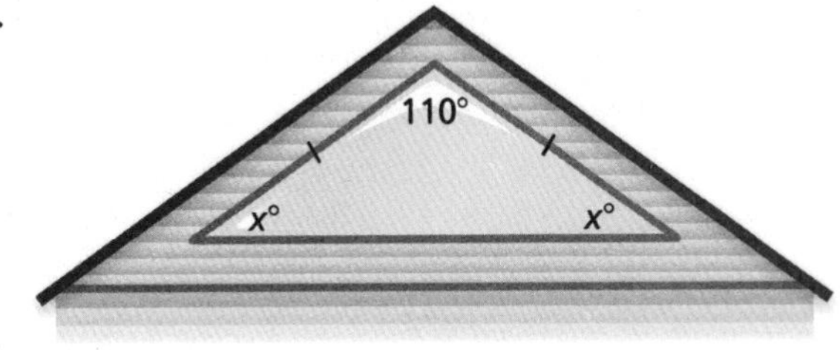

Name ____ Date __________

Practice (continued)

Tell whether a triangle can have the given angle measures. If not, change the first angle measure so that the angle measures form a triangle.

6. 25°, 64°, 91°

7. 55.5°, 94°, 31.5°

8. 85°, 64°, 30°

9. 33°, 140°, 12°

10. 99°, 53°, 28°

11. 79°, 54°, 47°

Name____ Date__________

12.4 Quadrilaterals

For use with Activity 12.4

Essential Question How can you classify quadrilaterals?

Quad means *four* and *lateral* means *side*. So, quadrilateral means a polygon with *four sides*.

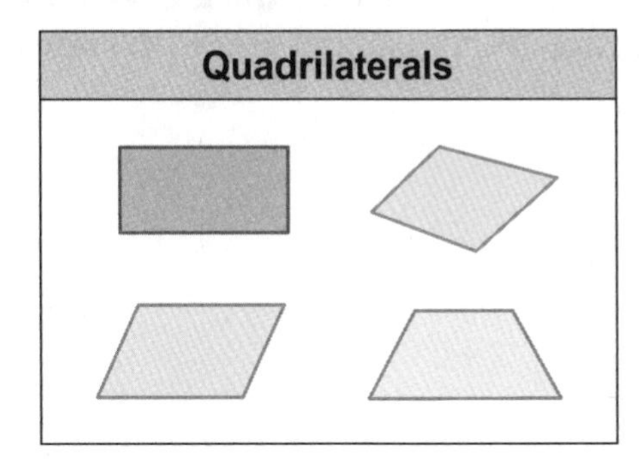

1 ACTIVITY: Using Descriptions to Form Quadrilaterals

Work with a partner. Use a geoboard to form a quadrilateral that fits the given description. Record your results on geoboard dot paper.

a. Form a quadrilateral with exactly one pair of parallel sides.

b. Form a quadrilateral with four congruent sides and four right angles.

c. Form a quadrilateral with four right angles that is *not* a square.

d. Form a quadrilateral with four congruent sides that is *not* a square.

e. Form a quadrilateral with two pairs of congruent adjacent sides and whose opposite sides are *not* congruent.

f. Form a quadrilateral with congruent and parallel opposite sides that is *not* a rectangle.

Name ____ Date __________

2 ACTIVITY: Naming Quadrilaterals

Work with a partner. Match the names *square, rectangle, rhombus, parallelogram, trapezoid,* and *kite* with your 6 drawings in Activity 1.

3 ACTIVITY: Forming Quadrilaterals

Work with a partner. Form each quadrilateral on your geoboard. Then move *only one* vertex to create the new type of quadrilateral. Record your results below.

a. Trapezoid  Kite

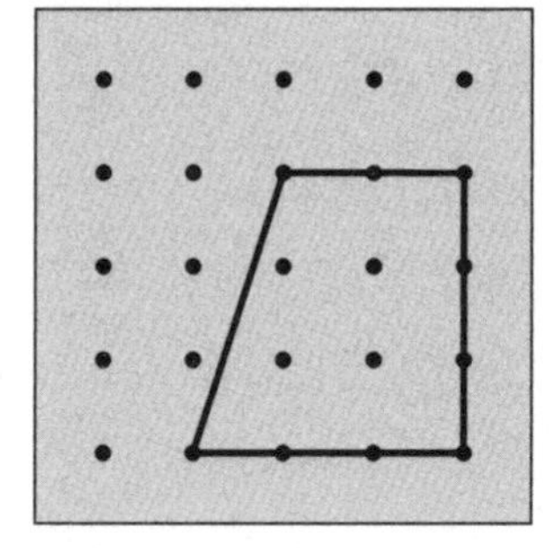

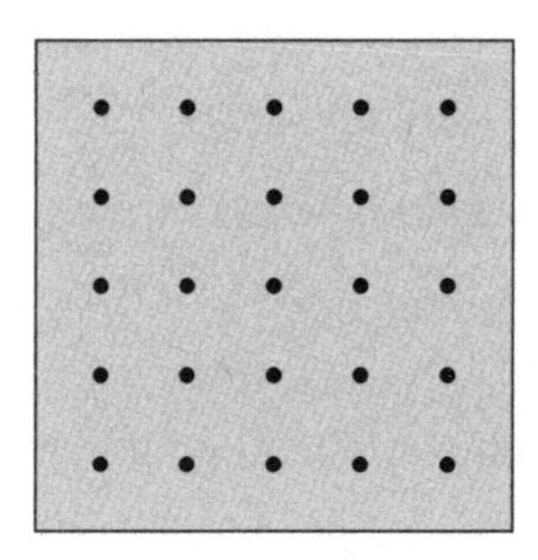

b. Kite  Rhombus (*not* a square)

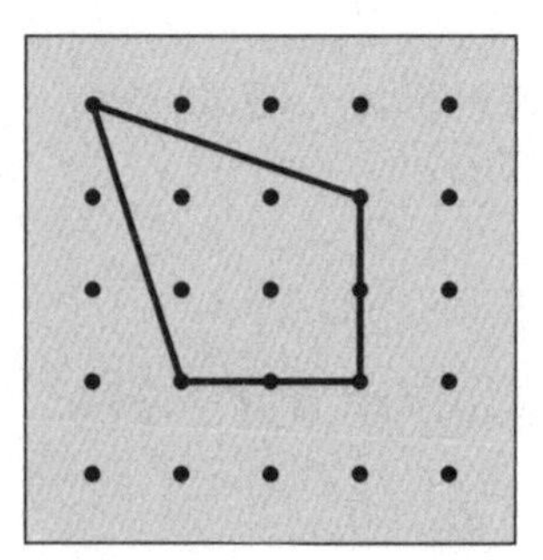

Name______________________________ Date__________

4 ACTIVITY: Using Technology to Draw Quadrilaterals

Work with a partner. Use geometry software to draw a quadrilateral that fits the given description.

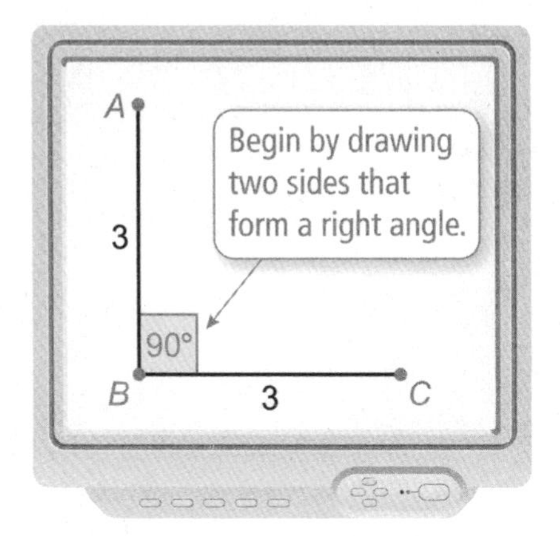

a. a square with a side length of 3 units

b. a rectangle with a width of 2 units and a length of 5 units

c. a parallelogram with side lengths of 6 units and 1 unit

d. a rhombus with a side length of 4 units

What Is Your Answer?

5. REASONING Measure the angles of each quadrilateral you formed in Activity 1. Record your results in a table. Include the sum of the angle measures. Then describe the pattern in the table and write a conclusion based on the pattern.

	$\angle 1$	$\angle 2$	$\angle 3$	$\angle 4$	$\angle 1 + \angle 2 + \angle 3 + \angle 4$
a.					
b.					
c.					
d.					
e.					
f.					

6. IN YOUR OWN WORDS How can you classify quadrilaterals? Explain using properties of sides and angles.

Name ____ Date __________

12.4 Practice

For use after Lesson 12.4

Classify the quadrilateral.

1.

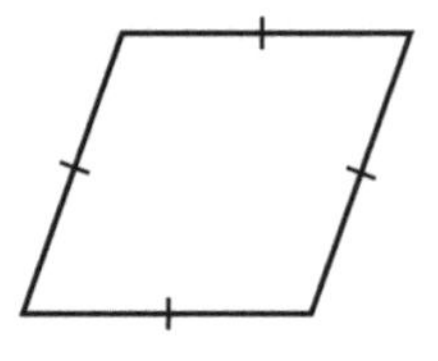

2.

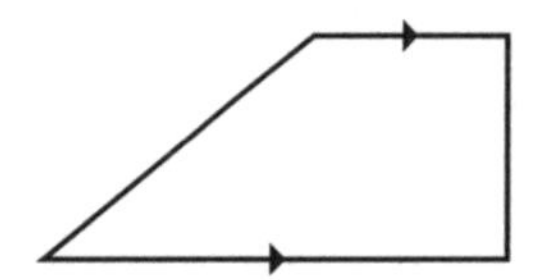

3.

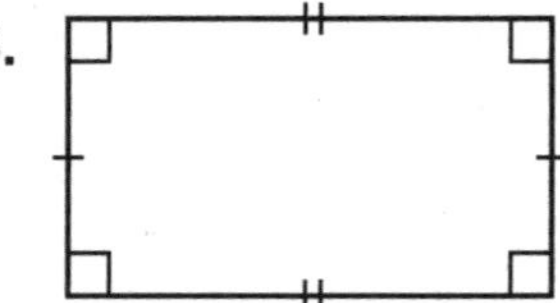

4.

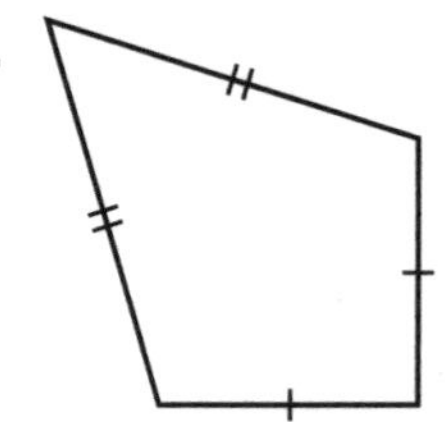

Find the value of *x*.

5.

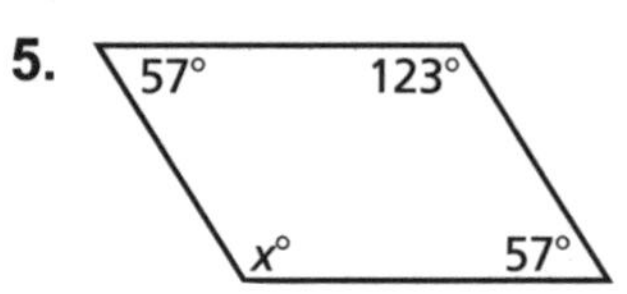

6.

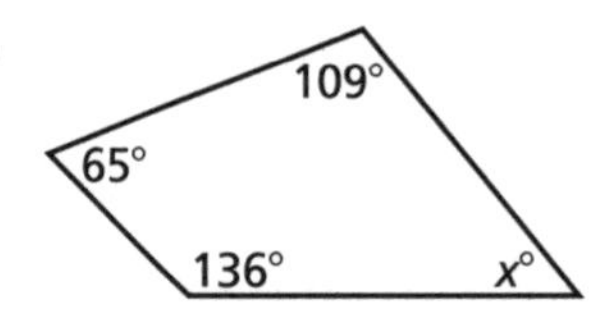

7. For a science fair, you are displaying your project on a trapezoidal piece of poster board. What is the measure of the missing angle?

132°

48° 48°

Name____ Date__________

12.5 Scale Drawings

For use with Activity 12.5

Essential Question How can you enlarge or reduce a drawing proportionally?

1 ACTIVITY: Comparing Measurements

Work with a partner. The diagram shows a food court at a shopping mall. Each centimeter in the diagram represents 40 meters.

a. Find the length and the width of the drawing of the food court.

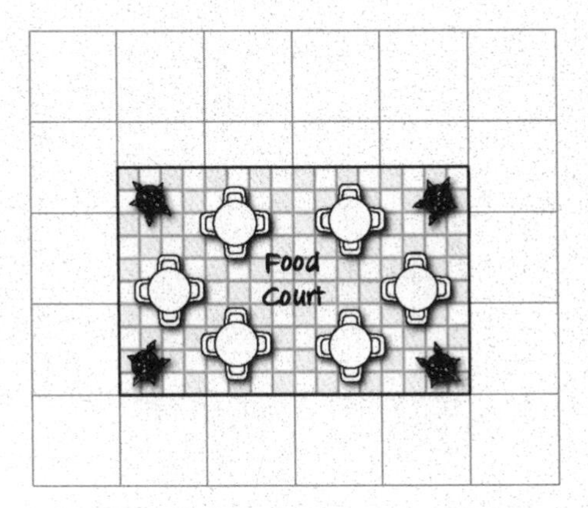

length: _________ cm width: _________ cm

b. Find the actual length and width of the food court. Explain how you found your answers.

length: _________ m width: _________ m

c. Find the ratios $\frac{\text{drawing length}}{\text{actual length}}$ and $\frac{\text{drawing width}}{\text{actual width}}$. What do you notice?

Name ______________________________ Date __________

12.5 Scale Drawings (continued)

2 ACTIVITY: Recreating a Drawing

Work with a partner. Draw the food court in Activity 1 on the grid paper so that each centimeter represents 20 meters.

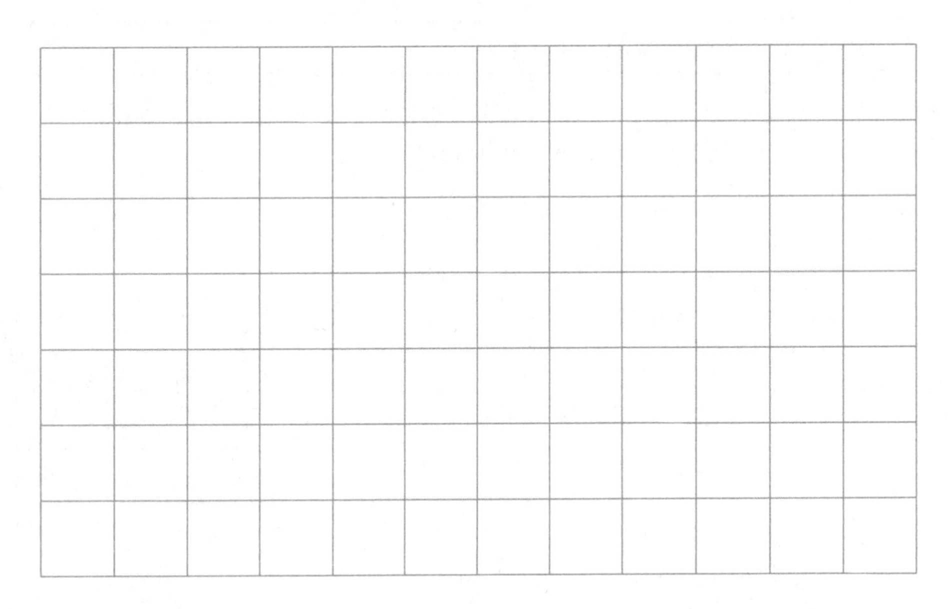

a. What happens to the size of the drawing?

b. Find the length and the width of your drawing. Compare these dimensions to the dimensions of the original drawing in Activity 1.

3 ACTIVITY: Comparing Measurements

Work with a partner. The diagram shows a sketch of a painting. Each unit in the sketch represents 8 inches.

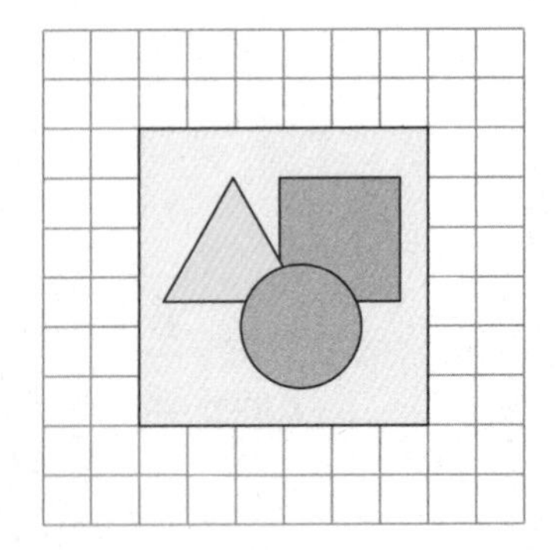

a. Find the length and the width of the sketch.

length: __________ units width: __________ units

b. Find the actual length and width of the painting. Explain how you found your answers.

length: __________ in. width: __________ in.

c. Find the ratios $\dfrac{\text{sketch length}}{\text{actual length}}$ and $\dfrac{\text{sketch width}}{\text{actual width}}$. What do you notice?

4 ACTIVITY: Recreating a Drawing

Work with a partner. Let each unit in the grid paper represent 2 feet. Now sketch the painting in Activity 3 onto the grid paper.

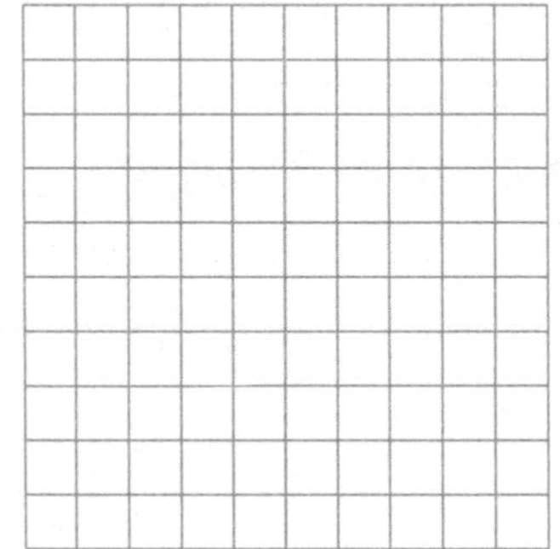

a. What happens to the size of the sketch?

b. Find the length and the width of your sketch. Compare these dimensions to the dimensions of the original sketch in Activity 3.

What Is Your Answer?

5. **IN YOUR OWN WORDS** How can you enlarge or reduce a drawing proportionally?

6. Complete the table for both the food court and the painting.

	Actual Object	Original Drawing	Your Drawing
Perimeter			
Area			

Compare the measurements in each table. What conclusions can you make?

7. **RESEARCH** Look at some maps in your school library or on the Internet. Make a list of the different scales used on the maps.

8. When you view a map on the Internet, how does the scale change when you zoom out? How does the scale change when you zoom in?

Name ______________________________ Date __________

12.5 Practice

For use after Lesson 12.5

Find the missing dimension. Use the scale factor 1 : 8.

Item	Model	Actual
1. Statue	Height: 168 in.	Height ________ ft
2. Painting	Width: ________ cm	Width: 200 m
3. Alligator	Height: ________ in.	Height: 6.4 ft
4. Train	Length: 36.5 in.	Length: ________ ft

5. The diameter of the moon is 2160 miles. A model has a scale of 1 in. : 150 mi. What is the diameter of the model?

6. A map has a scale of 1 in. : 4 mi.

a. You measure 3 inches between your house and the movie theater. How many miles is it from your house to the movie theater?

b. It is 17 miles to the mall. How many inches is that on the map?

Name______________________________ Date__________

Chapter 13 Fair Game Review

Identify the basic shapes in the figure.

1.

2.

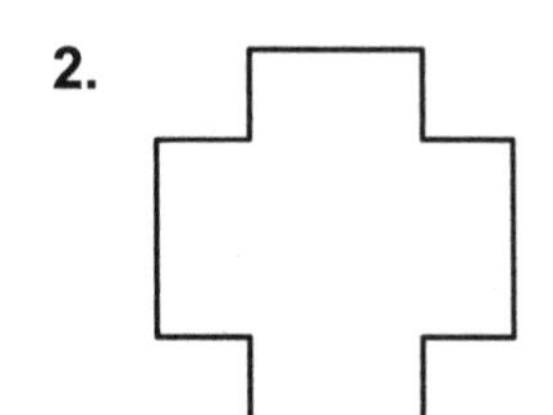

3.

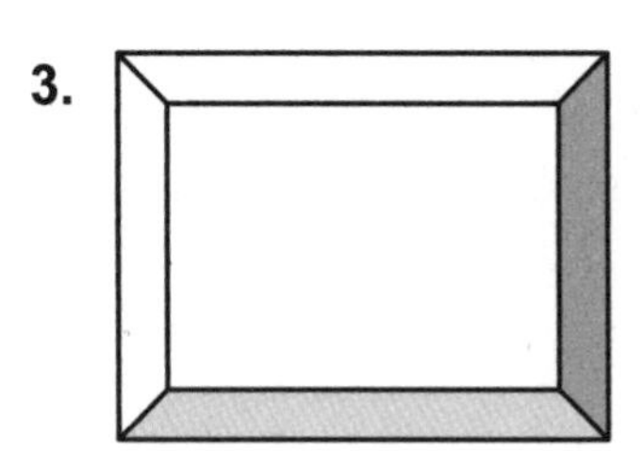

4.

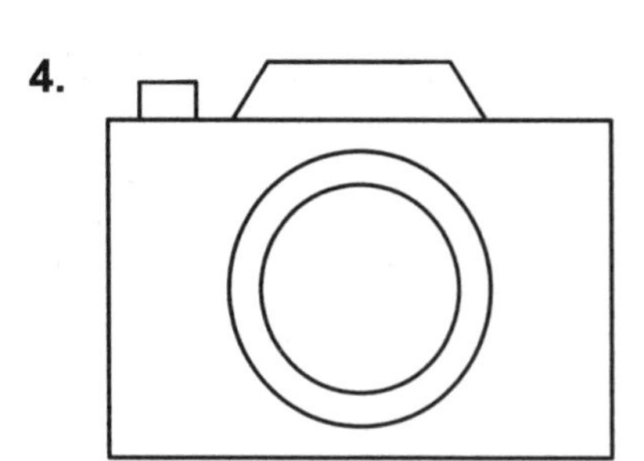

5. Identify the basic shapes that make up the top of your teacher's desk.

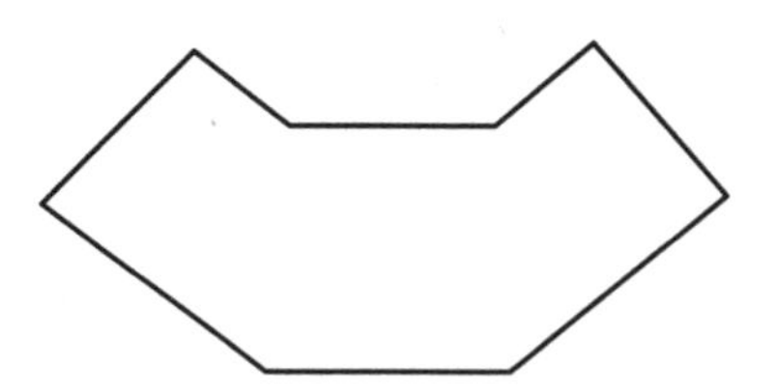

Name ______________________ Date __________

Fair Game Review (continued)

Evaluate the expression.

6. 7^2

7. 11^2

8. $4(5)^2$

9. $7 \bullet 10^2$

10. $4(4 + 2)^2$

11. $5(6 + 3)^2$

12. $6(8 + 3)^2 + 2 \bullet 9$

13. $4(12)^2 - (6 + 4)$

14. A kilometer is 10^3 meters. You run a 5-kilometer race. How many meters do you run?

Name____ Date__________

13.1 Circles and Circumference

For use with Activity 13.1

Essential Question How can you find the circumference of a circle?

Archimedes was a Greek mathematician, physicist, engineer, and astronomer.

Archimedes discovered that in any circle the ratio of circumference to diameter is always the same. Archimedes called this ratio pi, or π (a letter from the Greek alphabet).

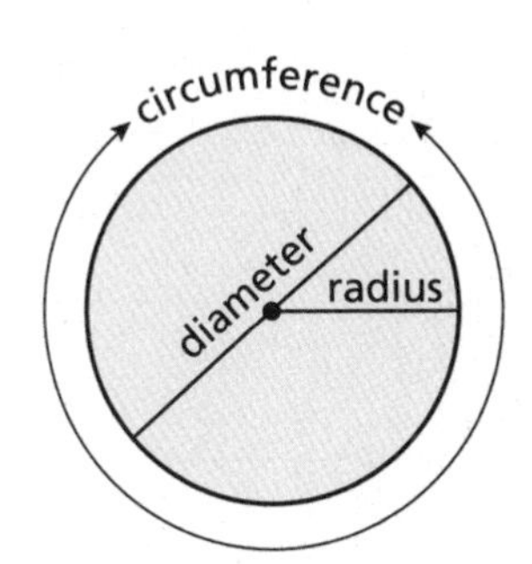

$$\pi = \frac{\text{circumference}}{\text{diameter}}$$

In Activities 1 and 2, you will use the same strategy Archimedes used to approximate π.

1 ACTIVITY: Approximating Pi

Work with a partner. Record your results in the first row of the table on the next page.

- Measure the perimeter of the large square in millimeters.
- Measure the diameter of the circle in millimeters.
- Measure the perimeter of the small square in millimeters.
- Calculate the ratios of the two perimeters to the diameter.
- The average of these two ratios is an approximation of π.

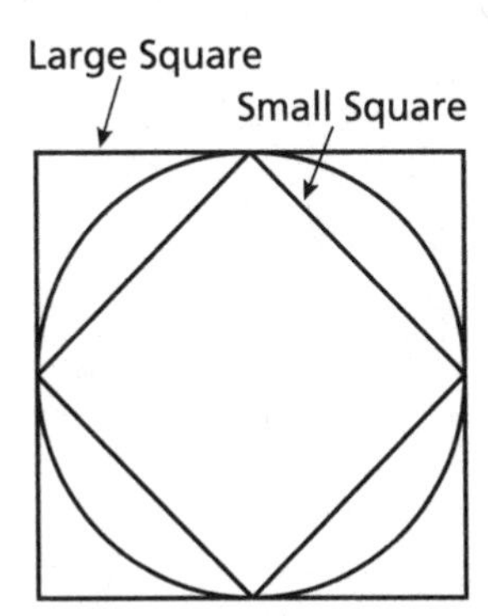

Name ____ Date __________

Sides	Large Perimeter	Diameter of Circle	Small Perimeter	$\frac{\text{Large Perimeter}}{\text{Diameter}}$	$\frac{\text{Small Perimeter}}{\text{Diameter}}$	Average of Ratios
4						
6						
8						
10						

2 ACTIVITY: Approximating Pi

Continue your approximation of pi. Complete the table above using a hexagon (6 sides), an octagon (8 sides), and a decagon (10 sides).

a. 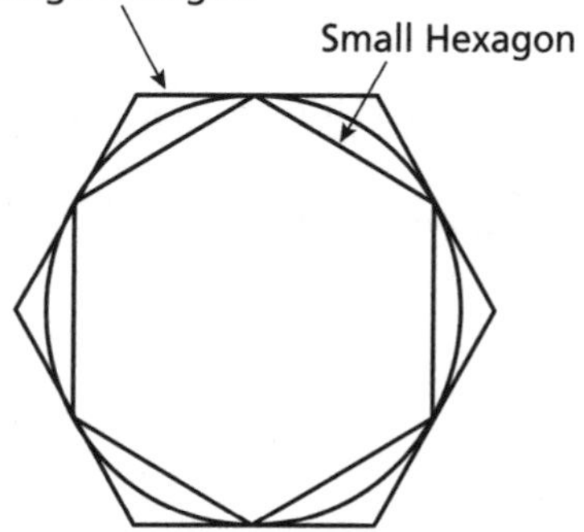

b. 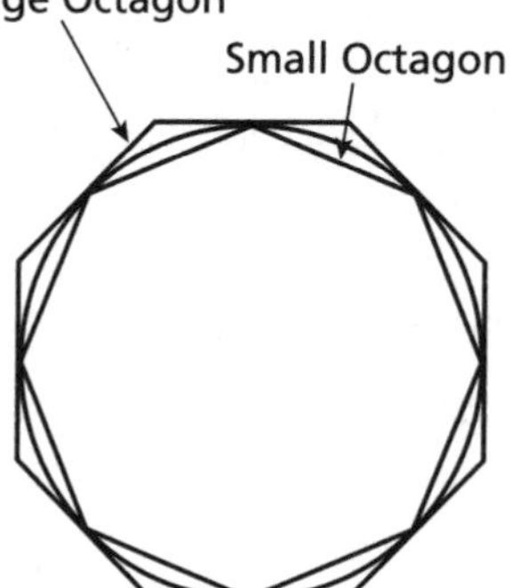

c. 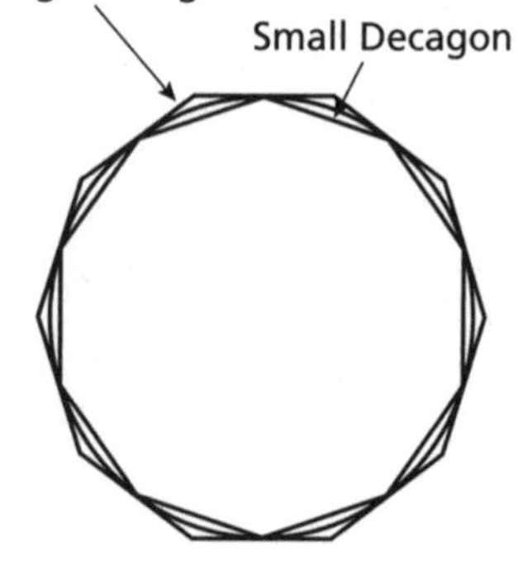

d. From the table, what can you conclude about the value of π? Explain your reasoning.

e. Archimedes calculated the value of π using polygons with 96 sides. Do you think his calculations were more or less accurate than yours?

What Is Your Answer?

3. IN YOUR OWN WORDS Now that you know an approximation for pi, explain how you can use it to find the circumference of a circle. Write a formula for the circumference C of a circle whose diameter is d.

4. CONSTRUCTION Use a compass to draw three circles. Use your formula from Question 3 to find the circumference of each circle.

Name Roselynn Sim Date ________

13.1 Practice

For use after Lesson 13.1

50k + 5 4 + 16k

1. Find the diameter of the circle.

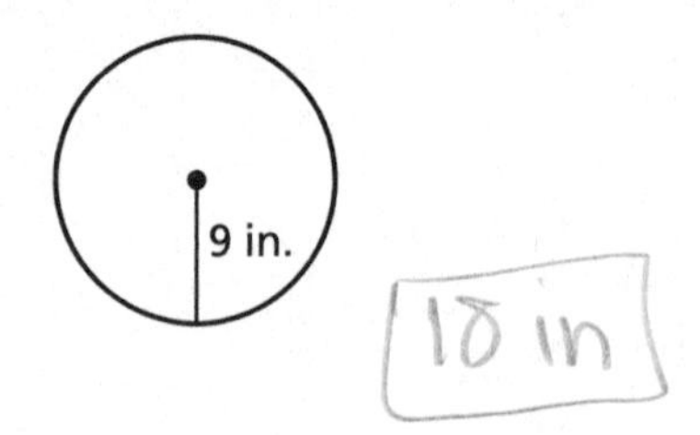

18 in

2. Find the radius of the circle.

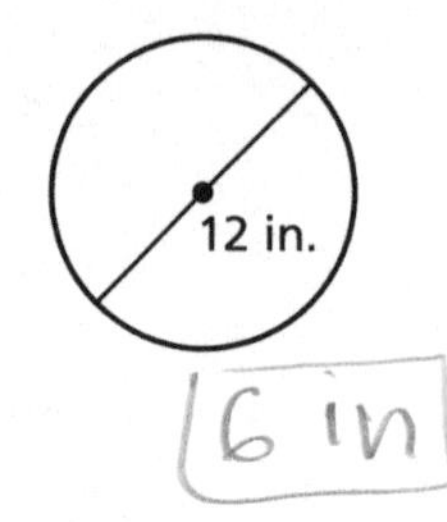

6 in

$2\frac{3}{4}$ 2.75

Find the circumference of the circle. Use 3.14 or $\frac{22}{7}$ for π.

3.

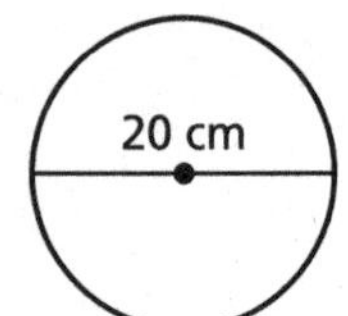

C = 3.14 • 20 = 62.8 cm

4.

C = 3.14 • 28 = 87.92 in

5.

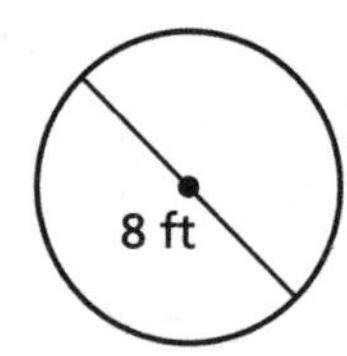

C = 3.14 • 8 = 25.12 ft

Find the perimeter of the semicircular region.

6.

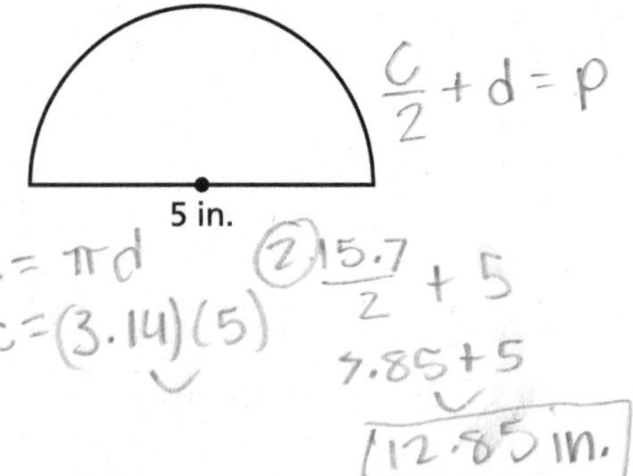

$\frac{C}{2} + d = P$

(1) C = πd
C = (3.14)(5)

(2) $\frac{15.7}{2} + 5$

7.85 + 5

12.85 in.

7.

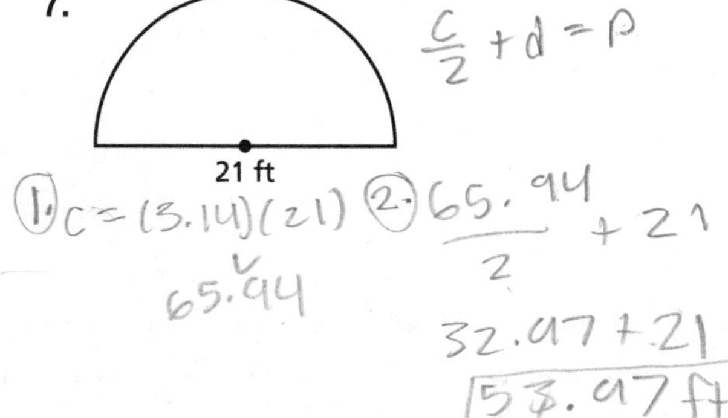

$\frac{C}{2} + d = P$

(1) C = (3.14)(21)
65.94

(2) $\frac{65.94}{2} + 21$

32.97 + 21

53.97 ft

8. A simple impact crater on the moon has a diameter of 15 kilometers. A complex impact crater has a radius of 30 kilometers. How much greater is the circumference of the complex impact crater than the simple impact crater?

The complex impact crater is 141.3 kilometers greater than the simple impact crater.

Simple: 15 — 15 • 3.14 = 47.1

Complex: 30 — 60 • 3.14 = 188.4

Name______________________________ Date__________

13.2 Perimeters of Composite Figures

For use with Activity 13.2

Essential Question How can you find the perimeter of a composite figure?

1 ACTIVITY: Finding a Pattern

Work with a partner. Describe the pattern of the perimeters. Use your pattern to find the perimeter of the tenth figure in the sequence. (Each small square has a perimeter of 4.)

a.

b.

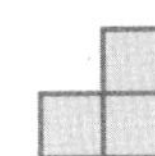
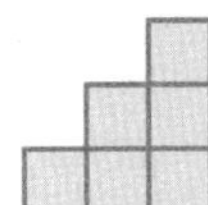
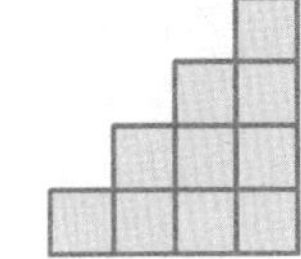

c.

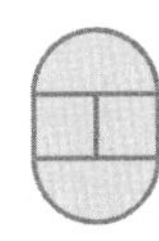
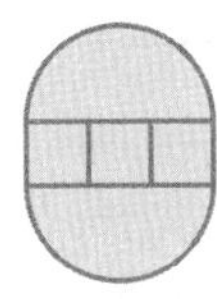
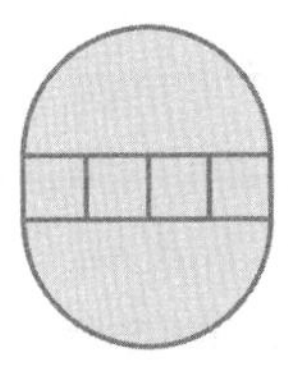

Name ______________________________ Date ________

2 ACTIVITY: Combining Figures

Work with a partner.

a. A rancher is constructing a rectangular corral and a trapezoidal corral, as shown. How much fencing does the rancher need to construct both corrals?

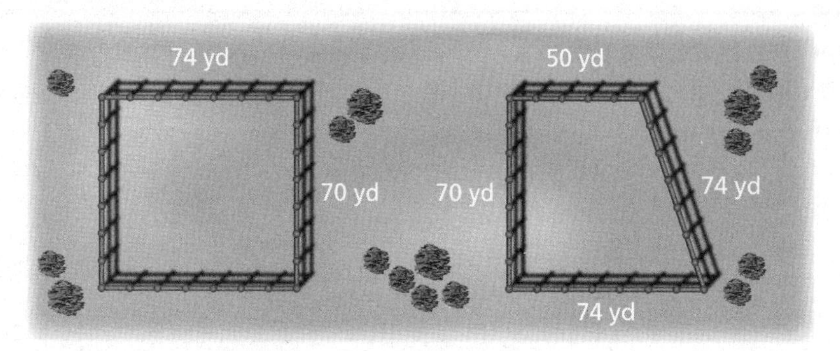

b. Another rancher is constructing one corral by combining the two corrals above, as shown. Does this rancher need more or less fencing? Explain your reasoning.

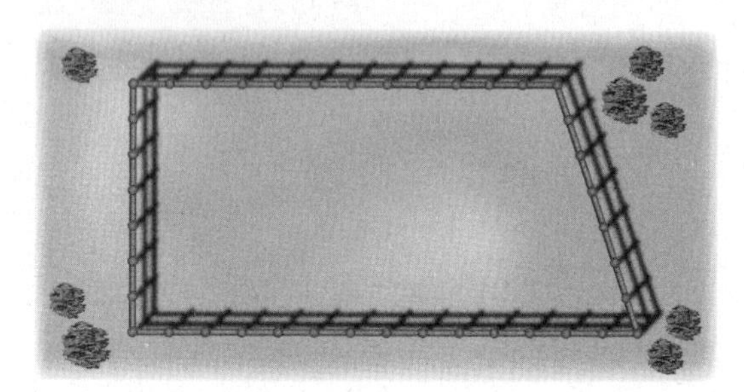

c. How can the rancher in part (b) combine the two corrals to use even less fencing?

3 ACTIVITY: Submitting a Bid

Work with a partner. You want to bid on a tiling contract. You will be supplying and installing the tile that borders the swimming pool shown on the next page. In the figure, each grid square represents 1 square foot.

- **Your cost for the tile is $4 per linear foot.**
- **It takes about 15 minutes to prepare, install, and clean each foot of tile.**

a. How many tiles do you need for the border?

b. Write a bid for how much you will charge to supply and install the tile. Include what you want to charge as an hourly wage. Estimate what you think your profit will be.

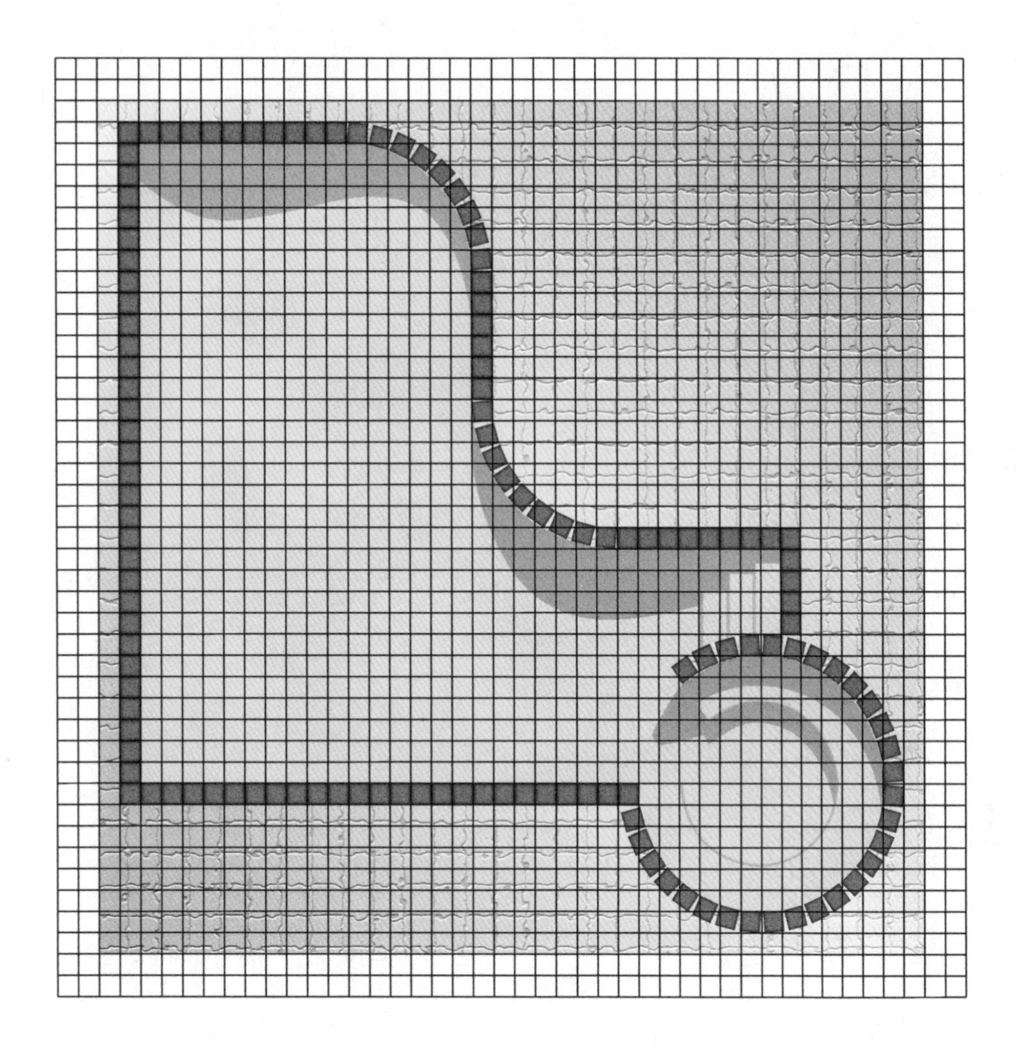

What Is Your Answer?

4. IN YOUR OWN WORDS How can you find the perimeter of a composite figure? Use a semicircle, a triangle, and a parallelogram to draw a composite figure. Label the dimensions. Find the perimeter of the figure.

Name ______________________________ Date __________

13.2 Practice

For use after Lesson 13.2

Estimate the perimeter of the figure.

1.

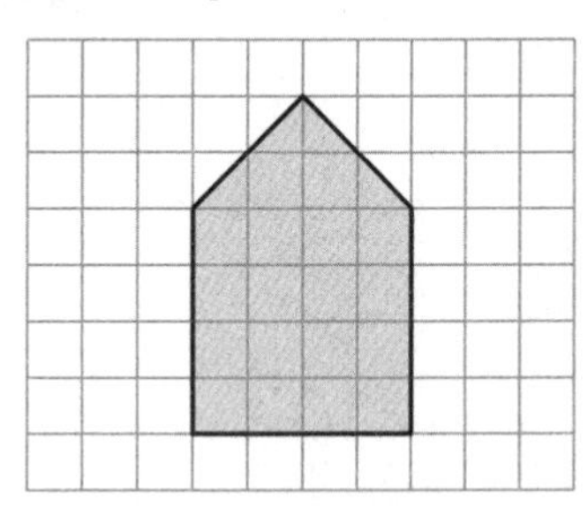

2.

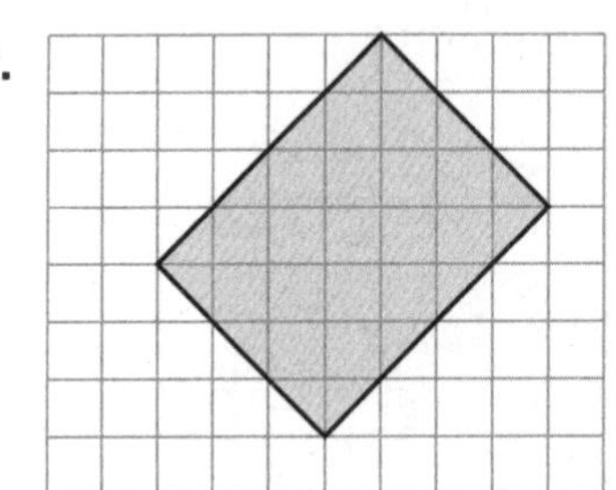

Find the perimeter of the figure.

3.

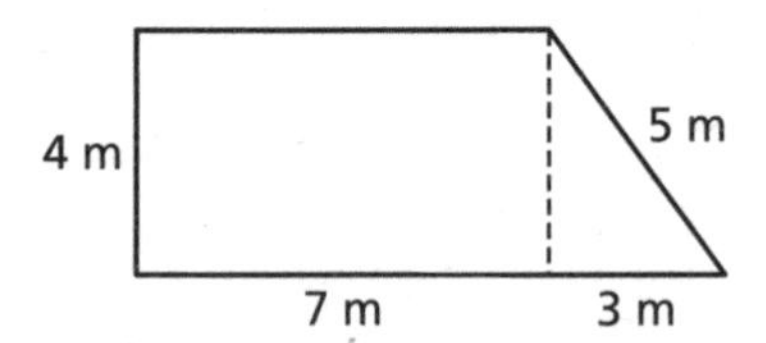

4.

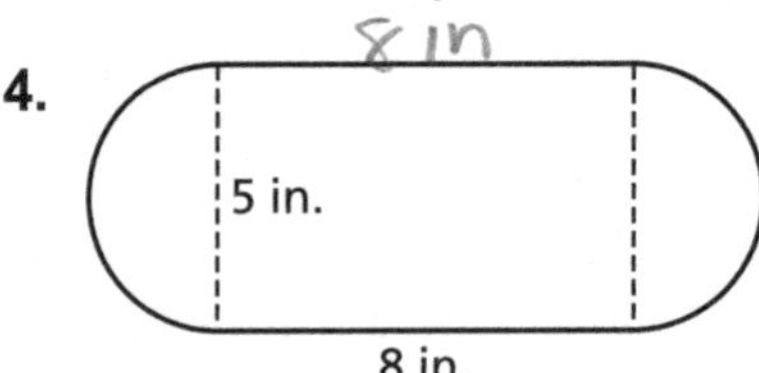

5. You are having a swimming pool installed.

a. Find the perimeter of the swimming pool.

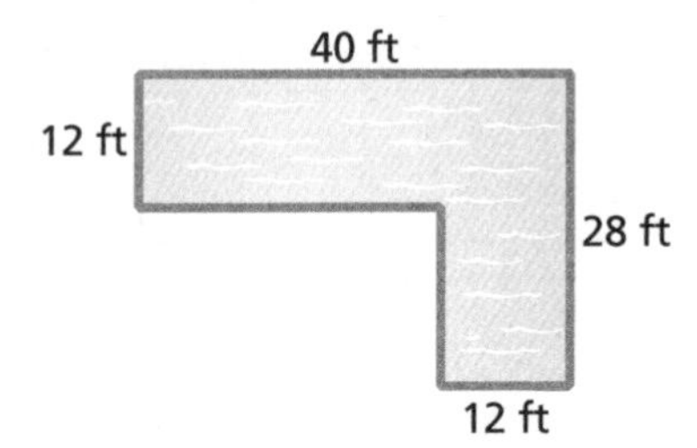

b. Tiling costs $15 per yard. How much will it cost to put tiles along the edge of the pool?

Name____ Date__________

13.3 Areas of Circles

For use with Activity 13.3

Essential Question How can you find the area of a circle?

1 ACTIVITY: Estimating the Area of a Circle

Work with a partner. Each square in the grid is 1 unit by 1 unit.

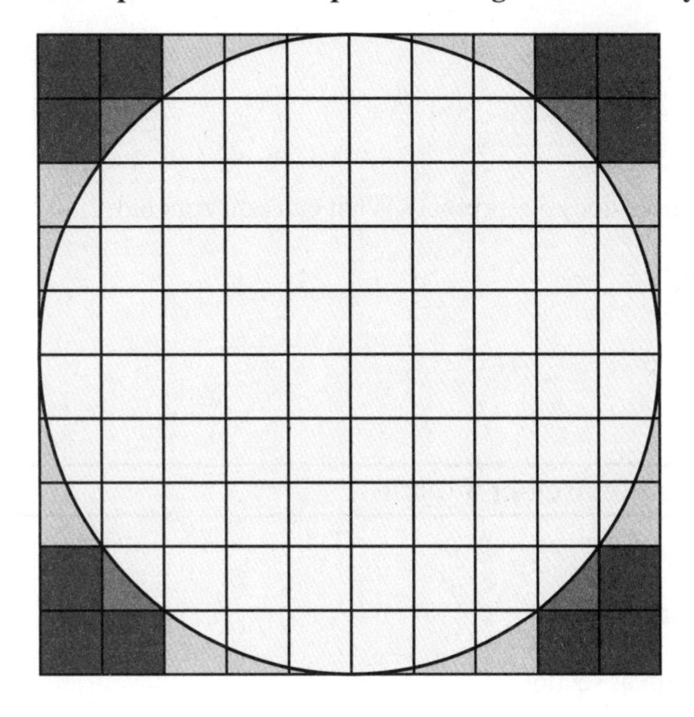

a. Find the area of the large 10-by-10 square.

b. Complete the table.

Region			
Area (square units)			

c. Use your results to estimate the area of the circle. Explain your reasoning.

Name ____ Date __________

d. Fill in the blanks. Explain your reasoning.

Area of large square = ______ • 5^2 square units

Area of circle ≈ ______ • 5^2 square units

e. What dimension of the circle does 5 represent? What can you conclude?

2 ACTIVITY: Approximating the Area of a Circle

Work with a partner.

a. Draw a circle. Label the radius as r.*

b. Divide the circle into 24 equal sections.

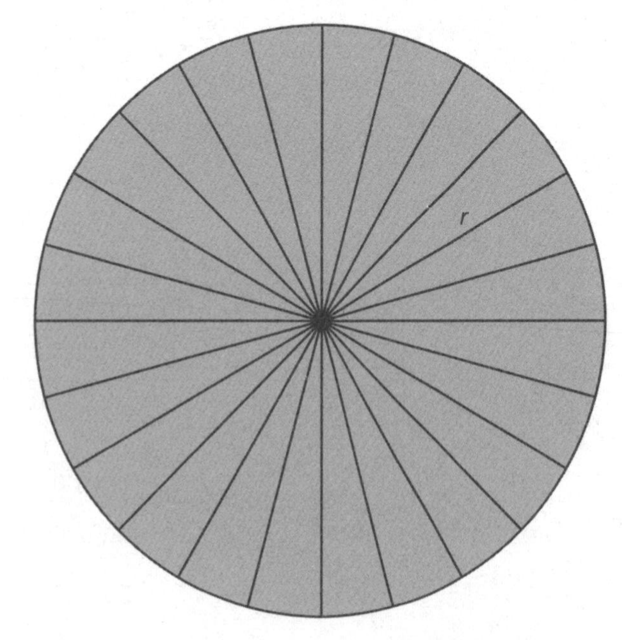

*Cut-outs are available in the back of the Record and Practice Journal.

13.3 Areas of Circles (continued)

c. Cut the sections apart. Then arrange them to approximate a parallelogram.

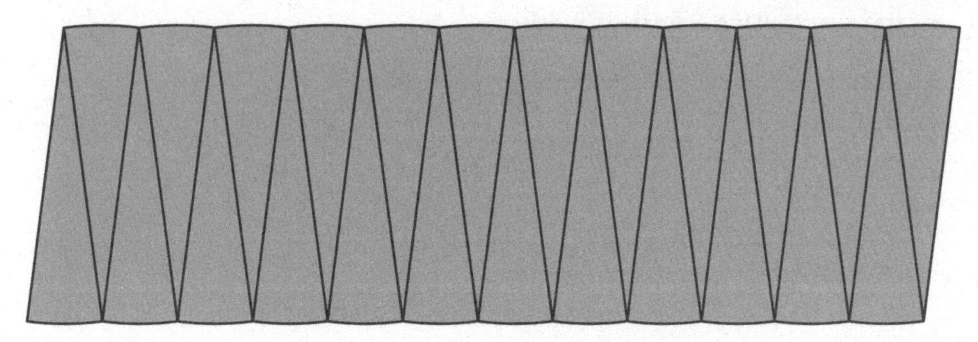

d. What is the approximate height and base of the parallelogram?

e. Find the area of the parallelogram. What can you conclude?

What Is Your Answer?

3. IN YOUR OWN WORDS How can you find the area of a circle?

4. Write a formula for the area of a circle with radius r. Find an object that is circular. Use your formula to find the area.

Name ____ Date __________

13.3 Practice

For use after Lesson 13.3

Find the area of the circle. Use 3.14 or $\frac{22}{7}$ for π.

1. 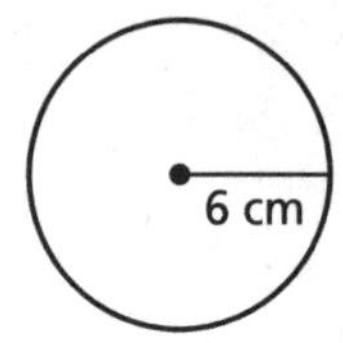

2. 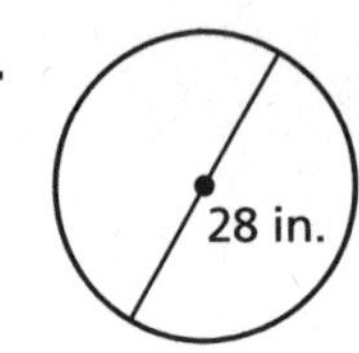

Find the area of the semicircle.

3. 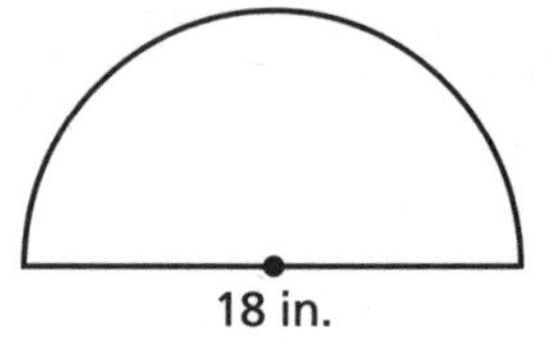

4. 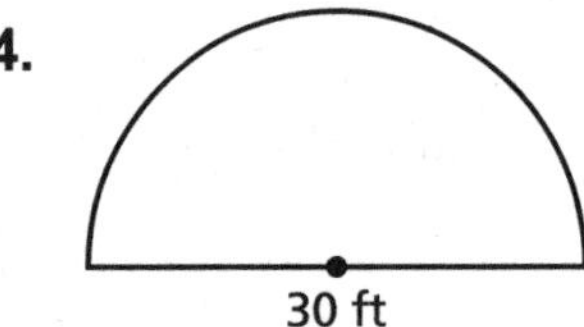

5. An FM radio station signal travels in a 40-mile radius. An AM radio station signal travels in a 4-mile radius. How much more area does the FM station cover than the AM station?

Name____ Date__________

13.4 Areas of Composite Figures

For use with Activity 13.4

Essential Question How can you find the area of a composite figure?

1 ACTIVITY: Estimating Area

Work with a partner.

a. Choose a state. On grid paper, draw a larger outline of the state.

b. Use your drawing to estimate the area (in square miles) of the state.

c. Which state areas are easy to find? Which are difficult? Why?

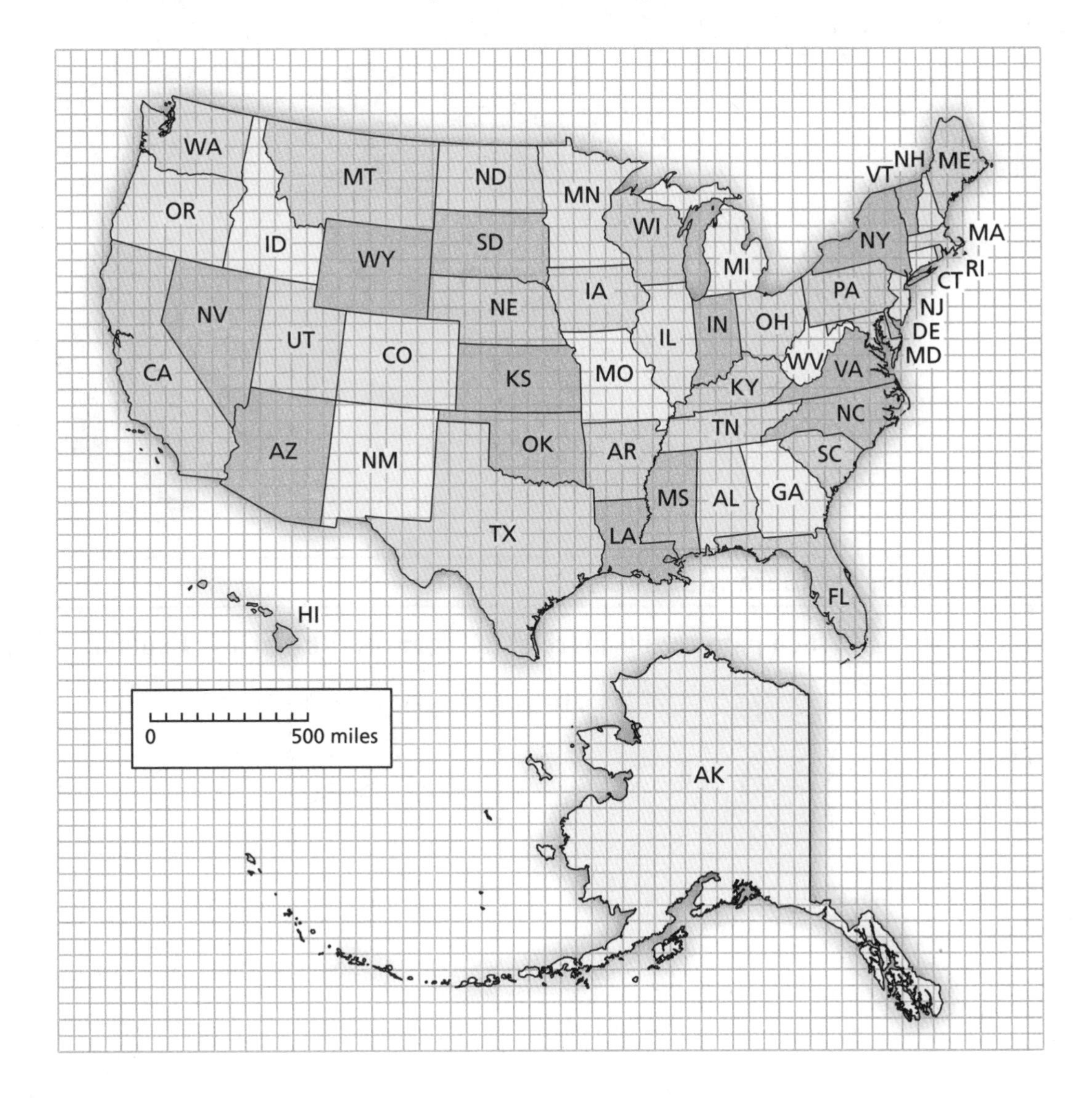

Name ____ Date __________

13.4 Areas of Composite Figures (continued)

2 ACTIVITY: Estimating Areas

Work with a partner. The completed puzzle has an area of 150 square centimeters.*

a. Estimate the area of each puzzle piece.

b. Check your work by adding the six areas. Why is this a check?

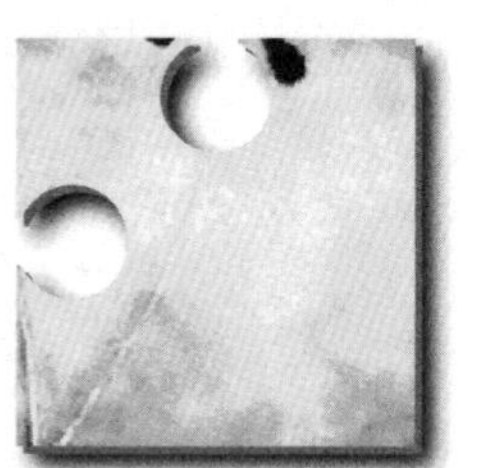

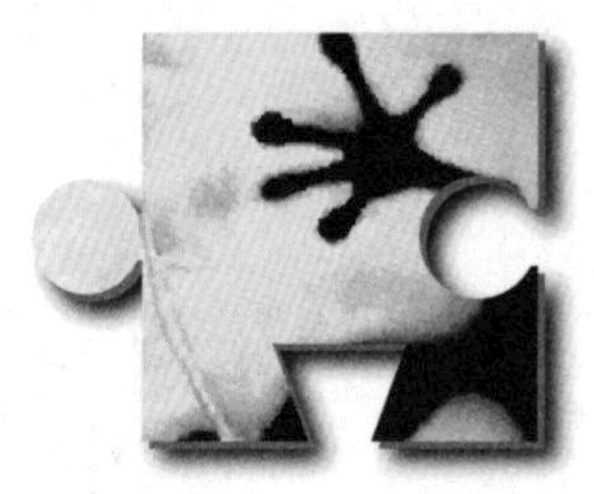

3 ACTIVITY: Filling a Square with Circles

Work with a partner. Which pattern fills more of the square with circles? Explain.

a.

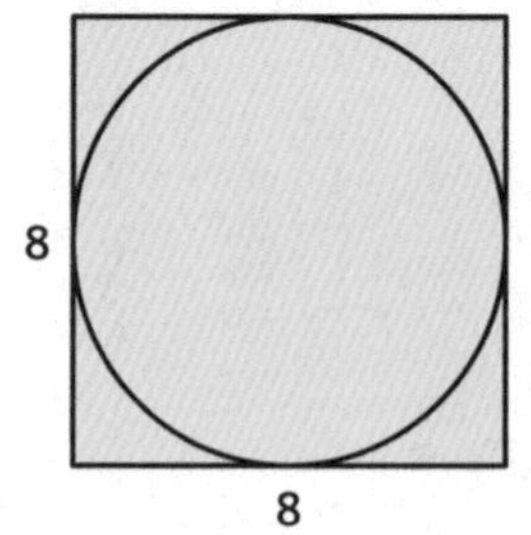

b.

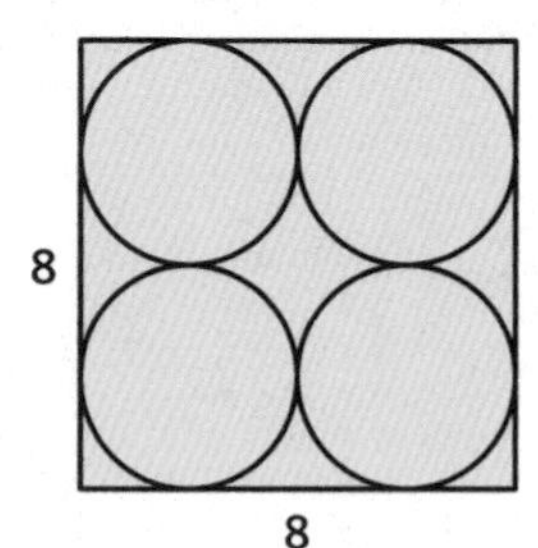

*Cut-outs are available in the back of the Record and Practice Journal.

13.4 Areas of Composite Figures (continued)

c.

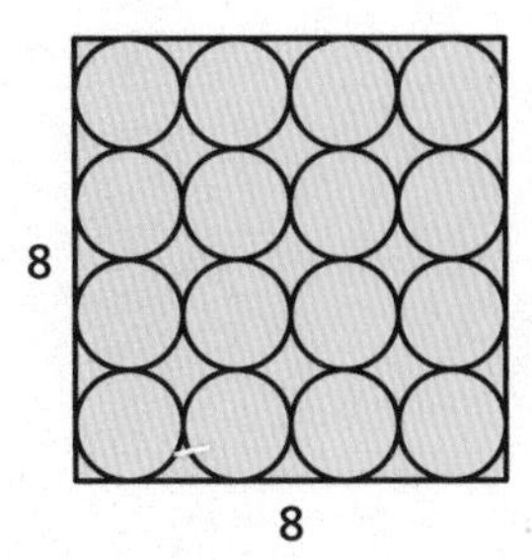

d.

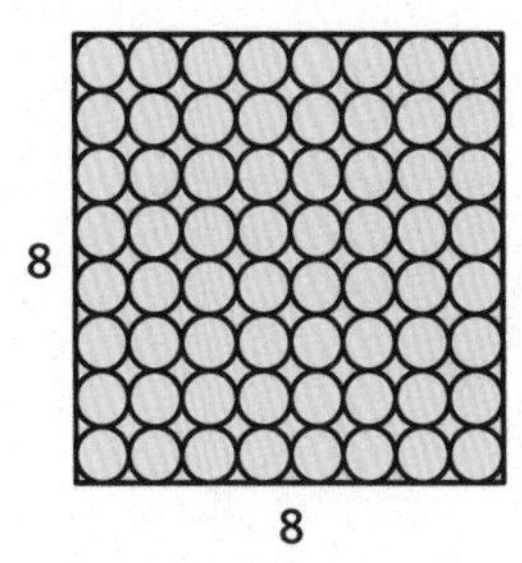

What Is Your Answer?

4. **IN YOUR OWN WORDS** How can you find the area of a composite figure?

5. Summarize the area formulas for all the basic figures you have studied. Draw a single composite figure that has each type of basic figure. Label the dimensions and find the total area.

Name ______________________________ Date __________

13.4 Practice
For use after Lesson 13.4

Find the area of the figure.

1. 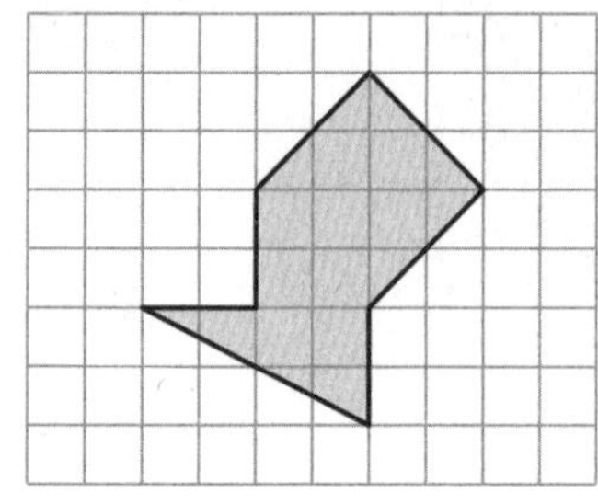

2. 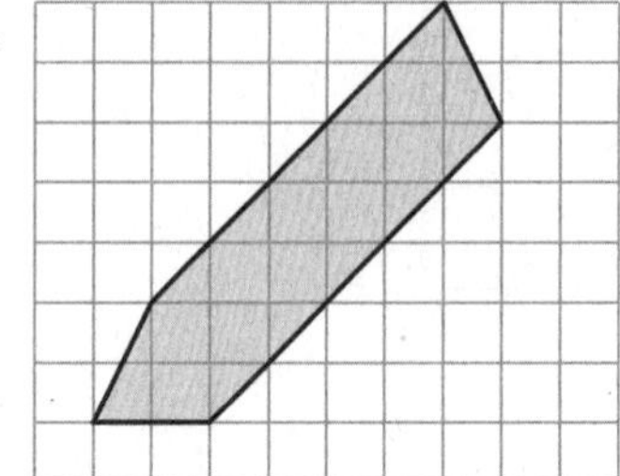

18 12

Find the area of the figure.

3. 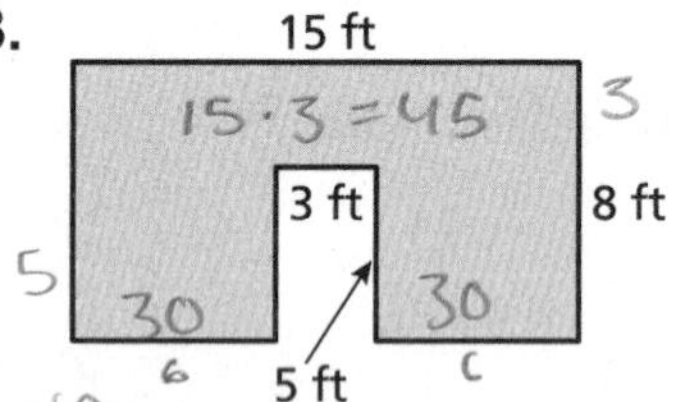

1. Identify shapes
   3 rectangles
2. Formula
   A = L·W
3. Dimensions
4. Find each area
5. Add all areas

45+30+30

105 ft

4. 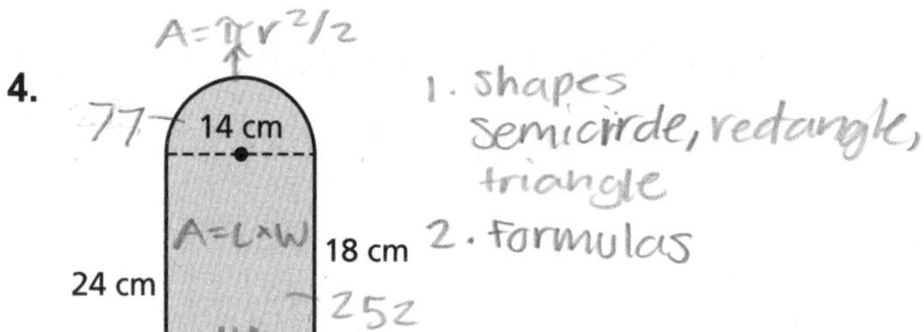

77 + 252 + 42

371 cm

5. The diagram shows the shape of the green of a miniature golf hole. What is the area of the green?

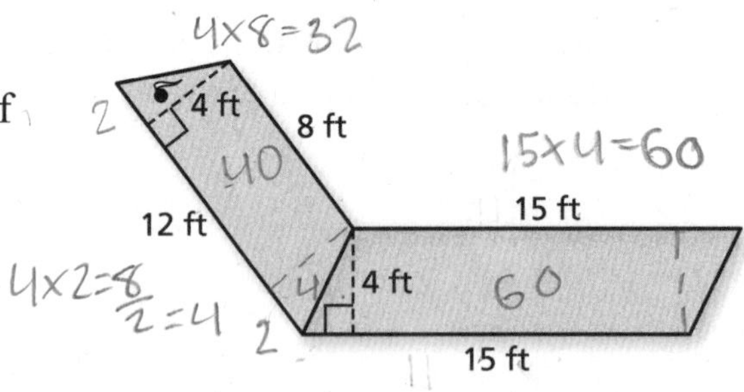

The area of the green is 100 ft.

(I can't tell what the color is so I'm assuming it means the whole figure.)

32+4+4=40

40+60=100 ft

345

Name________________________________ Date__________

Chapter 14 Fair Game Review

Find the area of the square or rectangle.

1.

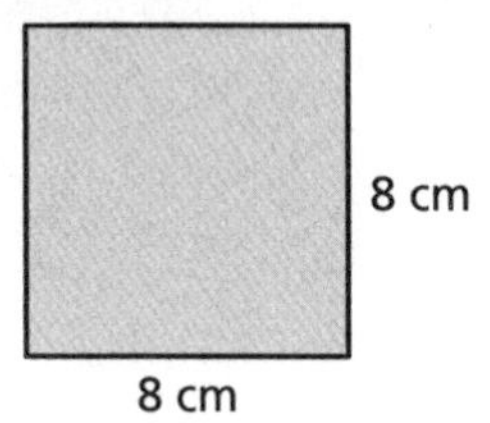

2.

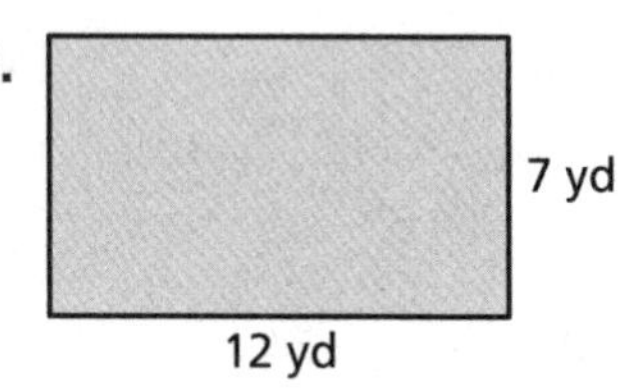

3.

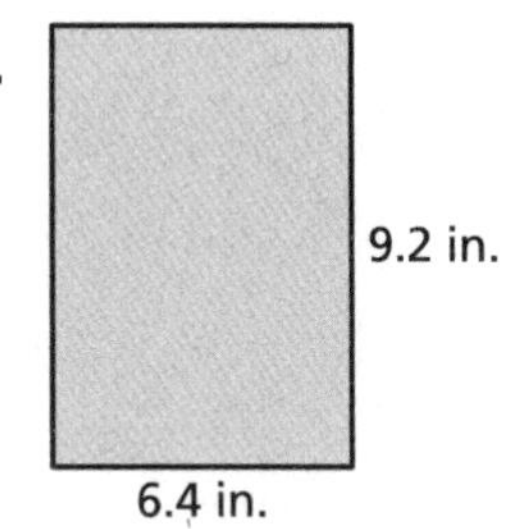

4.

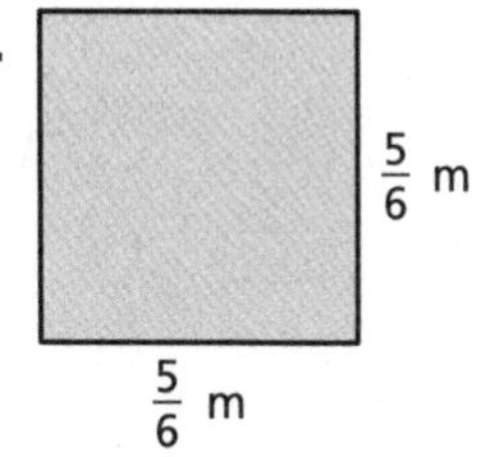

5.

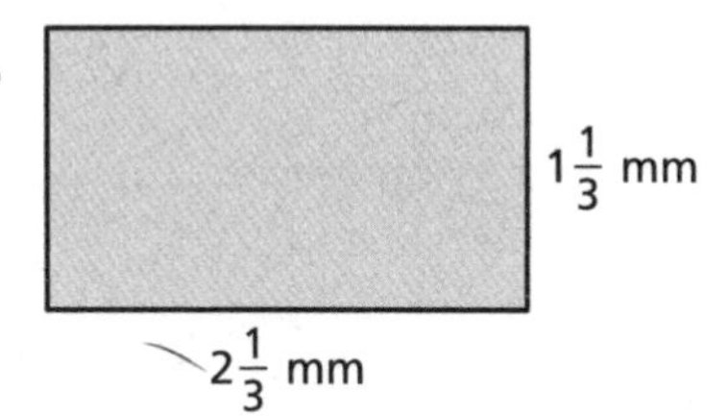

6.

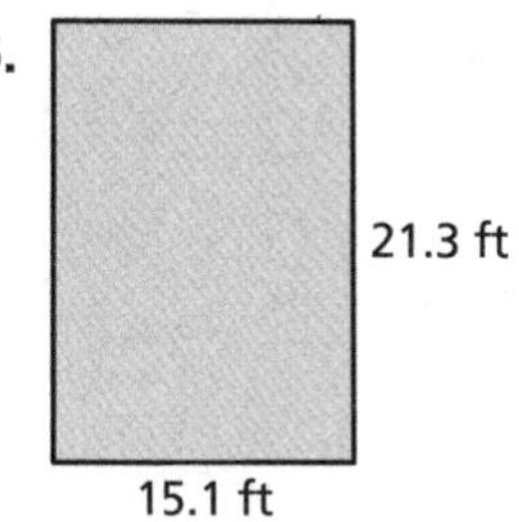

7. An artist buys a square canvas with a side length of 2.5 feet. What is the area of the canvas?

Name ______________________ Date ________

Chapter 14 Fair Game Review (continued)

Find the area of the triangle.

8.

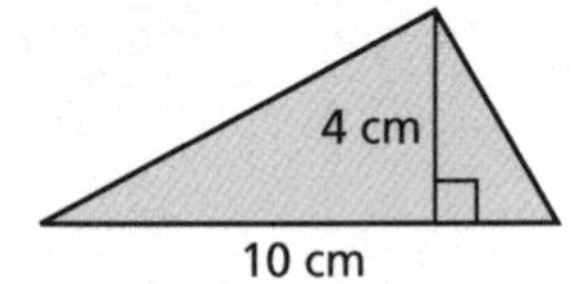

9.

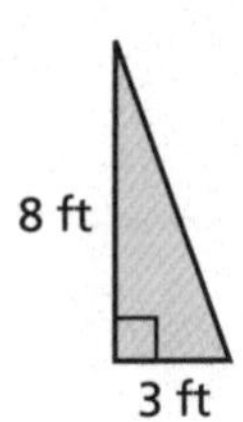

10.

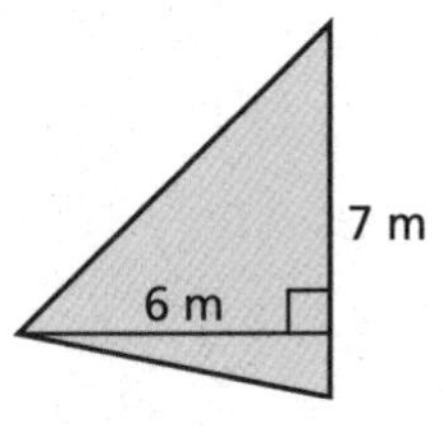

11.

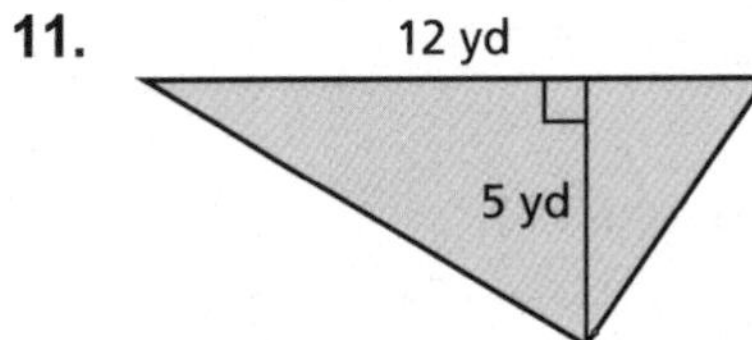

12.

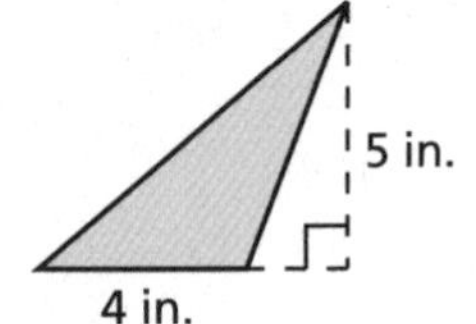

13.

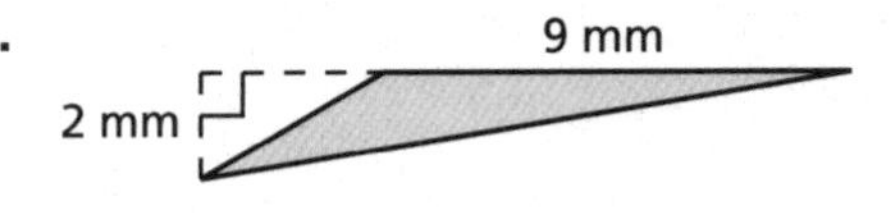

14. A spirit banner for a pep rally has the shape of a triangle. The base of the banner is 8 feet and the height is 6 feet. Find the area of the banner.

Name____ Date__________

14.1 Surface Areas of Prisms

For use with Activity 14.1

Essential Question How can you find the surface area of a prism?

1 ACTIVITY: Surface Area of a Rectangular Prism

Work with a partner. Use the net for a rectangular prism. Label each side as *h*, *w*, or *ℓ*. Then write a formula for the surface area of a rectangular prism.

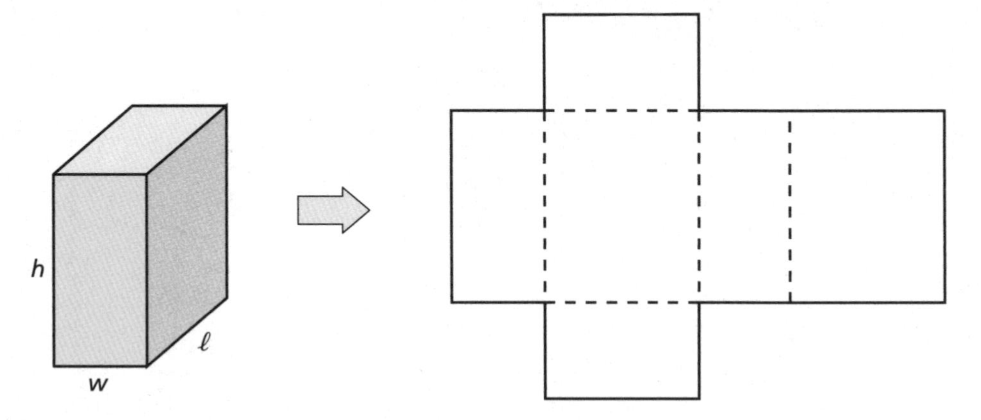

2 ACTIVITY: Surface Area of a Triangular Prism

Work with a partner.

a. Find the surface area of the solid shown by the net. Use a cut-out of the net.* Fold it to form a solid. Identify the solid.

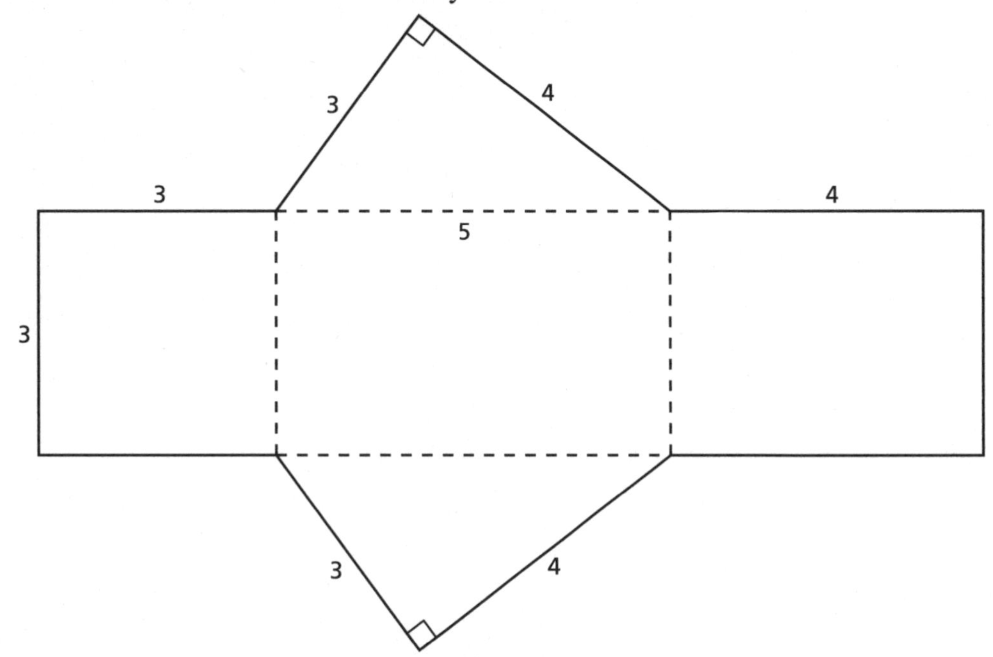

b. Which of the surfaces of the solid are bases? Why?

*Cut-outs are available in the back of the Record and Practice Journal.

Name ______________________________ Date __________

14.1 Surface Areas of Prisms (continued)

3 ACTIVITY: Forming Rectangular Prisms

Work with a partner.

- **Use 24 one-inch cubes to form a rectangular prism that has the given dimensions.**
- **Draw each prism.**
- **Find the surface area of each prism.**

a. $4 \times 3 \times 2$

b. $1 \times 1 \times 24$

c. $1 \times 2 \times 12$

d. $1 \times 3 \times 8$

e. $1 \times 4 \times 6$

f. $2 \times 2 \times 6$

g. $2 \times 4 \times 3$

What Is Your Answer?

4. Use your formula from Activity 1 to verify your results in Activity 3.

5. **IN YOUR OWN WORDS** How can you find the surface area of a prism?

6. **REASONING** When comparing ice blocks with the same volume, the ice with the greater surface area will melt faster. Which will melt faster, the bigger block or the three smaller blocks? Explain your reasoning.

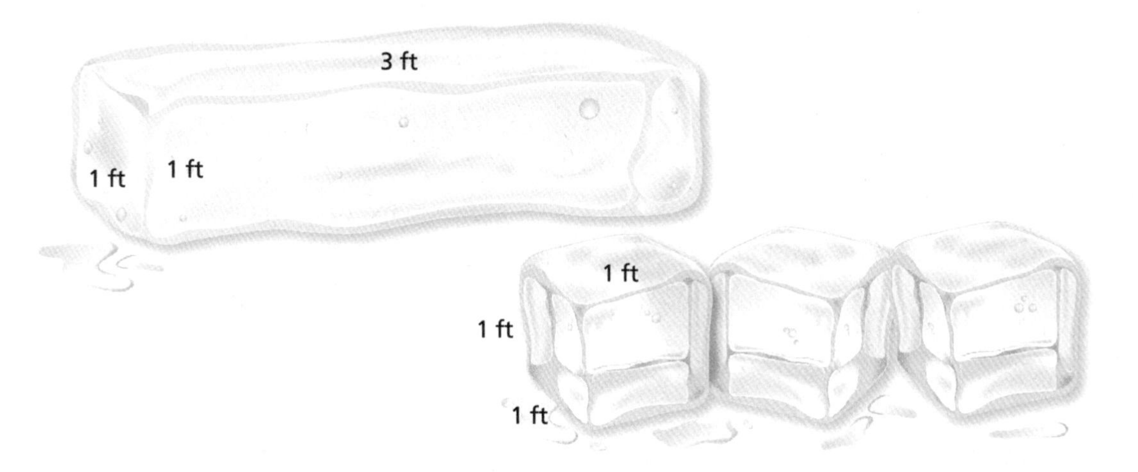

Name ______________________________ Date __________

14.1 Practice

For use after Lesson 14.1

Find the surface area of the prism.

1.

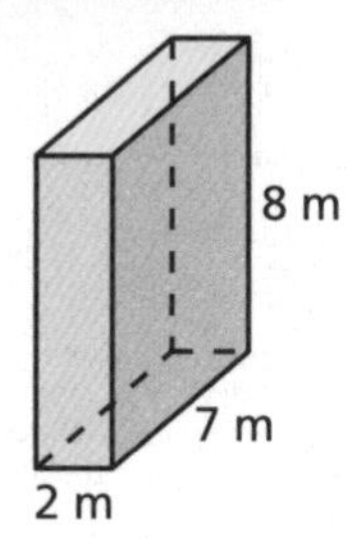

2.

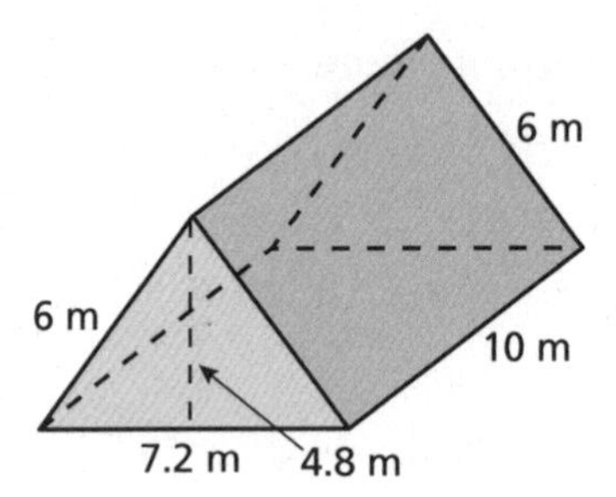

3.

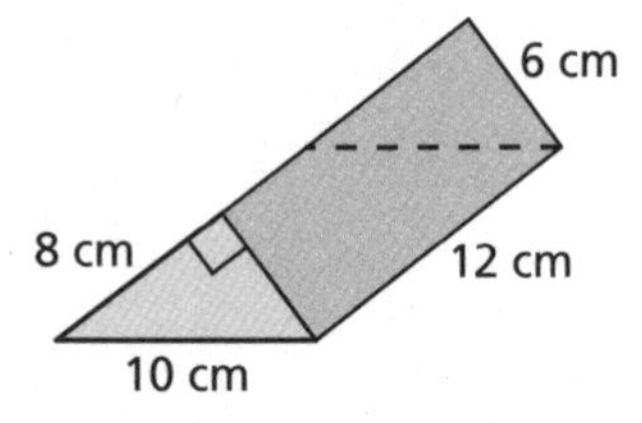

4.

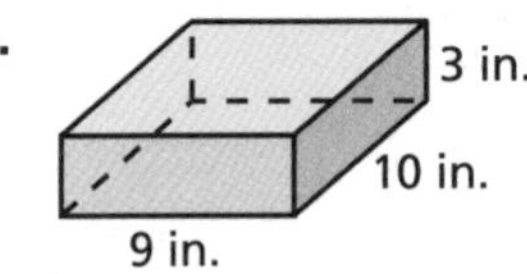

5. You buy a ring box as a birthday gift that is in the shape of a triangular prism. What is the least amount of wrapping paper needed to wrap the box?

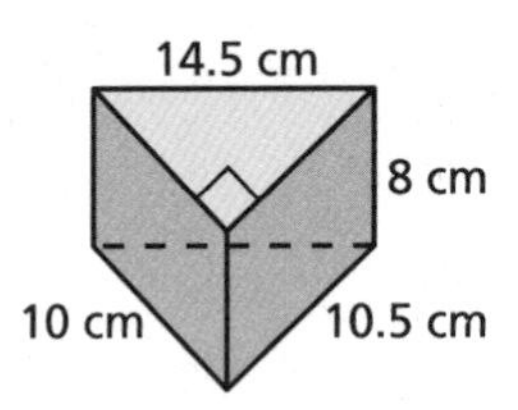

Name____ Date__________

14.2 Surface Areas of Pyramids

For use with Activity 14.2

Essential Question How can you find the surface area of a pyramid?

Even though many well-known pyramids have square bases, the base of a pyramid can be any polygon.

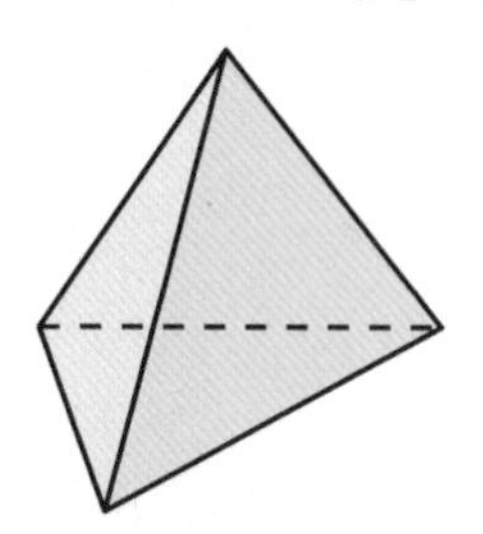

Triangular Base

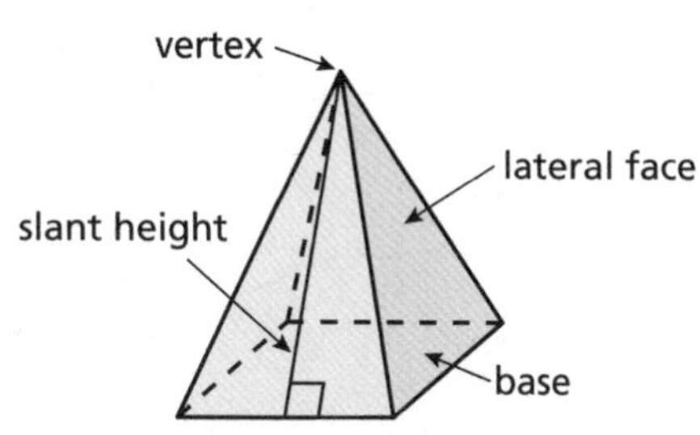

Square Base

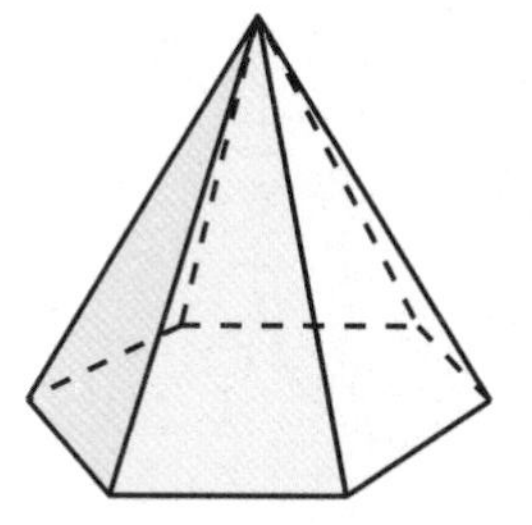

Hexagonal Base

1 ACTIVITY: Making a Scale Model

Work with a partner. Each pyramid has a square base.

- **Draw a net for a scale model of one of the pyramids. Describe your scale.**
- **Cut out the net and fold it to form a pyramid.**
- **Find the lateral surface area of the real-life pyramid.**

a. Cheops Pyramid in Egypt
Side = 230 m, Slant height ≈ 186 m

b. Muttart Conservatory in Edmonton
Side = 26 m, Slant height ≈ 27 m

c. Louvre Pyramid in Paris
Side = 35 m, Slant height ≈ 28 m

d. Pyramid of Caius Cestius in Rome
Side = 22 m, Slant height ≈ 29 m

Name ______________________ Date ________

2 ACTIVITY: Estimation

Work with a partner. There are many different types of gemstone cuts. Here is one called a brilliant cut.

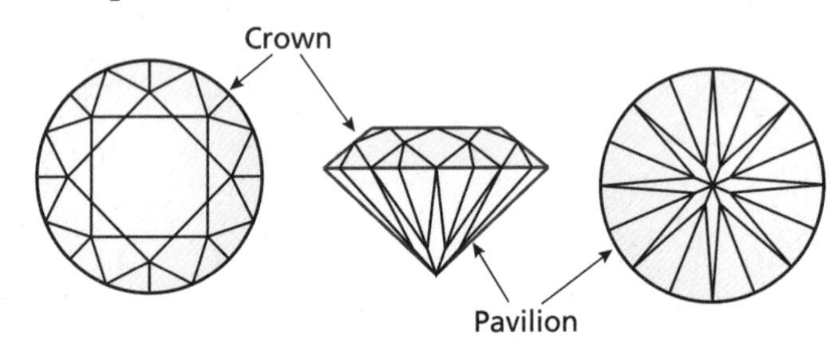

The size and shape of the pavilion can be approximated by an octagonal pyramid.

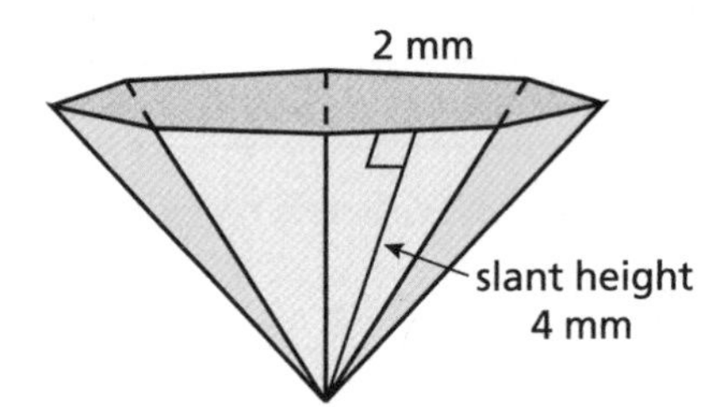

a. What does *octagonal* mean?

b. Draw a net for the pyramid.

c. Find the lateral surface area of the pyramid.

3 ACTIVITY: Comparing Surface Areas

Work with a partner. Both pyramids have the same side lengths of the base and the same slant heights.

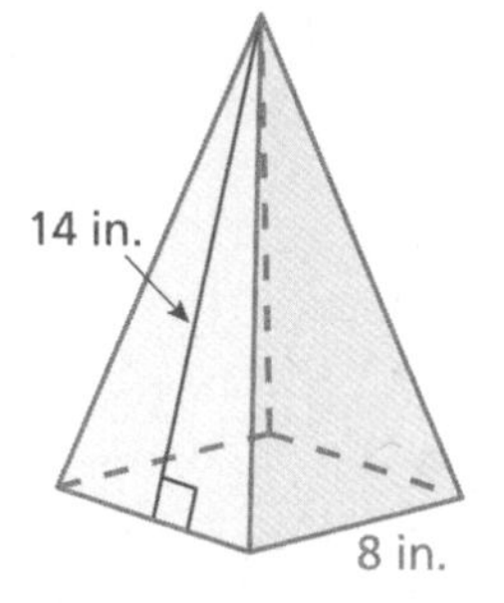

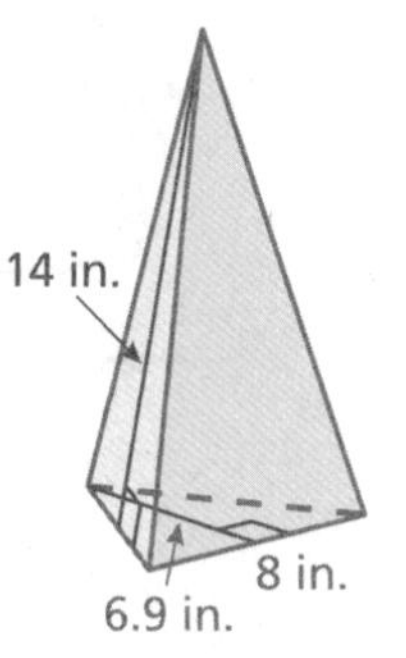

a. REASONING Without calculating, which pyramid has the greater surface area? Explain.

b. Verify your answer to part (a) by finding the surface area of each pyramid.

What Is Your Answer?

4. IN YOUR OWN WORDS How can you find the surface area of a pyramid? Draw a diagram with your explanation.

Name ______________________________ Date __________

14.2 Practice

For use after Lesson 14.2

Find the surface area of the regular pyramid.

1.

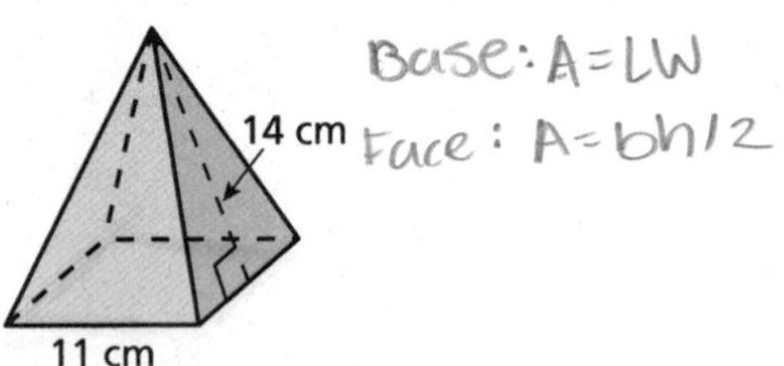

Base: A=LW
Face: A=bh/2

Base: (11)(11)=121
Face: (11)(14)/2=154/2=77
Add
121+77(4)=429 cm

2.

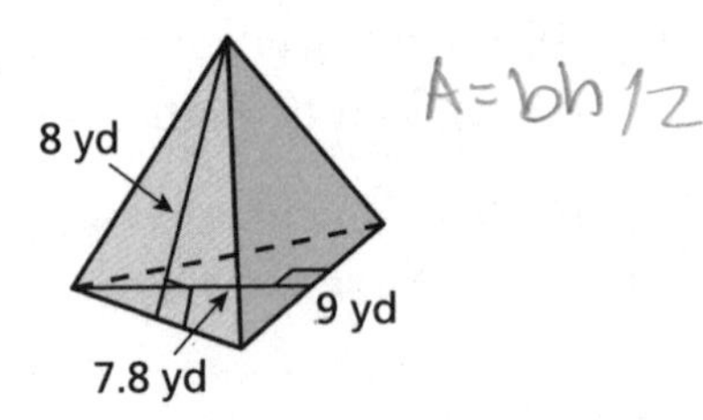

A=bh/2

Base: (7.8)(9)/2 =35.1
Face: (8)(9)/2=36
35.1 +36(3)=143.1yd

3.

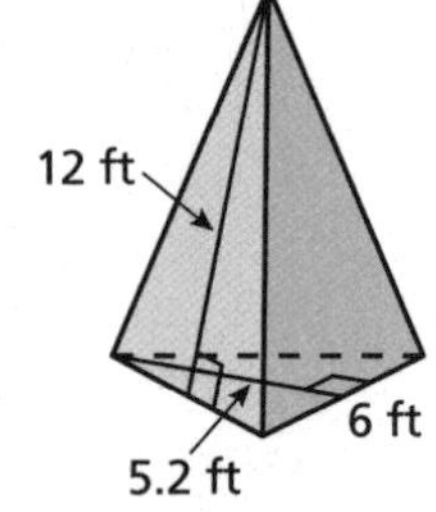

4.

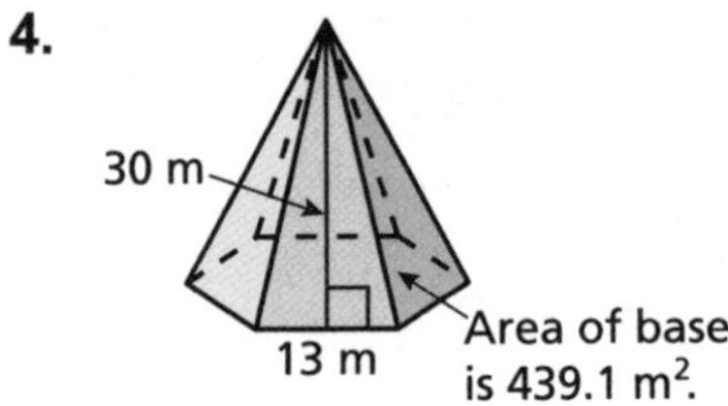

5. The surface area of a triangular pyramid is 305 square inches. The area of the base is 35 square inches. Each face has a base of 9 inches. What is the slant height?

Name____ Date__________

14.3 Surface Areas of Cylinders

For use with Activity 14.3

Essential Question How can you find the surface area of a cylinder?

A *cylinder* is a solid that has two parallel, identical circular bases.

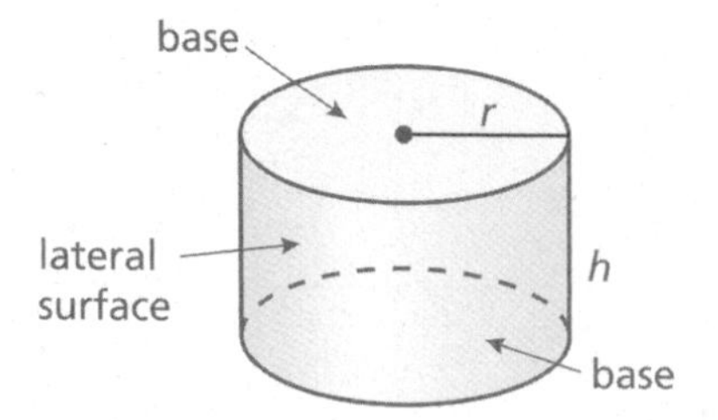

1 ACTIVITY: Finding Area

Work with a partner. Use a cardboard cylinder.

- **Talk about how you can find the area of the outside of the roll.**
- **Estimate the area using the methods you discussed.**
- **Use the roll and the scissors to find the actual area of the cardboard.**
- **Compare the actual area to your estimates.**

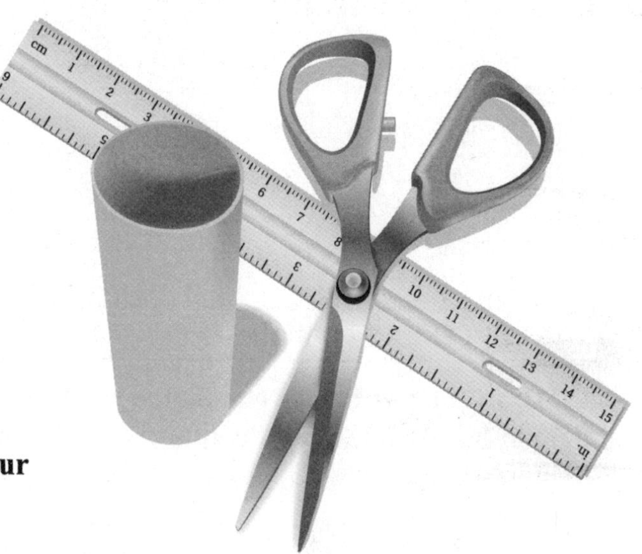

2 ACTIVITY: Finding Surface Area

Work with a partner.

- **Make a net for the can. Name the shapes in the net.**

Name ____ Date __________

- **Find the surface area of the can.**

- **How are the dimensions of the rectangle related to the dimensions of the can?**

3 ACTIVITY: Estimation

Work with a partner. From memory, estimate the dimensions of the real-life item in inches. Then use the dimensions to estimate the surface area of the item in square inches.

a.

b.

Name______________________________ Date__________

c.

d.

What Is Your Answer?

4. **IN YOUR OWN WORDS** How can you find the surface area of a cylinder? Give an example with your description. Include a drawing of the cylinder.

5. To eight decimal places, $\pi \approx 3.14159265$. Which of the following is closest to π?

a. 3.14 **b.** $\frac{22}{7}$ **c.** $\frac{355}{113}$

"To approximate $\pi \approx 3.141593$, I simply remember 1, 1, 3, 3, 5, 5."

"Then I compute $\frac{355}{113} \approx 3.141593$."

Name Roselynn Sim Date ________

14.3 Practice
For use after Lesson 14.3

Find the surface area of the cylinder. Round your answer to the nearest tenth.

SA = 2πr² + 2πrh

1.

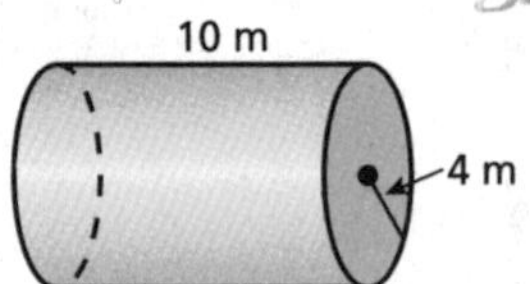

SA = 2(3.14)(4)² + 2(3.14)(4)(10)
100.48
= 351.7 m

2.

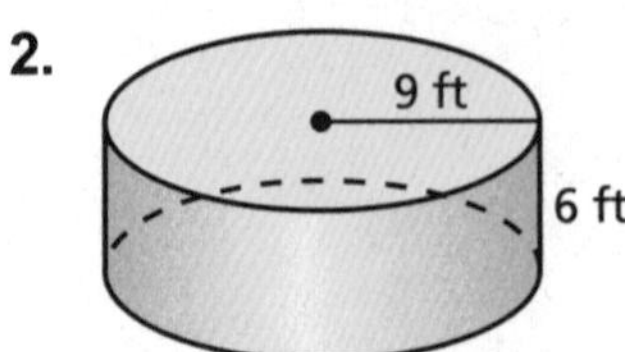

SA = 2(3.14)(9)² + 2(3.14)(6)(9)
508.68 + 339.12
847.8 ft

Find the lateral surface area of the cylinder. Round your answer to the nearest tenth.

3.

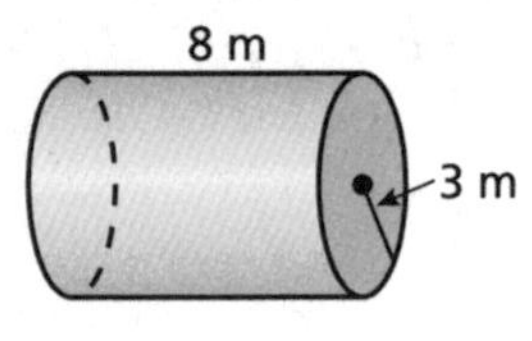

150.7 m

LSA = 2(3.14)(8)(3) = 150.72

4.

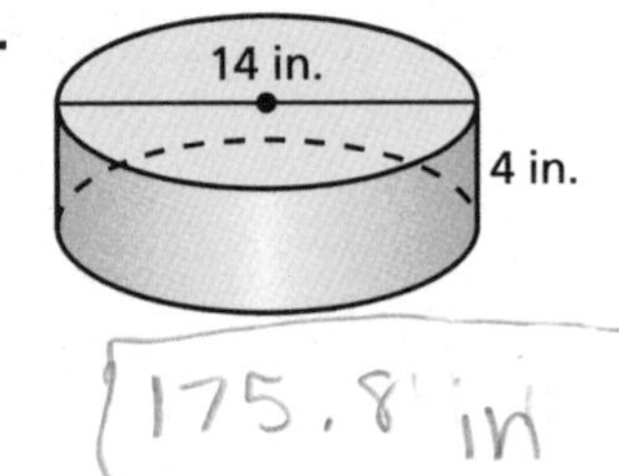

175.8 in

LSA = 2(3.14)(7)(4)

5. How much paper is used in the label for the can of cat food? Round your answer to the nearest whole number.

LSA = 2(3.14)(30)(24)

4,521 mm

Rosie

Name____ Date__________

14.4 Volumes of Prisms

For use with Activity 14.4

Essential Question How can you find the volume of a prism?

1 ACTIVITY: Pearls in a Treasure Chest

Work with a partner. A treasure chest is filled with valuable pearls. Each pearl is about 1 centimeter in diameter and is worth about $80.

Use the diagrams below to describe two ways that you can estimate the number of pearls in the treasure chest.

a.

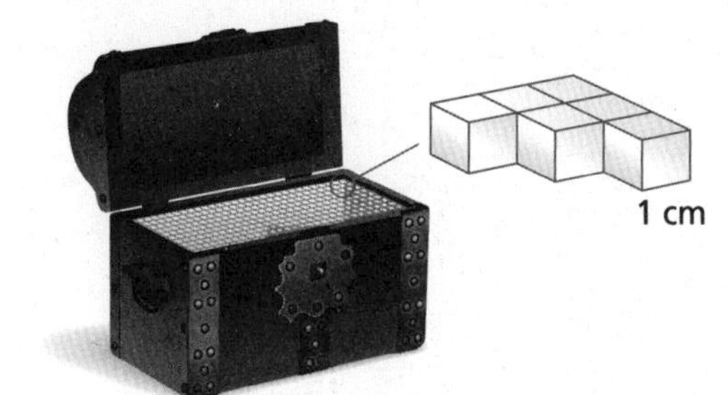

b.

c. Use the method in part (a) to estimate the value of the pearls in the chest.

Name ______________________ Date __________

2 ACTIVITY: Finding a Formula for Volume

Work with a partner. You know that the formula for the volume of a rectangular prism is $V = \ell wh$.

a. Write a formula that gives the volume in terms of the area of the base B and the height h.

b. Use both formulas to find the volume of each prism. Do both formulas give you the same volume?

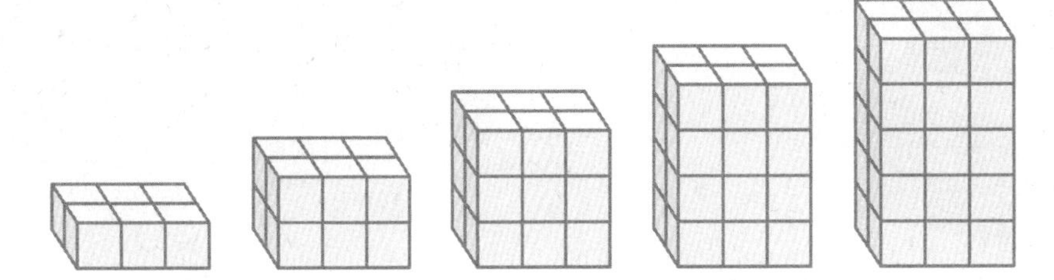

3 ACTIVITY: Finding a Formula for Volume

Work with a partner. Use the concept in Activity 2 to find a formula that gives the volume of any prism.

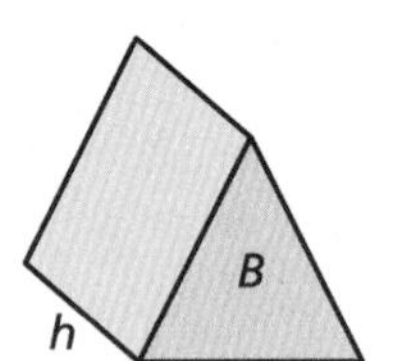

Triangular Prism

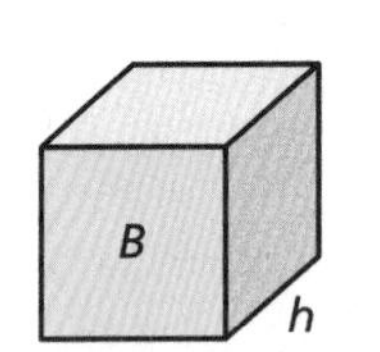

Rectangular Prism

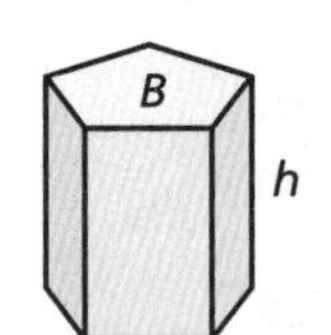

Pentagonal Prism

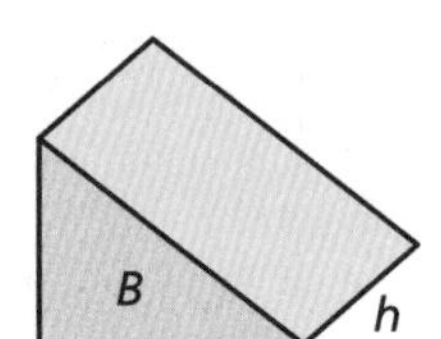

Triangular Prism

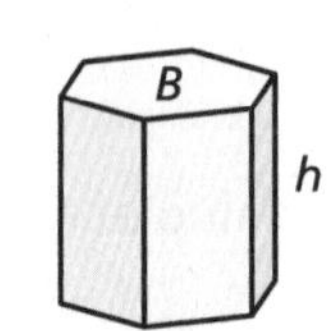

Hexagonal Prism

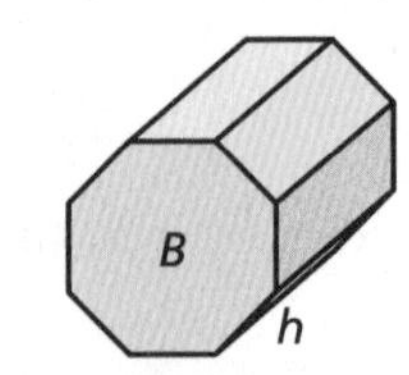

Octagonal Prism

Name________________________________ Date__________

4 ACTIVITY: Using a Formula

Work with a partner. A ream of paper has 500 sheets.

a. Does a single sheet of paper have a volume? Why or why not?

b. If so, explain how you can find the volume of a single piece of paper.

What Is Your Answer?

5. IN YOUR OWN WORDS How can you find the volume of a prism?

6. STRUCTURE Draw a prism that has a trapezoid as its base. Use your formula to find the volume of the prism.

Name ______________________________ Date __________

14.4 Practice
For use after Lesson 14.4

Find the volume of the prism.

1.

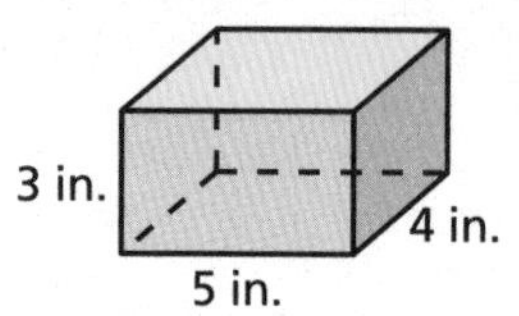

$L \times W \times H$

$3 \times 5 \times 4 = $ 60 in

2.

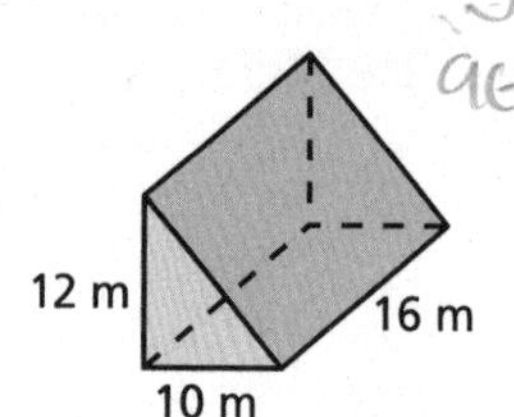

(60)(16)

960

Base → $A = \frac{bh}{2}$

$A = \frac{12(10)}{2}$

$A = 60$

$V = 960\ m^3$

3.

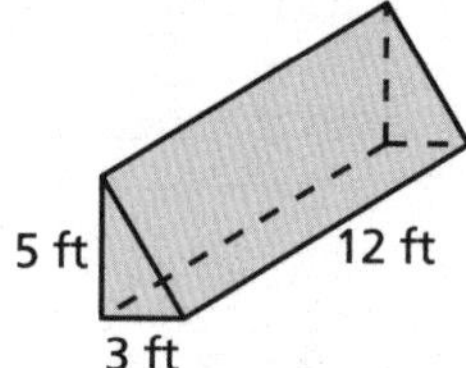

$A = \frac{(5)(3)}{2}$

$A = 7.5$

7.5(12)

90

$V = 90\ ft^3$

4.

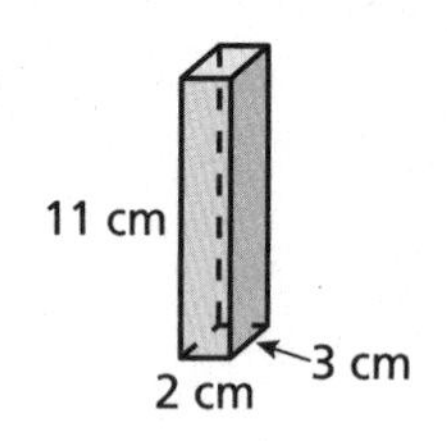

5. $B = 60\ \text{ft}^2$

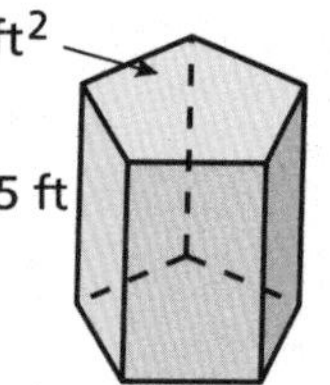

6. $B = 80\ \text{m}^2$

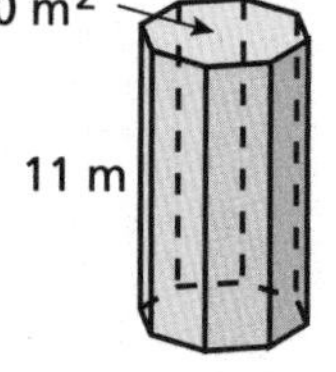

7. Each box is shaped like a rectangular prism. Which has more storage space? Explain.

Box 1 has more storage space because the volume is bigger.

Box 1

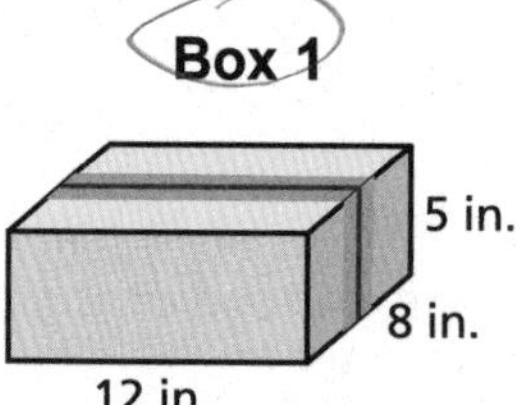

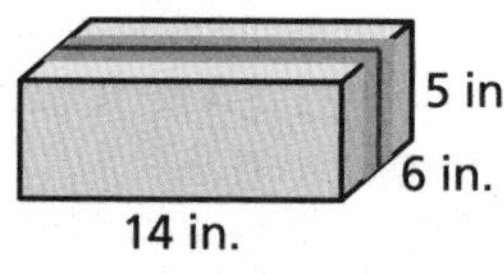

$480\ in^3$ $\quad$ $420\ in^3$

$12 \cdot 8 \cdot 5$ $\quad$ $14 \cdot 6 \cdot 5$

Name______________________________ Date__________

14.5 Volumes of Pyramids

For use with Activity 14.5

Essential Question How can you find the volume of a pyramid?

1 ACTIVITY: Finding a Formula Experimentally

Work with a partner.

- **Draw the two nets on cardboard and cut them out.***

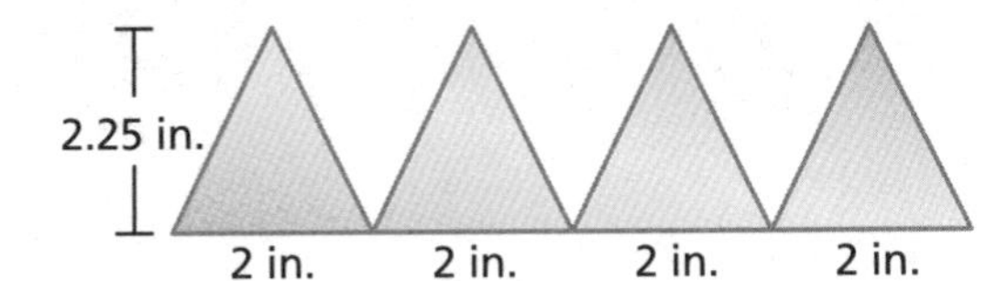

2 in. 2 in. 2 in.
2 in.
2 in.
2 in.

- **Fold and tape the nets to form an open square box and an open pyramid.**
- **Both figures should have the same size square base and the same height.**
- **Fill the pyramid with pebbles. Then pour the pebbles into the box. Repeat this until the box is full. How many pyramids does it take to fill the box?**

- **Use your result to find a formula for the volume of a pyramid.**

2 ACTIVITY: Comparing Volumes

Work with a partner. You are an archaeologist studying two ancient pyramids. What factors would affect how long it took to build each pyramid? Given similar conditions, which pyramid took longer to build? Explain your reasoning.

The Sun Pyramid in Mexico
Height: 246 ft
Base: 738 ft by 738 ft

Cheops Pyramid in Egypt
Height: about 480 ft
Base: about 755 ft by 755 ft

*Cut-outs are available in the back of the Record and Practice Journal.

Name ____ Date __________

3 ACTIVITY: Finding and Using a Pattern

Work with a partner.

- **Find the volumes of the pyramids.**

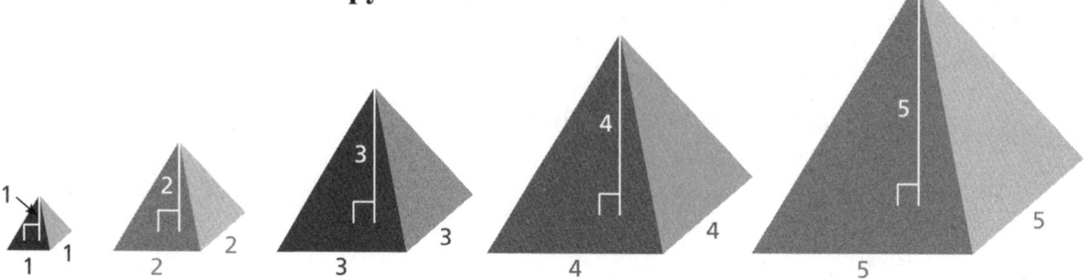

- **Organize your results in a table.**

Pyramid	Volume (cubic units)
1	
2	
3	
4	
5	

- **Describe the pattern.**

- **Use your pattern to find the volume of a pyramid with a side length and a height of 20.**

Name______________________________ Date__________

4 ACTIVITY: Breaking a Prism into Pyramids

Work with a partner. The rectangular prism can be cut to form three pyramids. Show that the sum of the volumes of the three pyramids is equal to the volume of the prism.

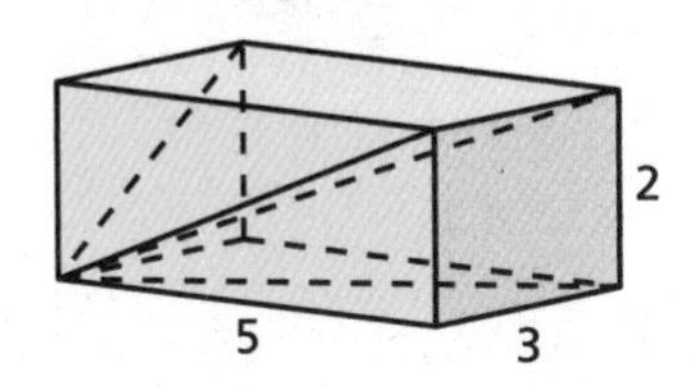

a. 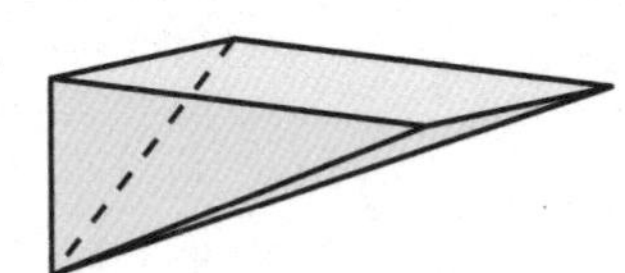

b. 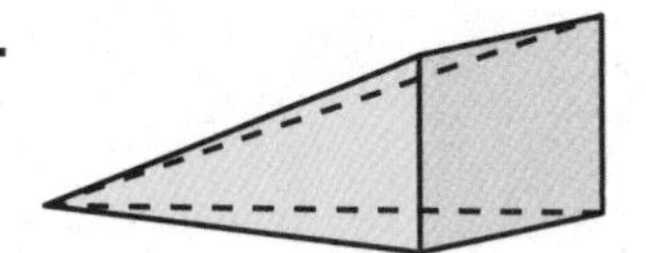

c. 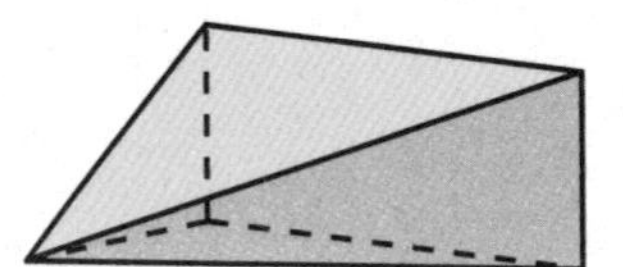

What Is Your Answer?

5. **IN YOUR OWN WORDS** How can you find the volume of a pyramid?

6. **STRUCTURE** Write a general formula for the volume of a pyramid.

Name ______________________________ Date __________

14.5 Practice
For use after Lesson 14.5

Find the volume of the pyramid.

1.

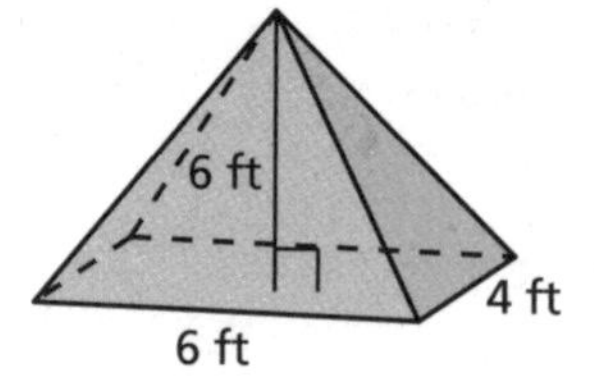

2.

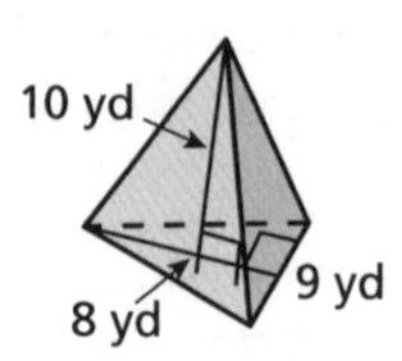

3.

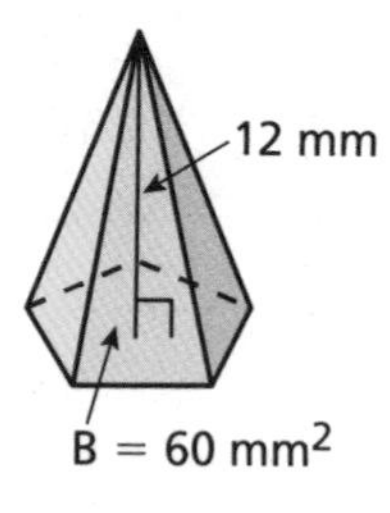

4.

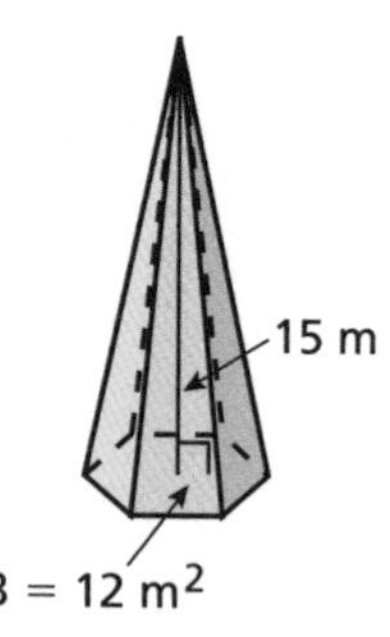

5. You create a simple tent in the shape of a pyramid. What is the volume of the tent?

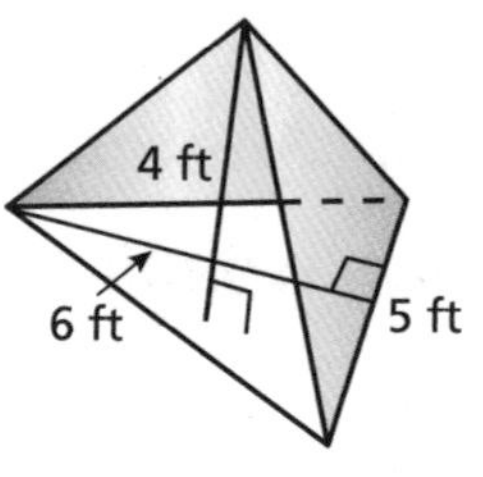

6. You work at a restaurant that has 20 tables. Each table has a set of salt and pepper shakers on it that are in the shape of square pyramids. How much salt do you need to fill all the salt shakers?

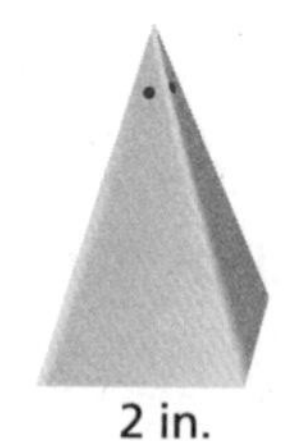

Name____ Date__________

Extension 14.5 Practice

For use after Extension 14.5

Describe the intersection of the plane and the solid.

1.

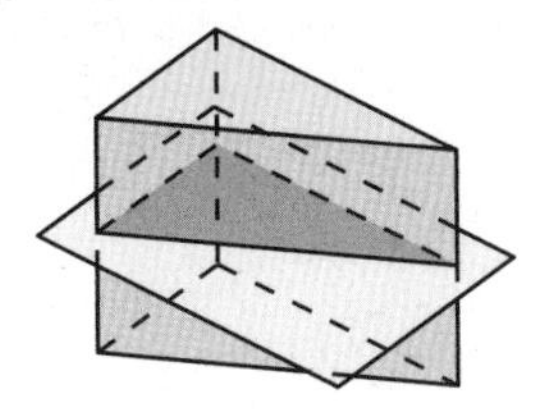

2.

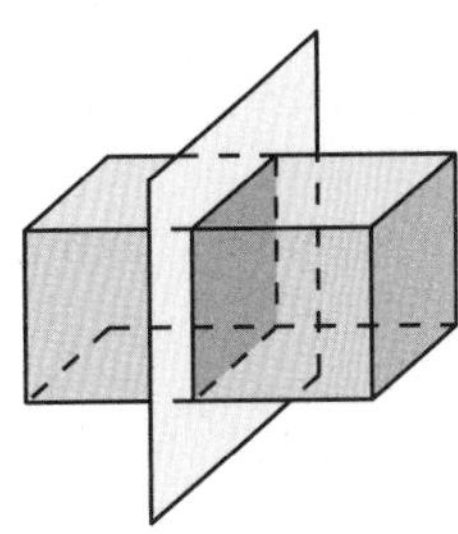

3.

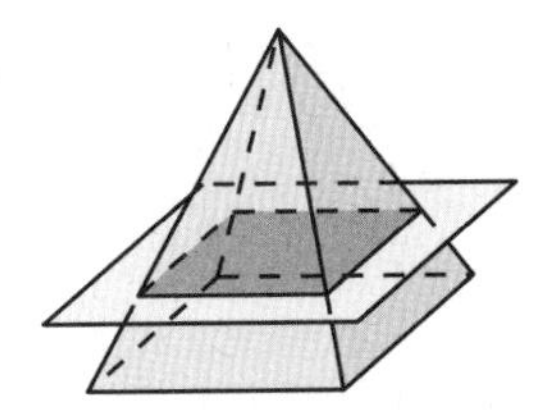

4.

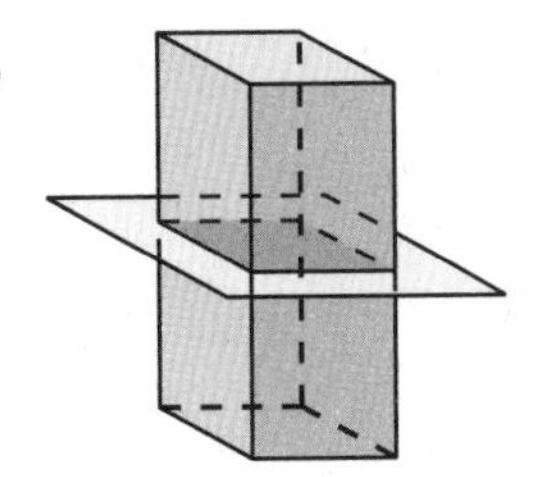

Describe the shape that is formed by the cut made in the food shown.

5.

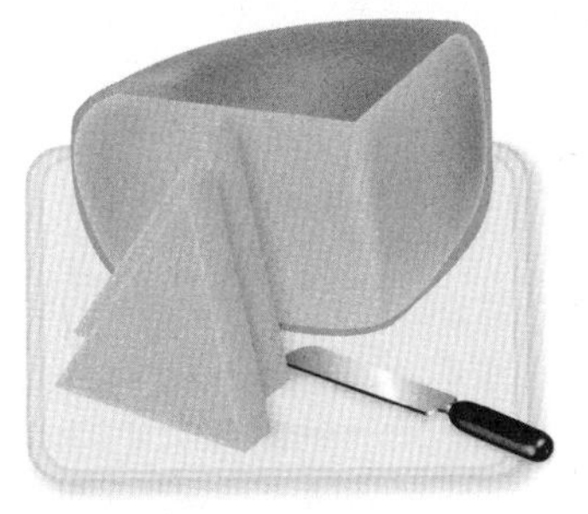

6.

Name ____ Date __________

Extension 14.5 **Practice (continued)**

Describe the intersection of the plane and the solid.

7.

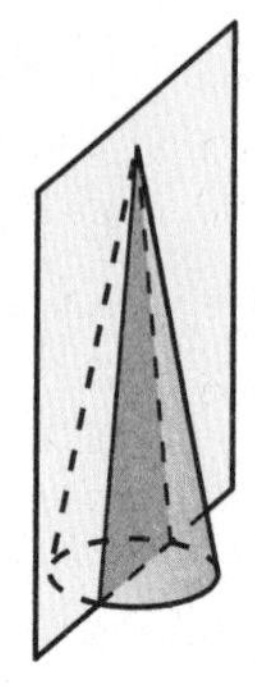

8.

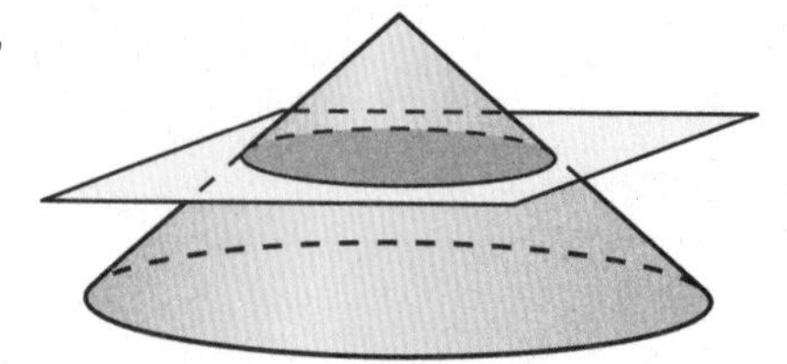

9.

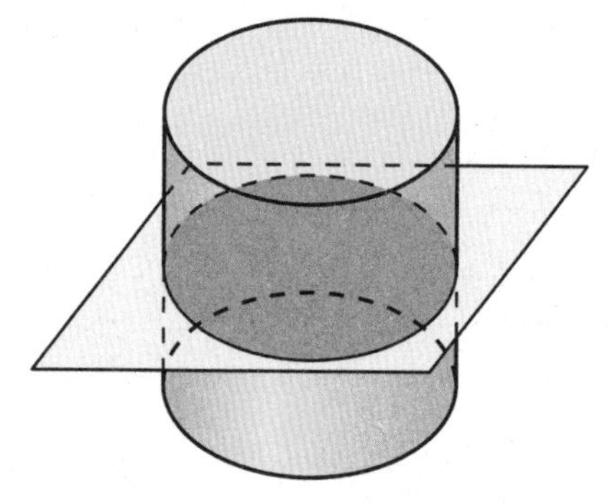

10.

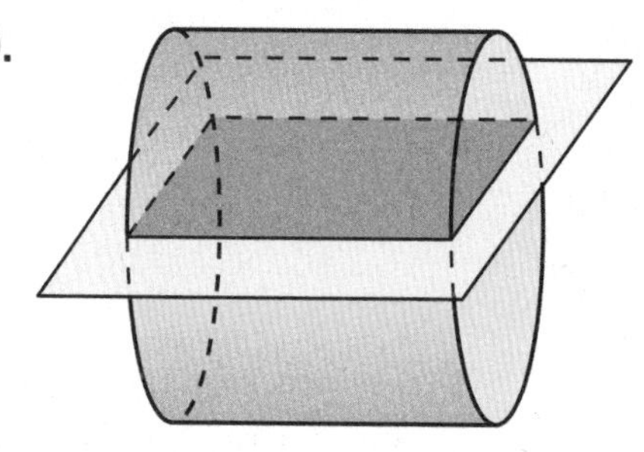

Describe the shape that is formed by the cut made in the food shown.

11.

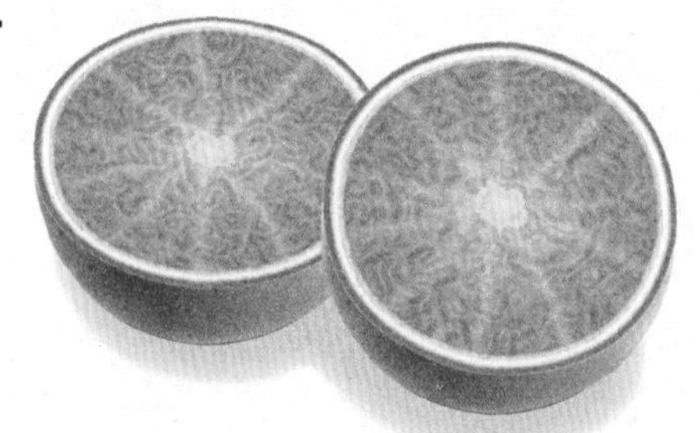

12.

Name____ Date__________

Fair Game Review

Write the ratio in simplest form.

1. bats to baseballs

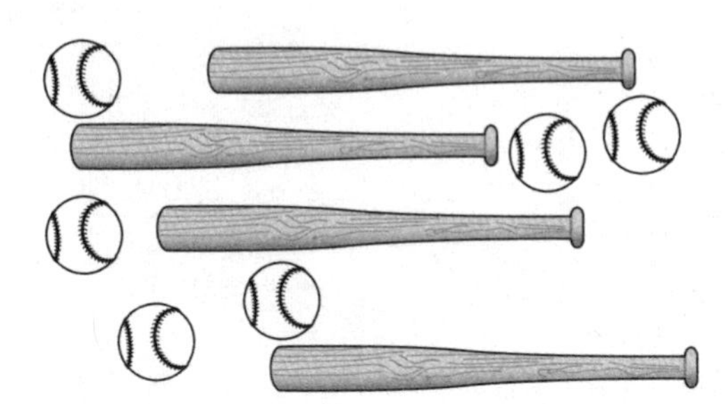

2. bows to gift boxes

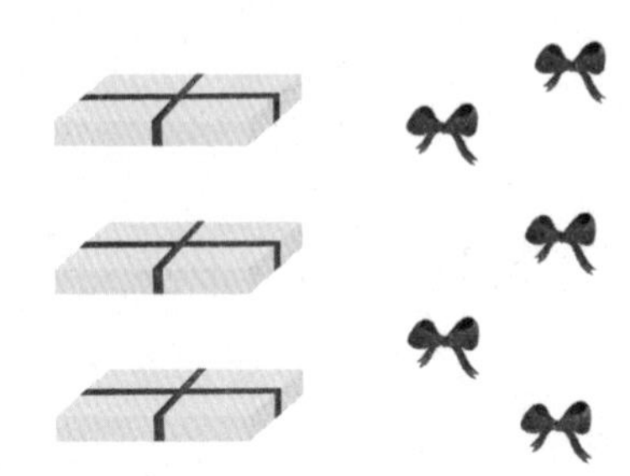

3. hammers to screwdrivers

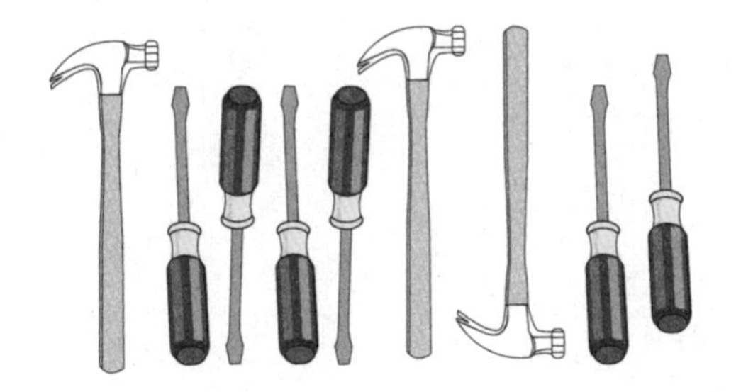

4. apples to bananas

5. There are 100 students in the sixth grade. There are 15 sixth-grade teachers. What is the ratio of teachers to students?

Name ______________________________ Date __________

Chapter 15

Fair Game Review (continued)

Write the ratio in simplest form.

6. golf balls to total number of balls

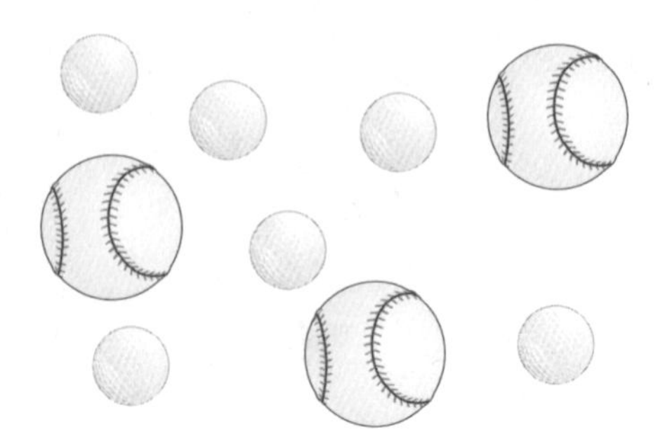

7. rulers to total pieces of equipment

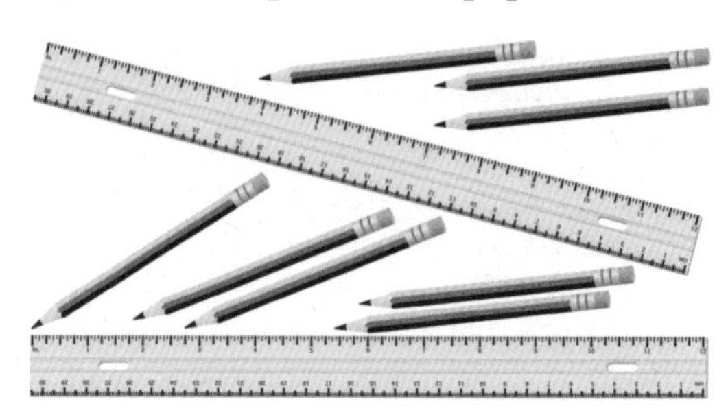

8. apples to total number of fruit

9. small fish to total number of fish

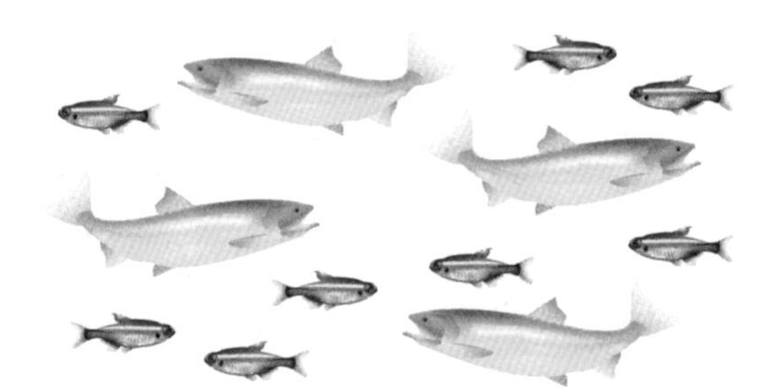

10. There are 24 flute players and 18 trumpet players in the band. Write the ratio of trumpet players to total number of trumpet players and flute players.

Name________________________________ Date__________

15.1 Outcomes and Events

For use with Activity 15.1

Essential Question In an experiment, how can you determine the number of possible results?

An *experiment* is an investigation or a procedure that has varying results. Flipping a coin, rolling a number cube, and spinning a spinner are all examples of experiments.

1 ACTIVITY: Conducting Experiments

Work with a partner.

a. You flip a dime.

There are __________ possible results.

Out of 20 flips, you think you will flip heads __________ times.

Flip a dime 20 times. Tally your results in a table. How close was your guess?

Flip	1	2	3	4	5	6	7	8	9	10	11	12	13	14	15	16	17	18	19	20
Result																				

b. You spin the spinner shown.

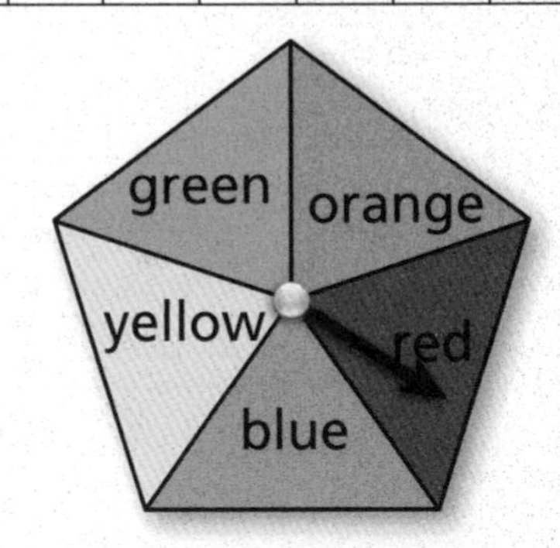

There are __________ possible results.

Out of 20 spins, you think you will spin orange __________ times.

Spin the spinner 20 times. Tally your results in a table. How close was your guess?

Spin	1	2	3	4	5	6	7	8	9	10	11	12	13	14	15	16	17	18	19	20
Result																				

c. You spin the spinner shown.

There are __________ possible results.

Out of 20 spins, you think you will spin a 4 __________ times.

Spin the spinner 20 times. Tally your results in a table. How close was your guess?

Spin	1	2	3	4	5	6	7	8	9	10	11	12	13	14	15	16	17	18	19	20
Result																				

Name ______________________________ Date __________

15.1 Outcomes and Events (continued)

2 ACTIVITY: Comparing Different Results

Work with a partner. Use the spinner in Activity 1(c).

a. Do you have a better chance of spinning an even number or a multiple of 4? Explain your reasoning.

b. Do you have a better chance of spinning an even number or an odd number? Explain your reasoning.

3 ACTIVITY: Rock Paper Scissors

Work with a partner.

a. Play Rock Paper Scissors 30 times. Tally your results in the table.

Rock *breaks* scissors.
Paper *covers* rock.
Scissors *cut* paper.

		Player A		
		Rock	Paper	Scissors
Player B	Rock			
	Paper			
	Scissors			

15.1 Outcomes and Events (continued)

b. How many possible results are there?

c. Of the possible results, in how many ways can Player A win? Player B win? the players tie?

d. Does one of the players have a better chance of winning than the other player? Explain your reasoning.

What Is Your Answer?

4. IN YOUR OWN WORDS In an experiment, how can you determine the number of possible results?

Name ______________________________ Date __________

15.1 Practice

For use after Lesson 15.1

A bag is filled with 4 red marbles, 3 blue marbles, 3 yellow marbles, and 2 green marbles. You randomly choose one marble from the bag. (a) Find the number of ways the event can occur. (b) Find the favorable outcomes of the event.

1. Choosing red

2. Choosing green

3. Choosing yellow

4. Choosing *not* blue

5. In order to figure out who will go first in a game, your friend asks you to pick a number between 1 and 25.

a. What are the possible outcomes?

b. What are the favorable outcomes of choosing an even number?

c. What are the favorable outcomes of choosing a number less than 20?

Name____ Date__________

15.2 Probability

For use with Activity 15.2

Essential Question How can you describe the likelihood of an event?

1 ACTIVITY: Black-and-White Spinner Game

Work with a partner. You work for a game company. You need to create a game that uses the spinner below.

a. Write rules for a game that uses the spinner. Then play it.

b. After playing the game, do you want to revise the rules? Explain.

c. **CHOOSE TOOLS** Using the center of the spinner as the vertex, measure the angle of each pie-shaped section. Is each section the same size? How do you think this affects the likelihood of spinning a given number?

d. Your friend is about to spin the spinner and wants to know how likely it is to spin a 3. How would you describe the likelihood of this event to your friend?

2 ACTIVITY: Changing the Spinner

Work with a partner. For each spinner, do the following.

- Measure the angle of each pie-shaped section.
- Tell whether you are more likely to spin a particular number. Explain your reasoning.
- Tell whether your rules from Activity 1 make sense for these spinners. Explain your reasoning.

a.

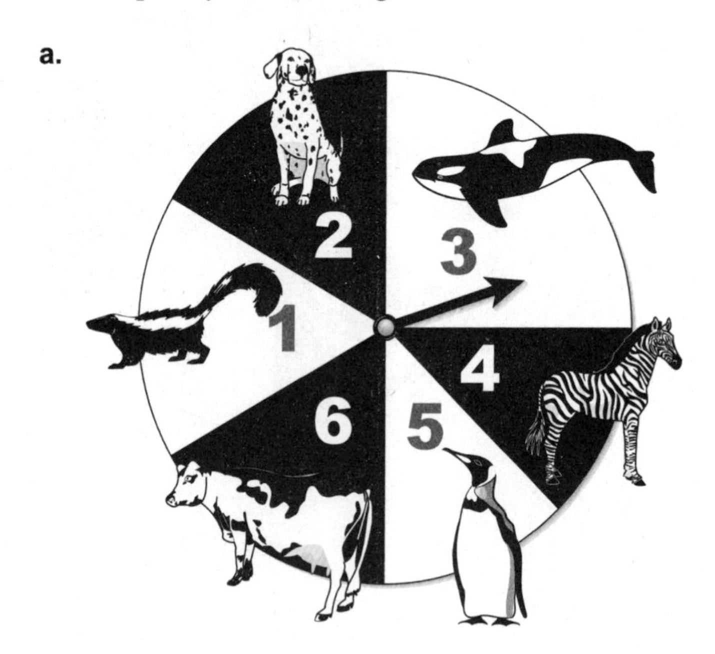

b.

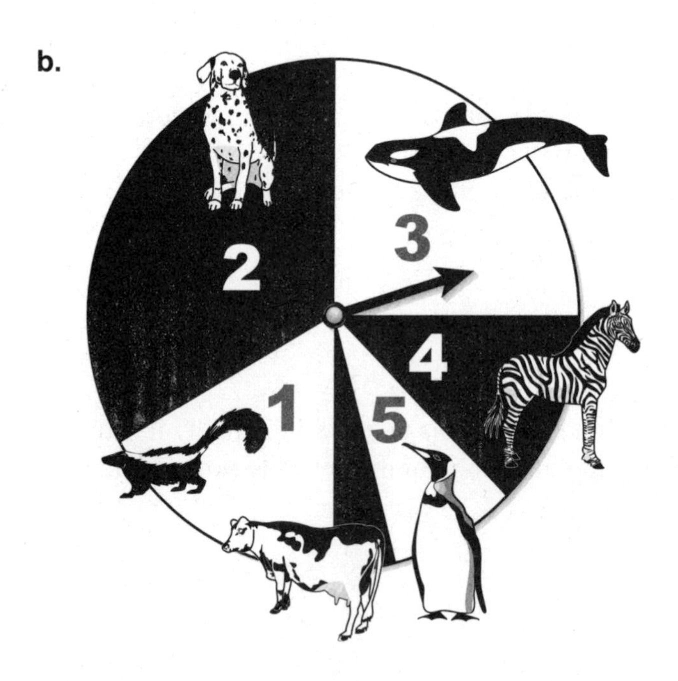

15.2 Probability (continued)

3 ACTIVITY: Is This Game Fair?

Work with a partner. Apply the following rules to each spinner in Activities 1 and 2. Is the game fair? Why or why not? If not, who has the better chance of winning?

- **Take turns spinning the spinner.**
- **If you spin an odd number, Player 1 wins.**
- **If you spin an even number, Player 2 wins.**

What Is Your Answer?

4. IN YOUR OWN WORDS How can you describe the likelihood of an event?

5. Describe the likelihood of spinning an 8 in Activity 1.

6. Describe a career in which it is important to know the likelihood of an event.

Name ______________________________ Date __________

15.2 Practice

For use after Lesson 15.2

Describe the likelihood of the event given its probability.

1. There is a 30% chance of snow tomorrow.

2. You solve a brain teaser 0.75 of the time.

You randomly choose one hat from 3 green hats, 4 black hats, 2 white hats, 2 red hats, and 1 blue hat. Find the probability of the event.

3. Choosing a red hat

4. Choosing a black hat

5. *Not* choosing a white hat

6. Choosing a blue hat

7. *Not* choosing a black hat

8. *Not* choosing a green hat

9. The probability that you draw a mechanical pencil from a group of 25 mechanical and wooden pencils is $\frac{3}{5}$. How many are mechanical pencils?

Name____ Date__________

15.3 Experimental and Theoretical Probability

For use with Activity 15.3

Essential Question How can you use relative frequencies to find probabilities?

When you conduct an experiment, the **relative frequency** of an event is the fraction or percent of the time that the event occurs.

$$\text{relative frequency} = \frac{\text{number of times the event occurs}}{\text{total number of times you conduct the experiment}}$$

1 ACTIVITY: Finding Relative Frequencies

Work with a partner.

a. Flip a quarter 20 times and record your results. Then complete the table. Are the relative frequencies the same as the probability of flipping heads or tails? Explain.

	Flipping Heads	Flipping Tails
Relative Frequency		

b. Compare your results with those of other students in your class. Are the relative frequencies the same? If not, why do you think they differ?

c. Combine all of the results in your class. Then complete the table again. Did the relative frequencies change? What do you notice? Explain.

d. Suppose everyone in your school conducts this experiment and you combine the results. How do you think the relative frequencies will change?

Name ____ Date __________

2 ACTIVITY: Using Relative Frequencies

Work with a partner. You have a bag of colored chips. You randomly select a chip from the bag and replace it. The table shows the number of times you select each color.

Red	Blue	Green	Yellow
24	12	15	9

a. There are 20 chips in the bag. Can you use the table to find the exact number of each color in the bag? Explain.

b. You randomly select a chip from the bag and replace it. You do this 50 times, then 100 times, and you calculate the relative frequencies after each experiment. Which experiment do you think gives a better approximation of the exact number of each color in the bag? Explain.

3 ACTIVITY: Conducting an Experiment

Work with a partner. You toss a thumbtack onto a table. There are two ways the thumbtack can land.

Point up

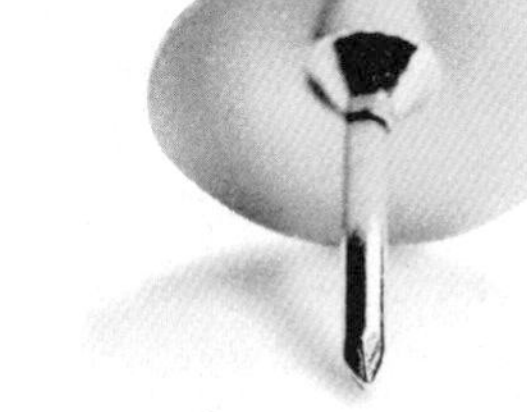

On its side

a. Your friend says that because there are two outcomes, the probability of the thumbtack landing point up must be $\frac{1}{2}$. Do you think this conclusion is true? Explain.

b. Toss a thumbtack onto a table 50 times and record your results. In a *uniform probability model*, each outcome is equally likely to occur. Do you think this experiment represents a uniform probability model? Explain.

Use the relative frequencies to complete the following.

$P(\text{point up}) =$ ________ $P(\text{on its side}) =$ ________

Name____ Date__________

What Is Your Answer?

4. **IN YOUR OWN WORDS** How can you use relative frequencies to find probabilities? Give an example.

5. Your friend rolls a number cube 500 times. How many times do you think your friend will roll an odd number? Explain your reasoning.

6. In Activity 2, your friend says, "There are no orange-colored chips in the bag." Do you think this conclusion is true? Explain.

7. Give an example of an experiment that represents a uniform probability model.

8. Tell whether you can use each spinner to represent a uniform probability model. Explain your reasoning.

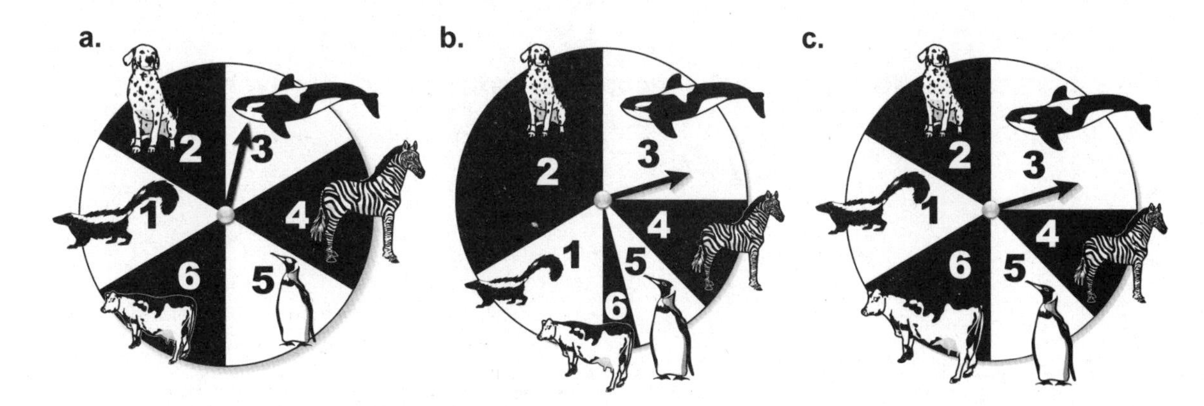

Name ______________________________ Date __________

15.3 Practice

For use after Lesson 15.3

Use the bar graph to find the experimental probability of the event.

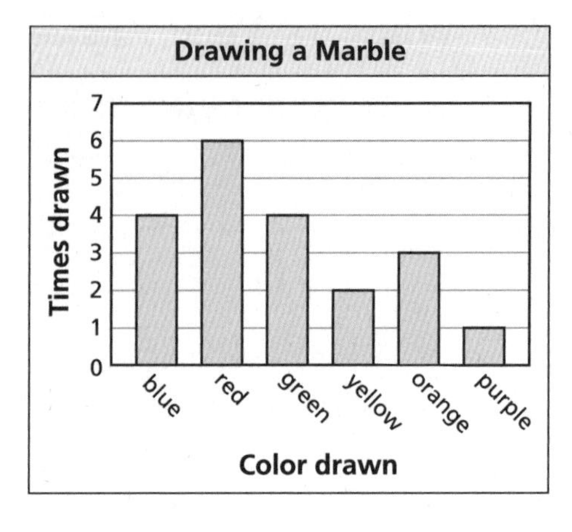

1. Drawing red

2. Drawing orange

3. Drawing *not* yellow

4. Drawing a color with more than 4 letters in its name

5. There are 25 students' names in a hat. You choose 5 names. Three are boys' names and two are girls' names. How many of the 25 names would you expect to be boys' names?

Use a number cube to determine the theoretical probability of the event.

6. Rolling a 2

7. Rolling a 5

8. Rolling an even number

9. Rolling a number greater than 1

Name____ Date__________

15.4 Compound Events

For use with Activity 15.4

Essential Question How can you find the number of possible outcomes of one or more events?

1 ACTIVITY: Comparing Combination Locks

Work with a partner. You are buying a combination lock. You have three choices.

a. This lock has 3 wheels. Each wheel is numbered from 0 to 9.

The least three-digit combination possible is ______.

The greatest three-digit combination possible is ______.

How many possible combinations are there?

b. Use the lock in part (a).

There are ______ possible outcomes for the first wheel.

There are ______ possible outcomes for the second wheel.

There are ______ possible outcomes for the third wheel.

How can you use multiplication to determine the number of possible combinations?

c. This lock is numbered from 0 to 39. Each combination uses three numbers in a right, left, right pattern. How many possible combinations are there?

d. This lock has 4 wheels.

Wheel 1: 0–9 **Wheel 2:** A–J

Wheel 3: K–T **Wheel 4:** 0–9

How many possible combinations are there?

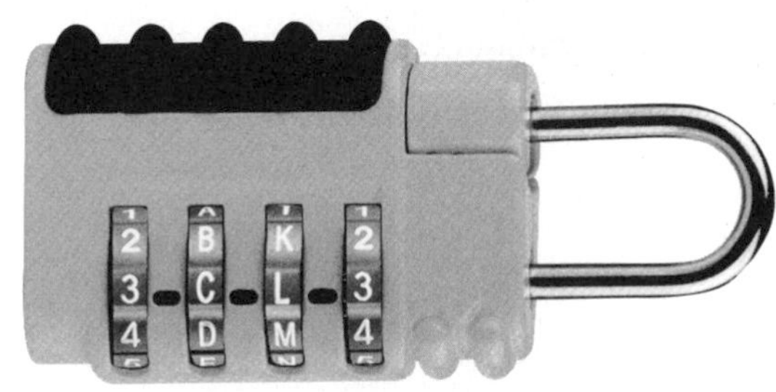

Name ______________________ Date ________

15.4 Compound Events (continued)

e. For which lock is it most difficult to guess the combination? Why?

2 ACTIVITY: Comparing Password Security

Work with a partner. Which password requirement is most secure? Explain your reasoning. Include the number of different passwords that are possible for each requirement.

a. The password must have four digits.

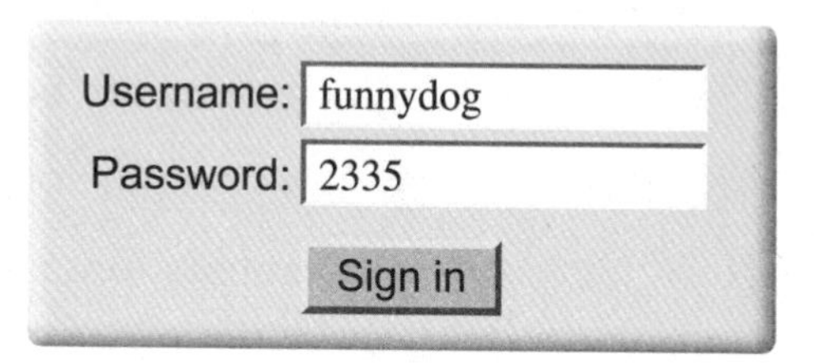

b. The password must have five digits.

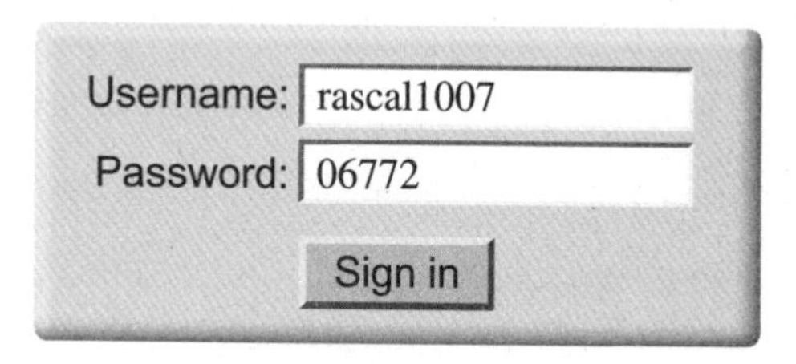

c. The password must have six letters.

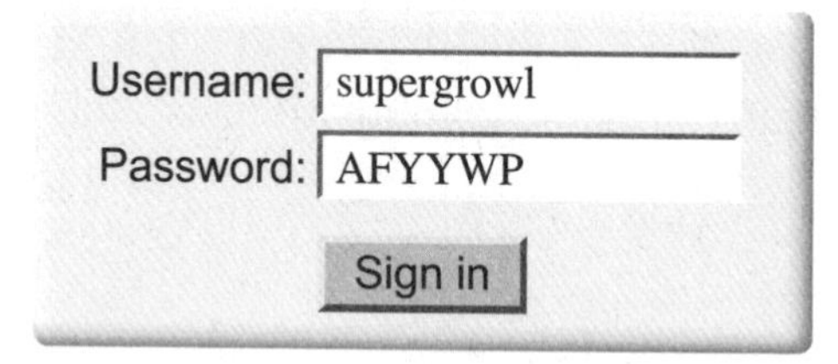

d. The password must have eight digits or letters.

Username: jupitermars
Password: 7TT3PX4W
Sign in

What Is Your Answer?

3. **IN YOUR OWN WORDS** How can you find the number of possible outcomes of one or more events?

4. **SECURITY** A hacker uses a software program to guess the passwords in Activity 2. The program checks 600 passwords per minute. What is the greatest amount of time it will take the program to guess each of the four types of passwords?

   a. four digits

   b. five digits

   c. six letters

   d. eight digits or letters

Name ______________________________ Date __________

15.4 Practice

For use after Lesson 15.4

1. Use a tree diagram to find the total number of possible outcomes.

Bed Sheets	
Size	Twin, Twin XL, Full, Queen, King
Style	Solid, Patterned

Use the Fundamental Counting Principle to find the total number of possible outcomes.

2.

Photos	
Size	Wallet, 4 by 6, 5 by 7, 8 by 10, 11 by 14, 16 by 20
Finish	Matte, Glossy
Edits	Red eye, Black and white, Crop

3.

Laptops	
Hard Drive	250 GB, 320 GB, 500 GB
Style	HD, LCD
Color	Black, White, Red, Blue, Pink, Green, Purple

You spin the spinner and flip a coin. Find the probability of the events.

4. Spinning a 2 and flipping tails

5. Spinning a 7 and flipping heads

6. *Not* spinning a 4 and flipping tails

Name_______________________________ Date__________

15.5 Independent and Dependent Events

For use with Activity 15.5

Essential Question What is the difference between dependent and independent events?

1 ACTIVITY: Drawing Marbles from a Bag (With Replacement)

Work with a partner. You have three marbles in a bag. There are two green marbles and one purple marble. Randomly draw a marble from the bag. Then put the marble back in the bag and draw a second marble.

a. Complete the tree diagram. Let G = Green and P = Purple. Find the probability that both marbles are green.

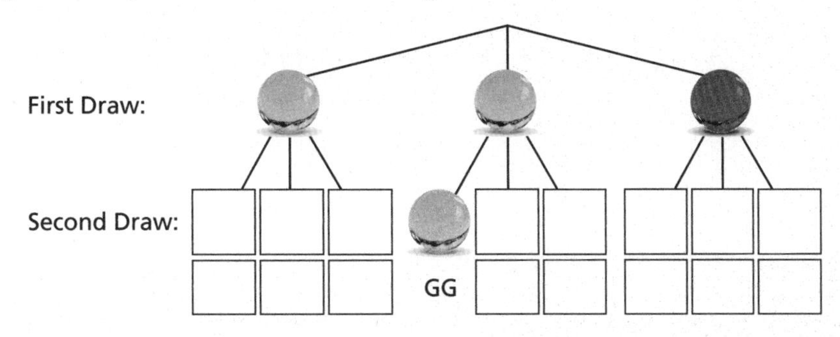

b. Does the probability of getting a green marble on the second draw *depend* on the color of the first marble? Explain.

2 ACTIVITY: Drawing Marbles from a Bag (Without Replacement)

Work with a partner. Using the same marbles from Activity 1, randomly draw two marbles from the bag.

a. Complete the tree diagram. Let G = Green and P = Purple. Find the probability that both marbles are green.

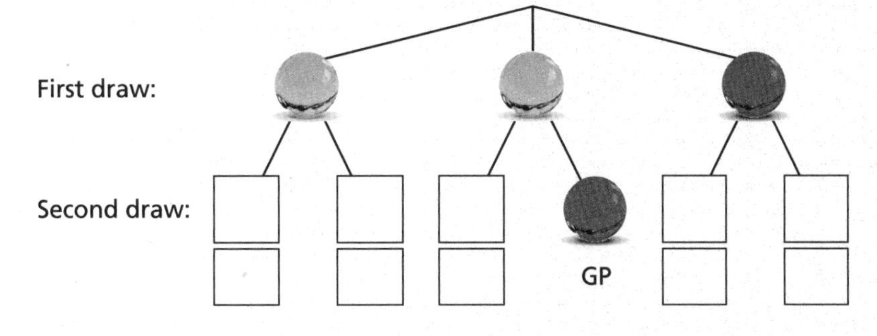

Is this event more likely than the event in Activity 1? Explain.

b. Does the probability of getting a green marble on the second draw *depend* on the color of the first marble? Explain.

Name ______________________ Date ________

3 ACTIVITY: Conducting an Experiment

Work with a partner. Conduct two experiments using two green marbles (G) and one purple marble (P).

a. In the first experiment, randomly draw one marble from the bag. Put it back. Draw a second marble. Repeat this 36 times. Record each result. Make a bar graph of your results.

GG	
GP	
PP	

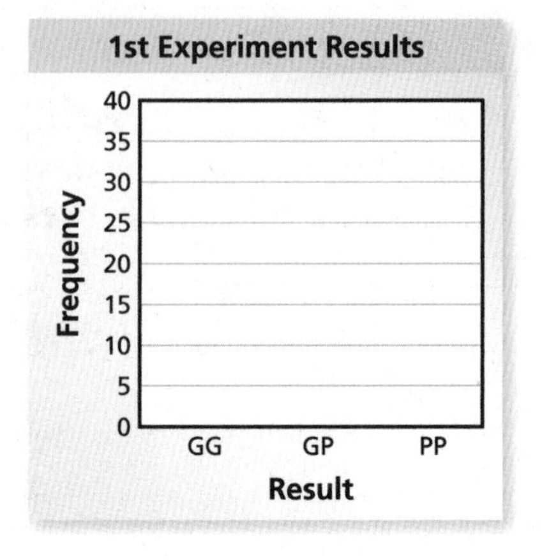

b. In the second experiment, randomly draw two marbles from the bag 36 times. Record each result. Make a bar graph of your results.

GG	
GP	
PP	

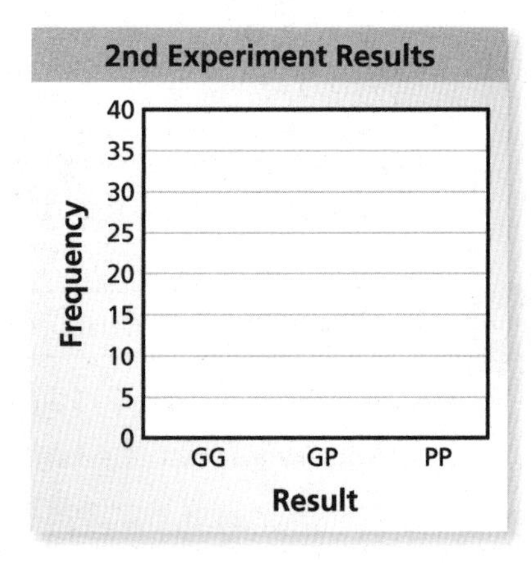

c. For each experiment, estimate the probability of drawing two green marbles.

d. Which experiment do you think represents *dependent events*? Which represents *independent events*? Explain your reasoning.

What Is Your Answer?

4. **IN YOUR OWN WORDS** What is the difference between *dependent* and *independent* events? Describe a real-life example of each.

In Questions 5–7, tell whether the events are *independent* or *dependent*. Explain your reasoning.

5. You roll a 5 on a number cube and spin blue on a spinner.

6. Your teacher chooses one student to lead a group, and then chooses another student to lead another group.

7. You spin red on one spinner and green on another spinner.

8. In Activities 1 and 2, what is the probability of drawing a green marble on the first draw? on the second draw? How do you think you can use these two probabilities to find the probability of drawing two green marbles?

Name ______________________________ Date __________

15.5 Practice

For use after Lesson 15.5

You roll a number cube twice. Find the probability of the events.

1. Rolling a 3 twice

2. Rolling an even number and a 5

3. Rolling an odd number and a 2 or a 4

4. Rolling a number less than 6 and a 3 or a 1

You randomly choose a letter from a hat with the letters A through J. Without replacing the first letter, you choose a second letter. Find the probability of the events.

5. Choosing an H and then a D

6. Choosing a consonant and then an E or an I

7. Choosing a vowel and then an F

8. Choosing a vowel and then a consonant

9. You have 3 clasp bracelets, 4 watches, and 5 stretch bracelets. You randomly choose two from your jewelry box. What is the probability that you will choose 2 watches?

You flip a coin, and then roll a number cube twice. Find the probability of the event.

10. Flipping heads, rolling a 5, and rolling a 2

11. Flipping tails, rolling an odd number, and rolling a 4

12. Flipping tails, rolling a 6 or a 1, and rolling a 3

13. Flipping heads, *not* rolling a 2, and rolling an even number

Name____ Date__________

Practice

For use after Extension 15.5

1. You write a four-question survey. Each question has a *yes* or *no* answer. You have your friend answer the survey.

a. Design a simulation that you can use to model the answers.

b. Use your simulation to find the experimental probability that your friend answers *yes* to all four questions.

Practice (continued)

2. There is a 70% chance of snow today and tomorrow.

a. Design and use a simulation that generates 50 randomly generated numbers.

b. Find the experimental probability that it snows one of those days.

Name____ Date__________

15.6 Samples and Populations
For use with Activity 15.6

Essential Question How can you determine whether a sample accurately represents a population?

A **population** is an entire group of people or objects. A **sample** is a part of the population. You can use a sample to make an inference, or conclusion, about a population.

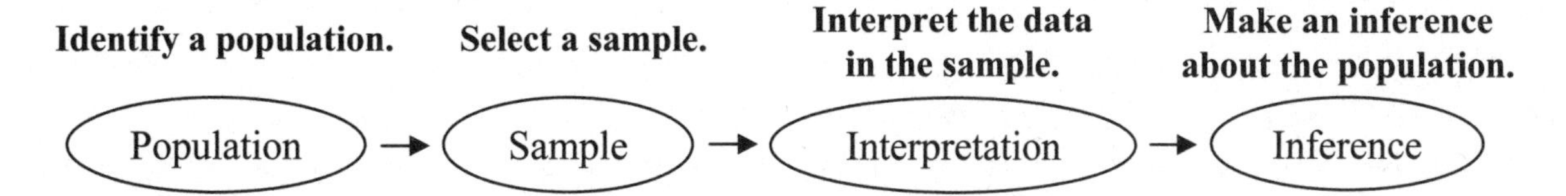

1 ACTIVITY: Identifying Populations and Samples

Work with a partner. Identify the population and the sample.

a.

The students in a school

The students in a math class

b.

The grizzly bears with GPS collars in a park

The grizzly bears in a park

c.

150 Quarters

All quarters in circulation

d.

All books in a library

10 fiction books in a library

Name ________________________________ Date __________

2 ACTIVITY: Identifying Random Samples

Work with a partner. When a sample is selected at random, each member of the population is equally likely to be selected. You want to know the favorite extracurricular activity of students at your school. Determine whether each method will result in a random sample. Explain your reasoning.

a. You ask members of the school band.

b. You publish a survey in the school newspaper.

c. You ask every eighth student who enters the school in the morning.

d. You ask students in your class.

3 ACTIVITY: Identifying Representative Samples

Work with a partner. A new power plant is being built outside a town. In each situation below and on the next page, residents of the town are asked how they feel about the new power plant. Determine whether each conclusion is valid. Explain your reasoning.

a. A local radio show takes calls from 500 residents. The table shows the results. The radio station concludes that most of the residents of the town oppose the new power plant.

New Power Plant	
For	70
Against	425
Don't know	5

Name______________________________ Date__________

b. A news reporter randomly surveys 2 residents outside a supermarket. The graph shows the results. The reporter concludes that the residents of the town are evenly divided on the new power plant.

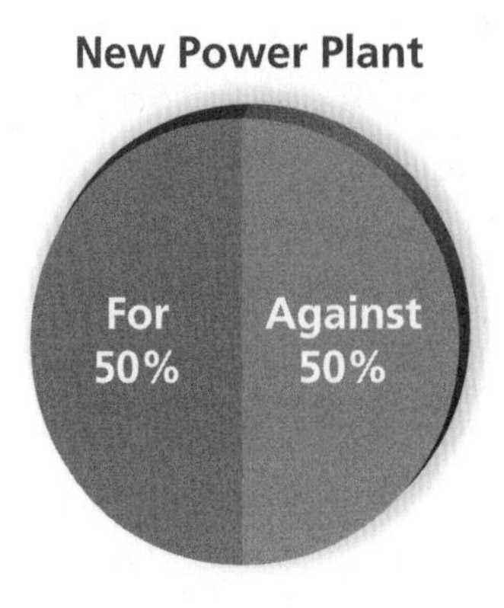

c. You randomly survey 250 residents at a shopping mall. The table shows the results. You conclude that there are about twice as many residents of the town against the new power plant than for the new power plant.

New Power Plant	
For	32%
Against	62%
Don't know	6%

What Is Your Answer?

4. IN YOUR OWN WORDS How can you determine whether a sample accurately represents a population?

5. RESEARCH Choose a topic that you would like to ask people's opinions about, and then write a survey question. How would you choose people to survey so that your sample is random? How many people would you survey? Conduct your survey and display your results. Would you change any part of your survey to make it more accurate? Explain.

6. Does increasing the size of a sample necessarily make the sample representative of a population? Give an example to support your explanation.

Name ______________________________ Date __________

15.6 Practice

For use after Lesson 15.6

Determine whether the sample is *biased* or *unbiased.* Explain.

1. You want to estimate the number of students in your school who want a football stadium to be built. You survey the first 20 students who attend a Friday night football game.

2. You want to estimate the number of students in your school who drive their own cars to school. You survey every 8th person who enters the cafeteria for lunch.

Determine whether the conclusion is valid. Explain.

3. You want to determine the number of city residents who want to have 38th Street repaved. You randomly survey 15 residents who live on 38th Street. Twelve want the street to be repaved and three do not. So, you conclude that 80% of city residents want the street to be repaved.

4. You want to determine how many students consider math to be their favorite school subject. You randomly survey 75 students. Thirty-three students consider math to be their favorite subject and forty-two do not. So, you conclude that 40% of students at your school consider math to be their favorite subject.

Name____ Date__________

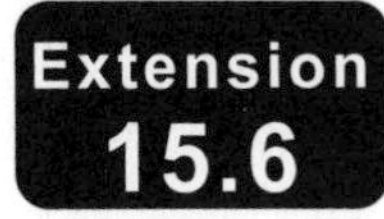

Generating Multiple Samples

For use with Extension 15.6

You have already used unbiased samples to make inferences about a population. In some cases, making an inference about a population from only one sample is not as precise as using multiple samples.

1 ACTIVITY: Using Multiple Random Samples

Work with a partner. You and a group of friends want to know how many students in your school listen to pop music. There are 840 students in your school. Each person in the group randomly surveys 20 students.

Step 1: The table shows your results. Make an inference about the number of students in your school who prefer pop music.

Favorite Type of Music			
Country	Pop	Rock	Rap
4	10	5	1

Step 2: The table shows Kevin's results. Use these results to make another inference about the number of students in your school who prefer pop music.

Favorite Type of Music			
Country	Pop	Rock	Rap
2	13	4	1

Compare the results of Steps 1 and 2.

Step 3: The table shows the results of three other friends. Use these results to make three more inferences about the number of students in your school who prefer pop music.

	Favorite Type of Music			
	Country	Pop	Rock	Rap
Steve	3	8	7	2
Laura	5	10	4	1
Ming	5	9	3	3

Name ______________________ Date __________

Extension 15.6 **Generating Multiple Samples (continued)**

Step 4: Describe the variation of the five inferences. Which one would you use to describe the number of students in your school who prefer pop music? Explain your reasoning.

Step 5: Show how you can use all five samples to make an inference.

Practice

1. Work with a partner. Mark 24 packing peanuts with either a red or a black marker. Put the peanuts into a paper bag. Trade bags with other students in the class.

a. Generate a sample by choosing a peanut from your bag six times, replacing the peanut each time. Record the number of times you choose each color. Repeat this process to generate four more samples. Organize the results in a table.

b. Use each sample to make an inference about the number of red peanuts in the bag. Then describe the variation of the five inferences. Make inferences about the numbers of red and black peanuts in the bag based on all the samples.

c. Take the peanuts out of the bag. How do your inferences compare to the population? Do you think you can make a more accurate prediction? If so, explain how.

2 ACTIVITY: Using Measures from Multiple Random Samples

Work with a partner. You want to know the mean number of hours students with part-time jobs work each week. You go to 8 different schools. At each school, you randomly survey 10 students with part-time jobs. Your results are shown at the right.

Hours Worked Each Week
1: 6, 8, 6, 6, 7, 4, 10, 8, 7, 8
2: 10, 4, 4, 6, 8, 6, 7, 12, 8, 8
3: 10, 9, 8, 6, 5, 8, 6, 6, 9, 10
4: 4, 8, 4, 4, 5, 4, 4, 6, 5, 6
5: 6, 8, 8, 6, 12, 4, 10, 8, 6, 12
6: 10, 10, 8, 9, 16, 8, 7, 12, 16, 14
7: 4, 5, 6, 6, 4, 5, 6, 6, 4, 4
8: 16, 20, 8, 12, 10, 8, 8, 14, 16, 8

Step 1: Find the mean of each sample.

Step 2: Make a box-and-whisker plot of the sample means.

Step 3: Use the box-and-whisker plot to estimate the actual mean number of hours students with part-time jobs work each week. How does your estimate compare to the mean of the entire data set?

3 ACTIVITY: Using a Simulation

Work with a partner. Another way to generate multiple samples of data is to use a simulation. Suppose 70% of all seventh graders watch reality shows on television.

Step 1: Design a simulation involving 50 packing peanuts by marking 70% of the peanuts with a certain color. Put the peanuts into a paper bag.

Step 2: Simulate choosing a sample of 30 students by choosing peanuts from the bag, replacing the peanut each time. Record the results. Repeat this process to generate eight more samples. How much variation do you expect among the samples? Explain.

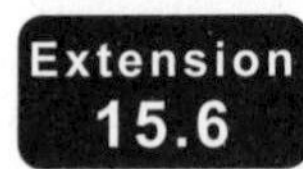

Generating Multiple Samples (continued)

Step 3: Display your results.

Practice

2. You want to know whether student-athletes prefer water or sports drinks during games. You go to 10 different schools. At each school, you randomly survey 10 student-athletes. The percents of student-athletes who prefer water are shown.

60% 70% 60% 50% 80% 70% 30% 70% 80% 40%

a. Make a box-and-whisker plot of the data.

b. Use the box-and-whisker plot to estimate the actual percent of student-athletes who prefer water. How does your estimate compare to the mean of the data?

3. Repeat Activity 2 using the medians of the samples.

4. In Activity 3, how do the percents in your samples compare to the actual percent of seventh graders who watch reality shows on television?

5. REASONING Why is it better to make inferences about a population based on multiple samples instead of only one sample? What additional information do you gain by taking multiple random samples? Explain.

Name____ Date__________

15.7 Comparing Populations

For use with Activity 15.7

Essential Question How can you compare data sets that represent two populations?

1 ACTIVITY: Comparing Two Data Distributions

Work with a partner. You want to compare the shoe sizes of male students in two classes. You collect the data shown in the table.

Male Students in Eighth-Grade Class														
7	9	8	$7\frac{1}{2}$	$8\frac{1}{2}$	10	6	$6\frac{1}{2}$	8	8	$8\frac{1}{2}$	9	11	$7\frac{1}{2}$	$8\frac{1}{2}$
Male Students in Sixth-Grade Class														
6	$5\frac{1}{2}$	6	$6\frac{1}{2}$	$7\frac{1}{2}$	$8\frac{1}{2}$	7	$5\frac{1}{2}$	5	$5\frac{1}{2}$	$6\frac{1}{2}$	7	$4\frac{1}{2}$	6	6

a. How can you display both data sets so that you can visually compare the measures of center and of variation? Make the data display you chose.

b. Describe the shape of each distribution.

c. Complete the table.

	Male Students in Eighth Grade Class	Male Students in Sixth Grade Class
Mean		
Median		
Mode		
Range		
Interquartile Range (IQR)		
Mean Absolute Deviation (MAD)		

d. Compare the measures of center for the data sets.

e. Compare the measures of variation for the data sets. Does one data set show more variation than the other? Explain.

f. Do the distributions overlap? How can you tell using the data display you chose in part (a)?

g. The double box-and-whisker plot below shows the shoe sizes of the members of two girls basketball teams. Can you conclude that at least one girl from each team has the same shoe size? Can you conclude that at least one girl from the Bobcats has a larger shoe size than one of the girls from the Tigers? Explain your reasoning.

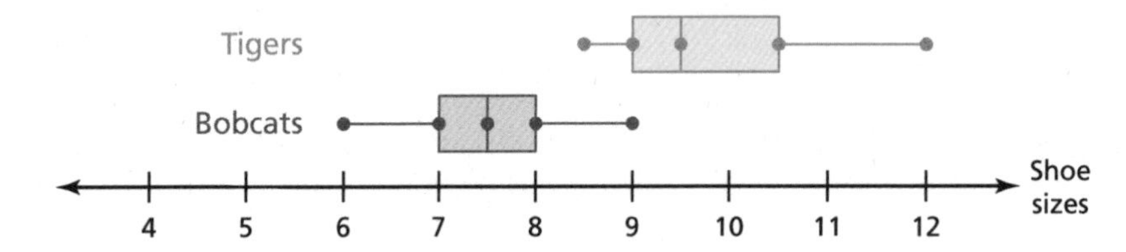

2 ACTIVITY: Comparing Two Data Distributions

Work with a partner. Compare the shapes of the distributions. Do the two data sets overlap? Explain. If so, use measures of center and the least and the greatest values to describe the overlap between the two data sets.

a.

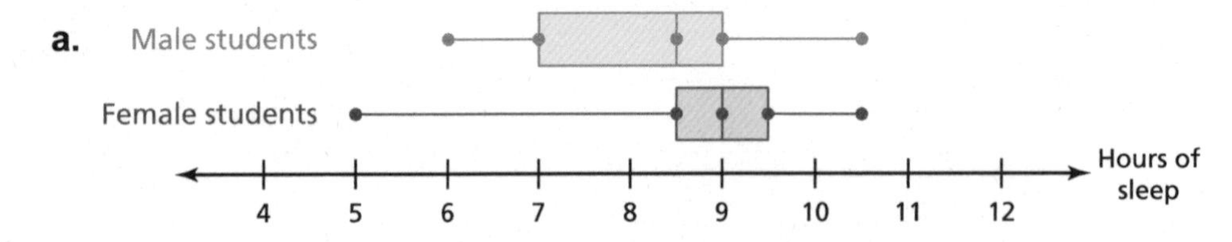

15.7 Comparing Populations (continued)

b.

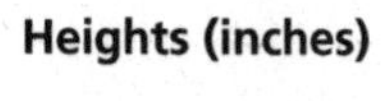

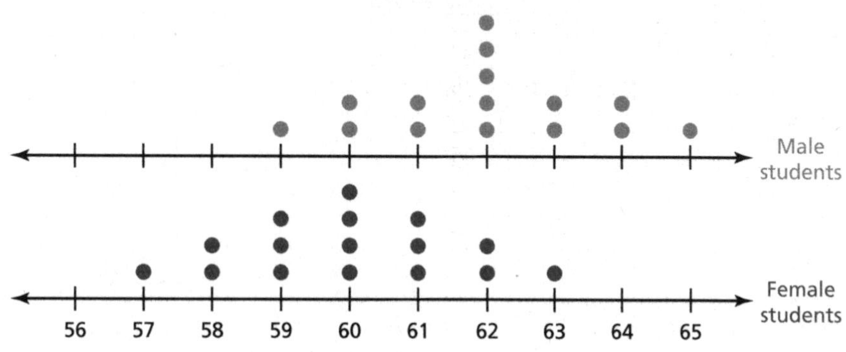

c. **Ages of People in Two Exercise Classes**

10:00 A.M. Class		8:00 P.M. Class
	1	8 9
	2	1 2 2 7 9 9
	3	0 3 4 5 7
9 7 3 2 2 2	4	0
7 5 4 3 1	5	
7 0 0	6	
0	7	

Key: $1 \mid 8 = 18$

What Is Your Answer?

3. **IN YOUR OWN WORDS** How can you compare data sets that represent two populations?

Name ____ Date __________

15.7 Practice
For use after Lesson 15.7

1. The dot plots show the quiz scores for two classes taught by the same teacher.

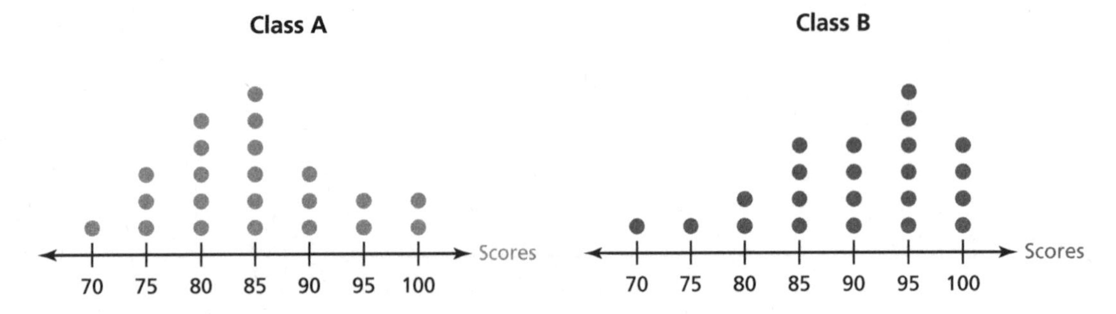

a. Compare the populations using measures of center and variation.

b. Express the difference in the measures of center as a multiple of the measure of variation.

2. The double box-and-whisker plot shows the number of song downloads a month by two seventh grade classes.

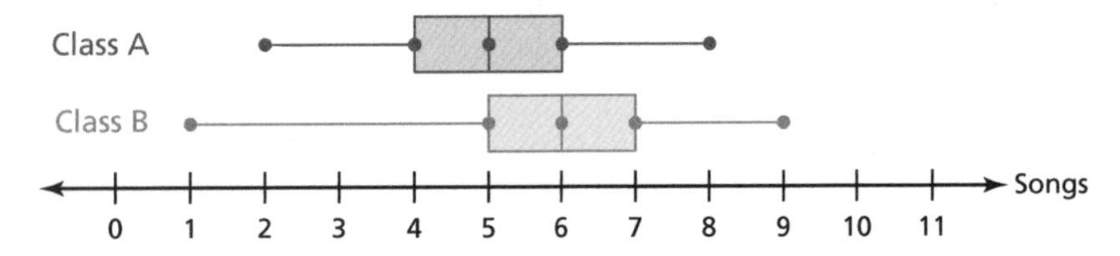

a. Compare the populations using measures of center and variation.

b. Express the difference in the measures of center as a multiple of the measure of variation.

Glossary

This student friendly glossary is designed to be a reference for key vocabulary, properties, and mathematical terms. Several of the entries include a short example to aid your understanding of important concepts.

Also available at *BigIdeasMath.com*:

- multi-language glossary
- vocabulary flash cards

Addition Property of Equality

Adding the same number to each side of an equation produces an equivalent equation.

$$\begin{aligned} x - 7 &= -6 \\ \underline{+7} \quad & \quad \underline{+7} \\ x &= 1 \end{aligned}$$

Addition Property of Inequality

When you add the same number to each side of an inequality, the inequality remains true.

$$\begin{aligned} x - 3 &> -10 \\ \underline{+3} \quad & \quad \underline{+3} \\ x &> -7 \end{aligned}$$

adjacent angles

Two angles that share a common side and have the same vertex

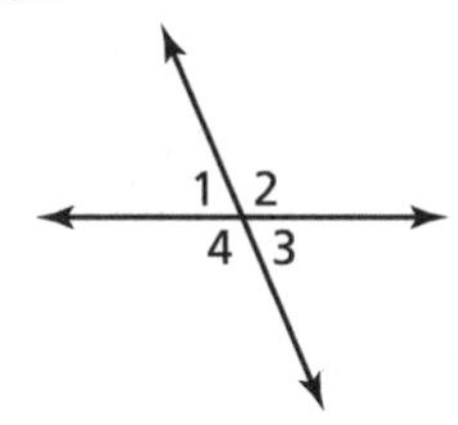

$\angle 1$ and $\angle 2$ are adjacent.

$\angle 2$ and $\angle 4$ are not adjacent.

angle

A figure formed by two rays with the same endpoint

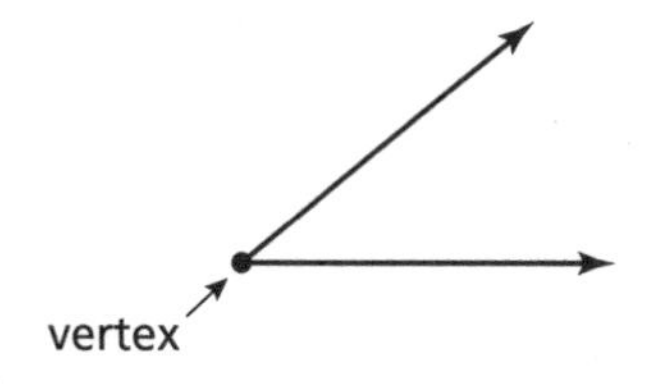

angle of rotation

The number of degrees a figure rotates

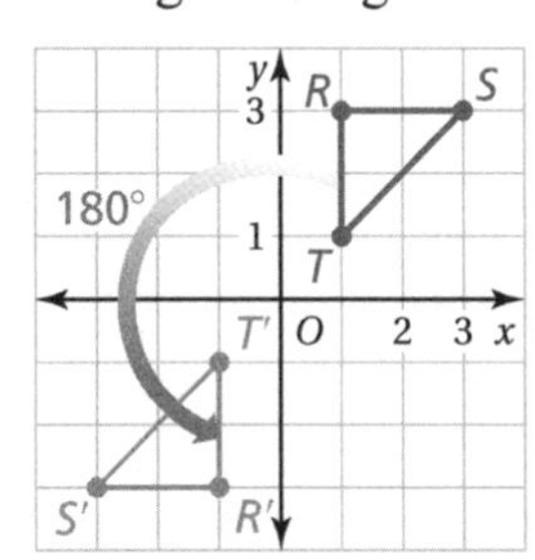

ΔRST has been rotated 180° to $\Delta R'S'T'$.

base (of a power)

The base of a power is the common factor.

See power.

biased sample

A sample that is not representative of a population; One or more parts of the population are favored over others.

You want to estimate the number of students in your school who like to play basketball. You survey 100 students at a basketball game.

center (of a circle)

The point inside a circle that is the same distance from all points on the circle

See circle.

center of dilation

A point with respect to which a figure is dilated

See dilation.

center of rotation

A point about which a figure is rotated

See rotation.

center of a sphere

The point inside a sphere that is the same distance from all points on the sphere

See sphere.

circle

The set of all points in a plane that are the same distance from a point called the center

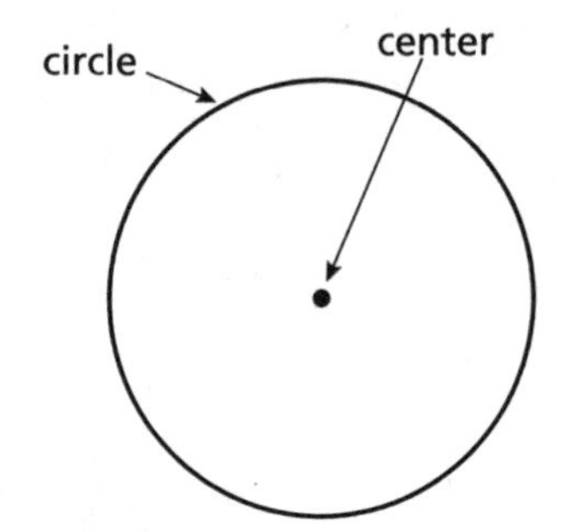

circumference

The distance around a circle

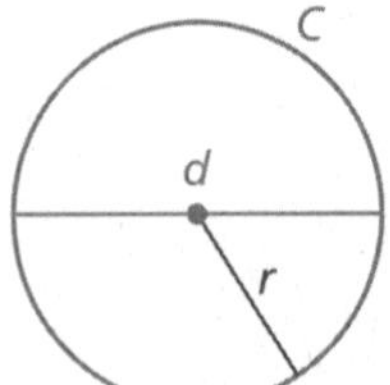

complementary angles

Two angles whose measures have a sum of 90°

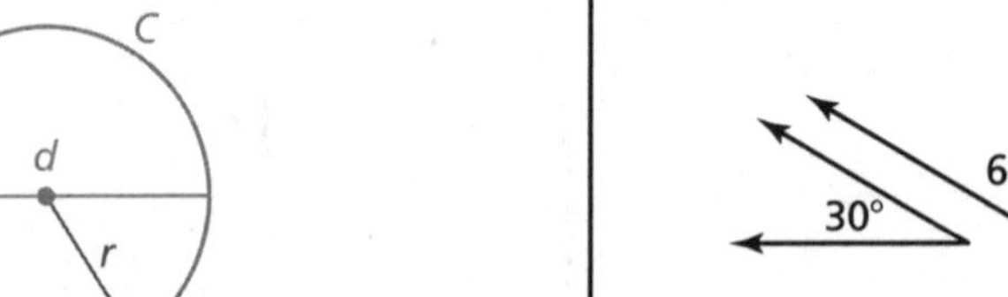

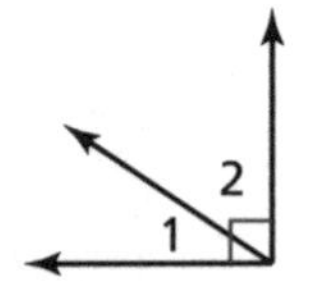

composite figure

A figure made up of triangles, squares, rectangles, semicircles, and other two-dimensional figures

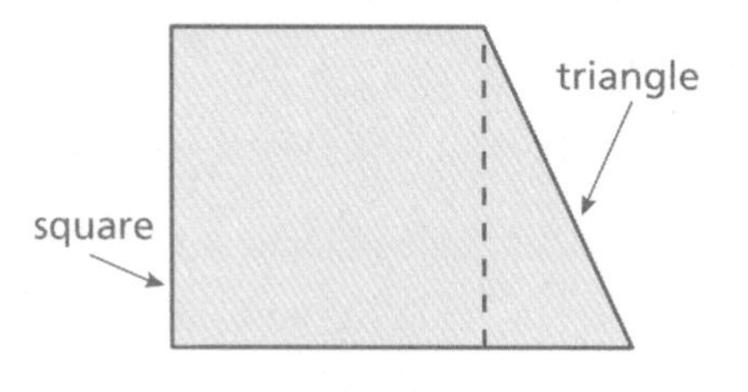

compound event

An event that consists of two or more events

Spinning a spinner and flipping a coin

concave polygon

A polygon in which at least one line segment connecting any two vertices lies outside the polygon

congruent angles

Angles that have the same measure

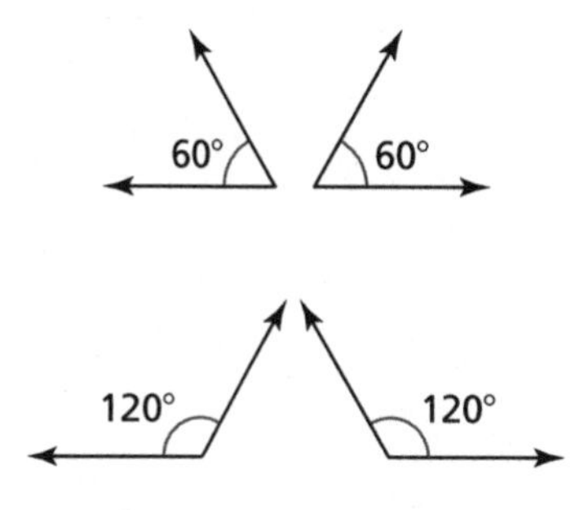

congruent figures

Figures that have the same size and the same shape

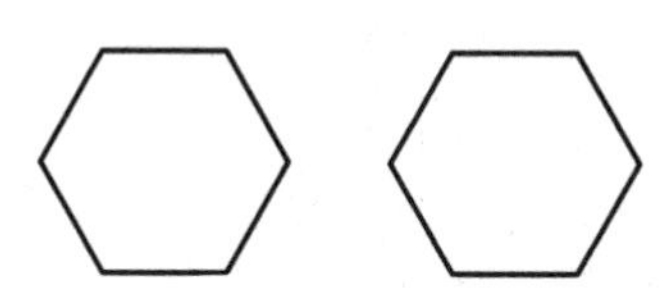

congruent sides

Sides that have the same length

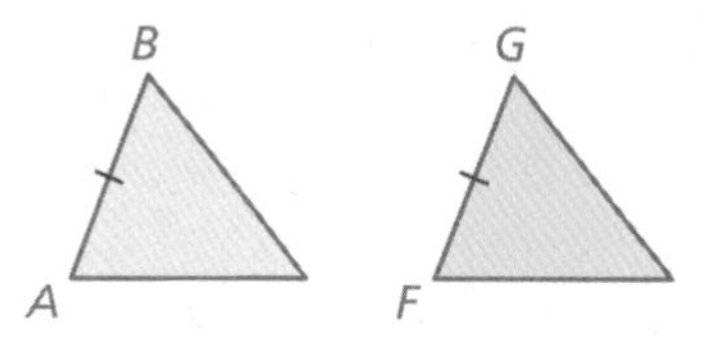

Side AB and side FG are congruent sides.

convex polygon

A polygon in which every line segment connecting any two vertices lies entirely inside the polygon

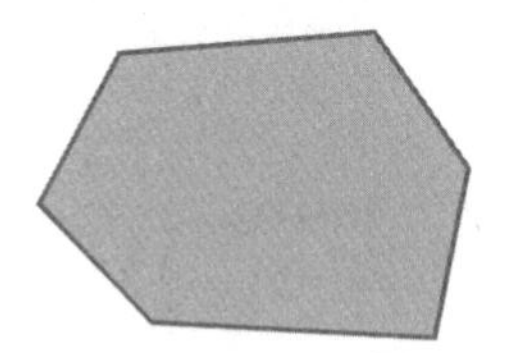

coordinate plane

A coordinate plane is formed by the intersection of a horizontal number line and a vertical number line.

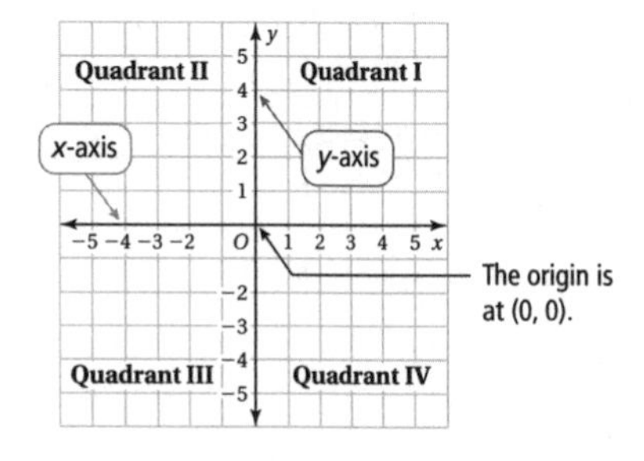

corresponding angles

Matching angles of two congruent figures

$\triangle ABC \cong \triangle DEF$

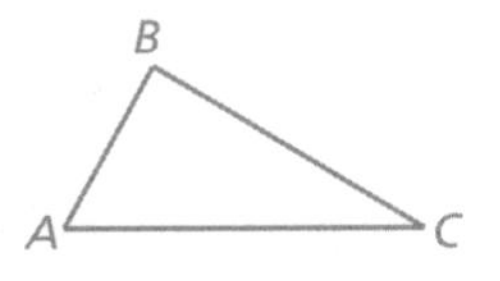

Corresponding angles: $\angle A$ and $\angle D$

$\angle B$ and $\angle E$

$\angle C$ and $\angle F$

corresponding sides

Matching sides of two congruent figures

$\triangle ABC \cong \triangle DEF$

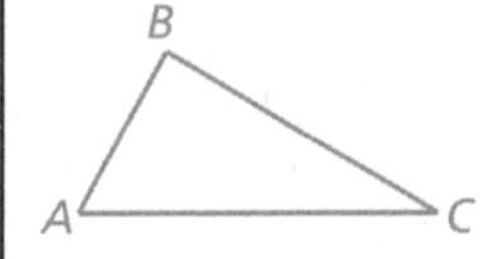

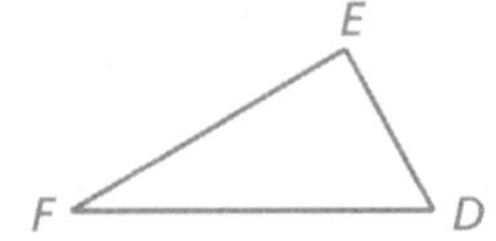

Corresponding sides: side AB and side DE

side BC and side EF

side AC and side DF

cross section

A two-dimensional shape formed by the intersection of a plane and a solid

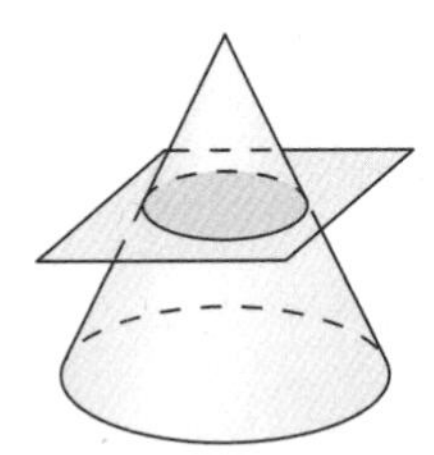

The intersection of the plane and the cone is a circle.

cube root

A number that, when multiplied by itself, and then multiplied by itself again, equals a given number

$\sqrt[3]{8} = 2$

$\sqrt[3]{-27} = -3$

degree

A unit used to measure angles

$90°, 45°, 32°$

dependent events

Two events such that the occurrence of one event affects the likelihood that the other event(s) will occur

A bag contains 3 red marbles and 4 blue marbles. You randomly draw a marble, do not replace it, then randomly draw another marble. The events "first marble is blue" and "second marble is red" are dependent events.

dependent variable

The variable whose value depends on the independent variable in an equation in two variables

In the equation $y = 5x - 8$, y is the dependent variable.

diameter (of a circle)

The distance across a circle through the center

See circumference.

dilation

A transformation in which a figure is made larger or smaller with respect to a fixed point called the center of dilation

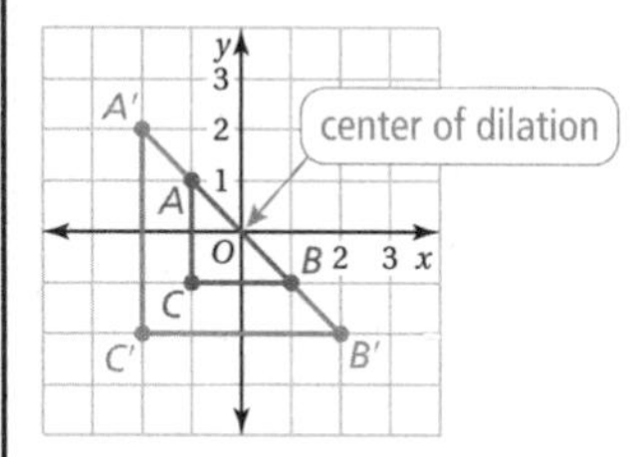

$A'B'C'$ is a dilation of ABC with respect to the origin. The scale factor is 2.

distance formula

The distance d between any two points (x_1, y_1) and (x_2, y_2) is given by the formula

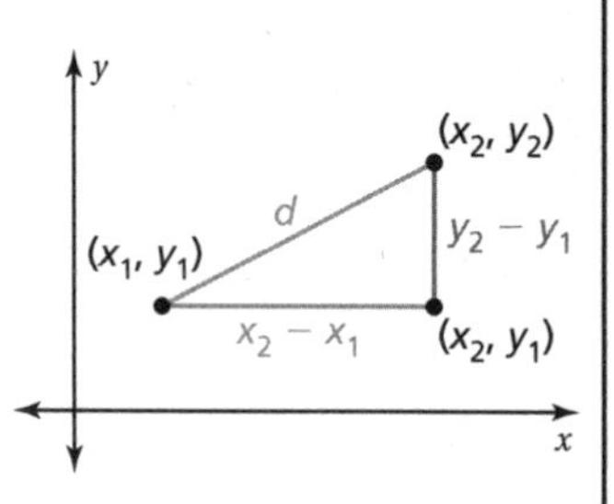

$$d = \sqrt{(x_2 - x_1)^2 + (y_2 - y_1)^2}.$$

Division Property of Equality

Dividing each side of an equation by the same number produces an equivalent equation.

$$4x = -40$$
$$\frac{4x}{4} = \frac{-40}{4}$$
$$x = -10$$

Division Property of Inequality

When you divide each side of an inequality by the same positive number, the inequality remains true. When you divide each side of an inequality by the same negative number, the direction of the inequality symbol must be reversed for the inequality to remain true.

$$4x > -12 \qquad -5x > 30$$
$$\frac{4x}{4} > \frac{-12}{4} \qquad \frac{-5x}{-5} < \frac{30}{-5}$$
$$x > -3 \qquad x < -6$$

enlargement

A dilation with a scale factor greater than 1

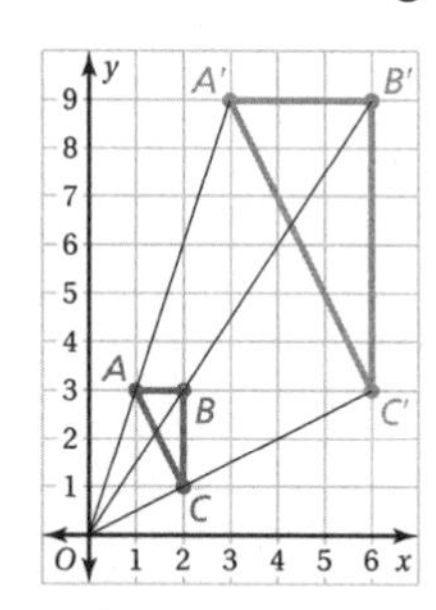

$A'B'C'$ is an enlargement of ABC.

equation

A mathematical sentence that uses an equal sign to show that two expressions are equal

$$4x = 16, a + 7 = 21$$

equivalent equations

Equations that have the same solutions

$$2x - 8 = 0 \text{ and } 2x = 8$$

evaluate (an algebraic expression)

Substitute a number for each variable in an algebraic expression. Then use the order of operations to find the value of the numerical expression.

Evaluate $3x + 5$ when $x = 6$.

$$3x + 5 = 3(6) + 5$$
$$= 18 + 5$$
$$= 23$$

event

A collection of one or more outcomes of an experiment

Flipping heads on a coin

experiment

An investigation or procedure that has varying results

Rolling a number cube

experimental probability

Probability that is based on repeated trials of an experiment

$$P(\text{event}) = \frac{\text{number of times the event occurs}}{\text{total number of trials}}$$

A basketball player makes 19 baskets in 28 attempts. The experimental probability that the player makes a basket is $\frac{19}{28}$, or about 68%.

exponent

The exponent of a power indicates the number of times a base is used as a factor.

See power.

expression

A mathematical phrase containing numbers, operations, and/or variables

$12 + 6, 18 + 3 \times 4,$

$8 + x, 6 \times a - b$

exterior angles

When two parallel lines are cut by a transversal, four exterior angles are formed on the outside of the parallel lines.

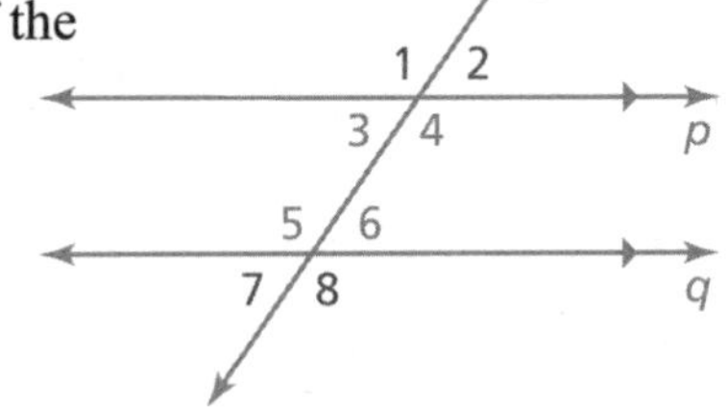

$\angle 3$, $\angle 4$, $\angle 5$, and $\angle 6$ are interior angles.

$\angle 1$, $\angle 2$, $\angle 7$, and $\angle 8$ are exterior angles.

exterior angles of a polygon

The angles outside a polygon that are adjacent to the interior angles

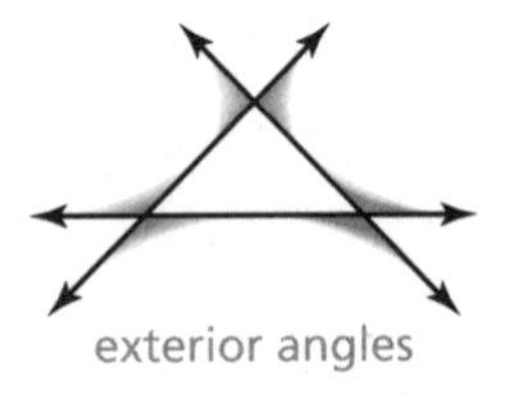

factor

When whole numbers other than zero are multiplied together, each number is a factor of the product.

$2 \times 3 \times 4 = 24$, so 2, 3, and 4 are factors of 24.

favorable outcomes

The outcomes of a specific event

When rolling a number cube, the favorable outcomes for the event "rolling an even number" are 2, 4, and 6.

function

A relation that pairs each input with exactly one output

The ordered pairs $(0, 1)$, $(1, 2)$, $(2, 4)$, and $(3, 6)$ represent a function.

Ordered Pairs	Input	Output
(0, 1)	0	1
(1, 2)	1	2
(2, 4)	2	4
(3, 6)	3	6

function rule

An equation that describes the relationship between inputs (independent variable) and outputs (dependent variable)

The function rule "The output is three less than the input" is represented by the equation $y = x - 3$.

Fundamental Counting Principle

An event M has m possible outcomes and event N has n possible outcomes. The total number of outcomes of event M followed by event N is $m \times n$.

You have 7 shirts, 5 pairs of pants, and 2 pairs of shoes. You can make $7 \times 5 \times 2 = 70$ different outfits.

graph of an inequality

A graph that shows all the solutions of an inequality on a number line

$x > -2$

−3 −2 −1 0 1 2 3

hemisphere

One-half of a sphere

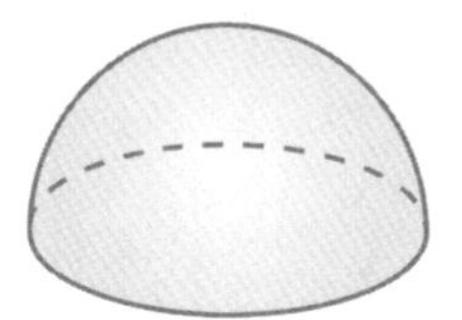

hypotenuse

The side of a right triangle that is opposite the right angle

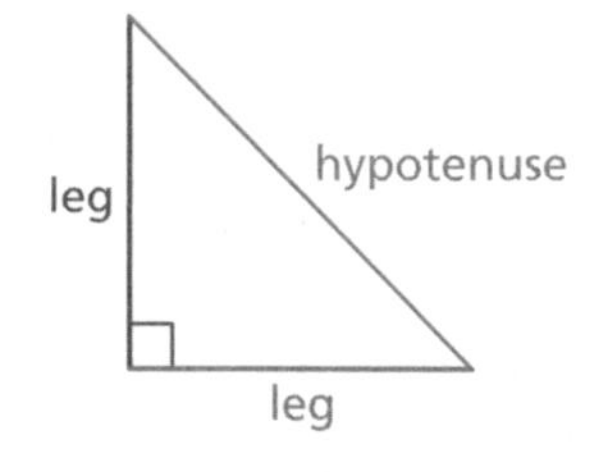

image

The new figure formed by a transformation

See translation, reflection, rotation, and dilation.

independent events

Two events such that the occurrence of one event does not affect the likelihood that the other event(s) will occur

You flip a coin and roll a number cube. The events "flipping tails" and "rolling a 4" are independent events.

independent variable

The variable representing the quantity that can change freely in an equation in two variables

In the equation $y = 5x - 8$, x is the independent variable.

indirect measurement

Indirect measurement uses similar figures to find a missing measure when it is difficult to find directly.

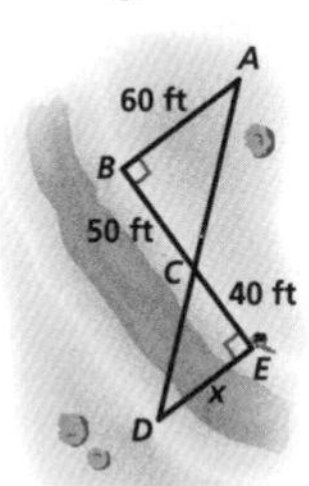

$$\frac{x}{60} = \frac{40}{50}$$

$$60 \bullet \frac{x}{60} = 60 \bullet \frac{40}{50}$$

$$x = 48$$

The distance across the river is 48 feet.

inequality

A mathematical sentence that compares expressions; It contains the symbols <, >, ≤, or ≥.

$x - 4 < 14,\ x + 5 \geq -12$

input

In a relation, inputs are associated with outputs.

inputs → (0, 0) (1, 2) (2, 4) ← outputs

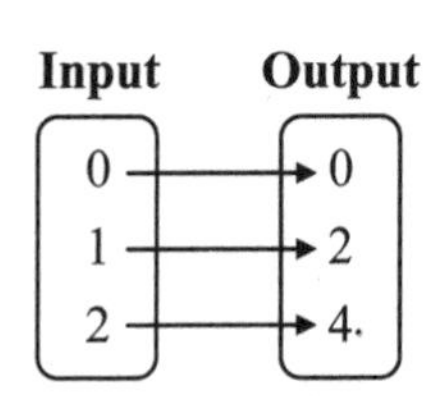

integers

The set of whole numbers and their opposites

$\ldots -3, -2, -1, 0, 1, 2, 3, \ldots$

interior angles

When two parallel lines are cut by a transversal, four interior angles are formed on the inside of the parallel lines.

See exterior angles.

interior angles of a polygon

The angles inside a polygon

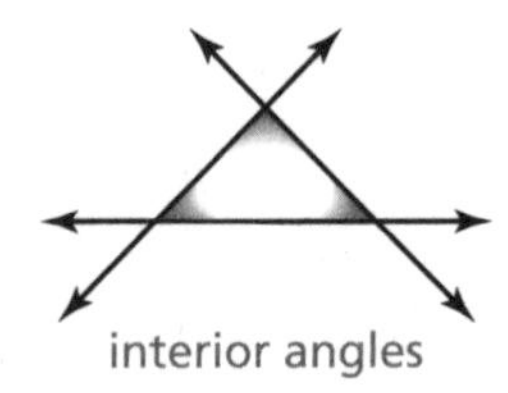

irrational number

A number that cannot be written as the ratio of two integers

$\pi, \sqrt{14}$

joint frequency

Each entry in a two-way table

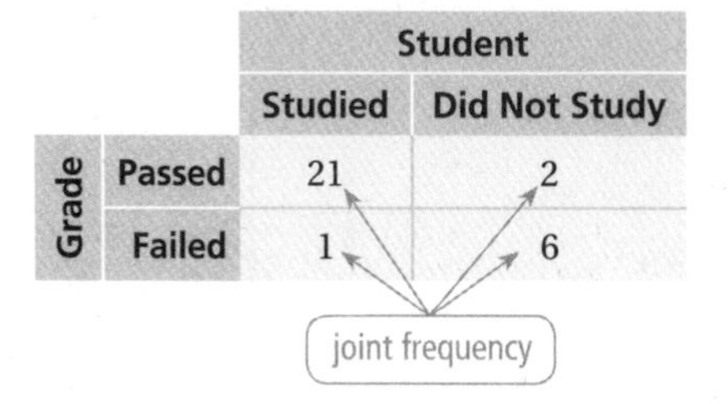

		Student	
		Studied	Did Not Study
Grade	Passed	21	2
	Failed	1	6

kite

A quadrilateral with two pairs of congruent adjacent sides and opposite sides that are not congruent

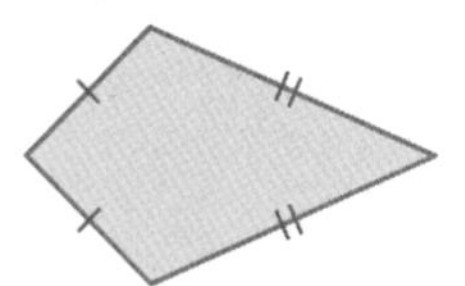

lateral surface area (of a prism)

The sum of the areas of the lateral faces of a prism

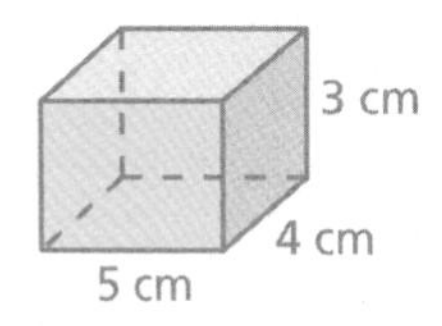

$$\begin{aligned}\text{Lateral surface area} &= 2(4)(3) + 2(5)(3)\\ &= 24 + 30 = 54\ \text{cm}^2\end{aligned}$$

legs

The two sides of a right triangle that form the right angle

See hypotenuse.

line of best fit

A precise line of fit that best models a set of data

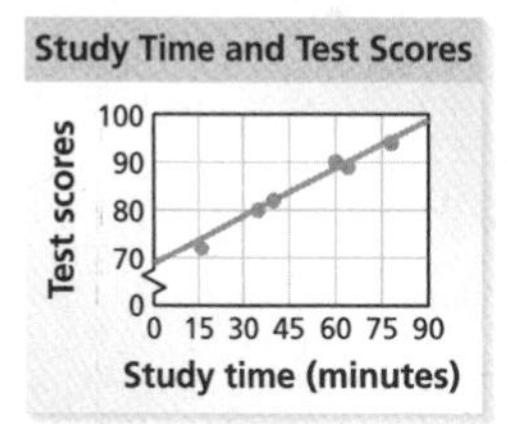

line of fit

A line drawn on a scatter plot close to most of the data points; It can be used to estimate data on a graph.

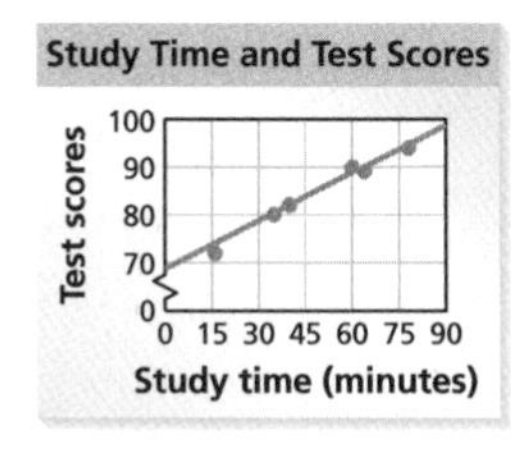

line of reflection

A line that a figure is reflected in to create a mirror image of the original figure

See reflection.

linear equation

An equation whose graph is a line

$y = x - 1$

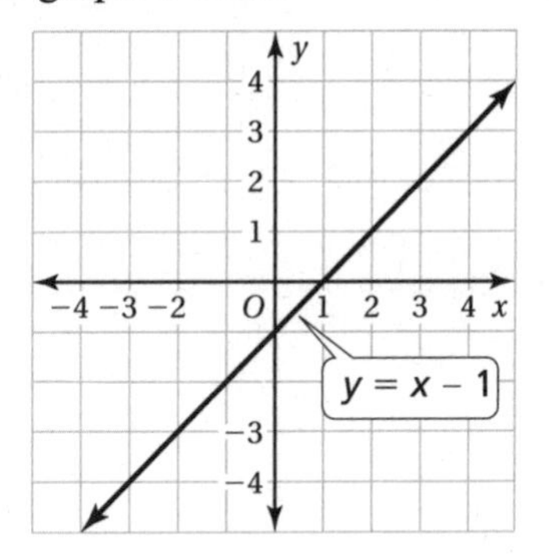

linear function

A function whose graph is a nonvertical line; a function that has a constant rate of change

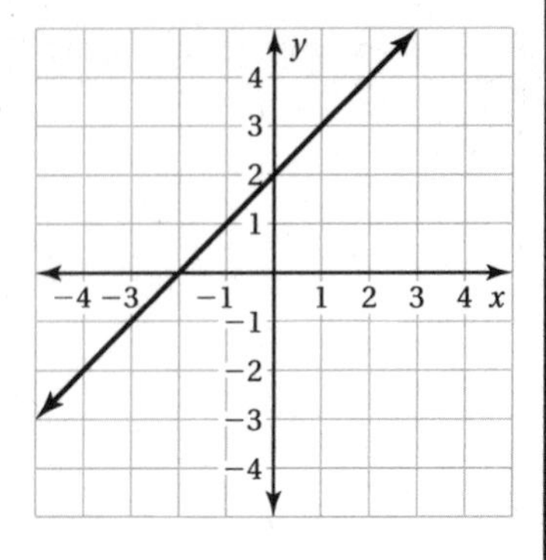

literal equation

An equation that has two or more variables

$$2y + 6x = 12$$

mapping diagram

A way to represent a relation

Input	Output
1	3
2	4
3	5
4	6

marginal frequencies

The sums of the rows and columns in a two-way table

		Age			
		12–13	14–15	16–17	Total
Student	Rides Bus	24	12	14	50
	Does Not Ride Bus	16	13	21	50
	Total	40	25	35	100

Multiplication Property of Equality

Multiplying each side of an equation by the same number produces an equivalent equation.

$$-\frac{2}{3}x = 8$$

$$-\frac{3}{2} \bullet \left(-\frac{2}{3}x\right) = -\frac{3}{2} \bullet 8$$

$$x = -12$$

Multiplication Property of Inequality

When you multiply each side of an inequality by the same positive number, the inequality remains true.

When you multiply each side of an inequality by the same negative number, the direction of the inequality symbol must be reversed for the inequality to remain true.

$$\frac{x}{2} < -9 \qquad \frac{x}{-6} < 3$$

$$2 \bullet \frac{x}{2} < 2 \bullet (-9) \qquad -6 \bullet \frac{x}{-6} > -6 \bullet 3$$

$$x < -18 \qquad x > -18$$

negative number

A number less than 0

$$-0.25, -10, -500$$

nonlinear function

A function that does not have a constant rate of change; a function whose graph is not a line

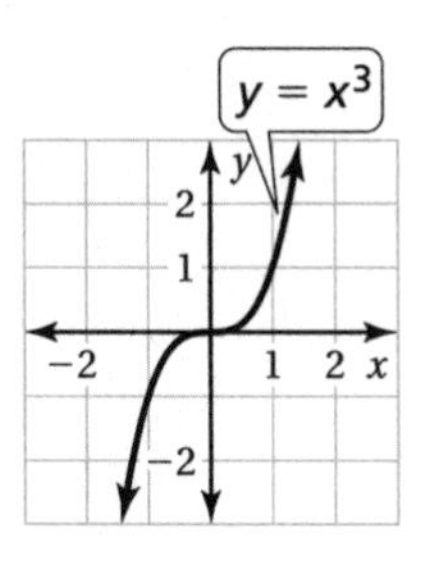

ordered pair

A pair of numbers (x, y) used to locate a point in a coordinate plane; The first number is the x-coordinate, and the second number is the y-coordinate.

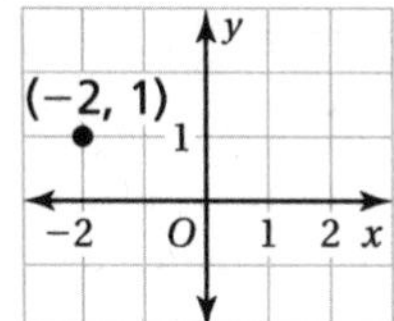

The x-coordinate of the point $(-2, 1)$ is -2, and the y-coordinate is 1.

origin

The point, represented by the ordered pair (0, 0), where the horizontal and the vertical number lines intersect in a coordinate plane

See coordinate plane.

outcomes

The possible results of an experiment

The outcomes of flipping a coin are heads and tails.

output

In a relation, inputs are associated with outputs.

See input.

parallel lines

Lines in the same plane that do not intersect; Nonvertical parallel lines have the same slope. All vertical lines are parallel.

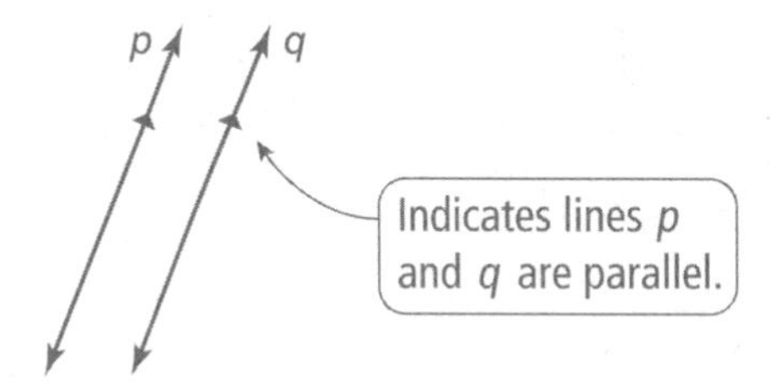

perfect cube

A number that can be written as the cube of an integer

$-27, 8, 125$

perfect square

A number with integers as its square roots

16, 25, 81

perpendicular lines

Lines in the same plane that intersect at right angles; Two nonvertical lines are perpendicular when the product of their slopes is -1. Vertical lines are perpendicular to horizontal lines.

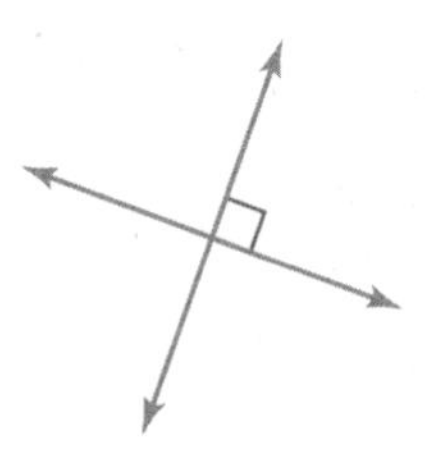

pi (π)

The ratio of the circumference of a circle to its diameter

The value of π can be approximated as 3.14 or $\frac{22}{7}$.

point-slope form

A linear equation written in the form $y - y_1 = m(x - x_1)$ is in point-slope form. The line passes through the point (x_1, y_1), and the slope of the line is m.

$$y - 1 = \frac{2}{3}(x + 6)$$

polygon

A closed figure in a plane that is made up of three or more line segments that intersect only at their endpoints

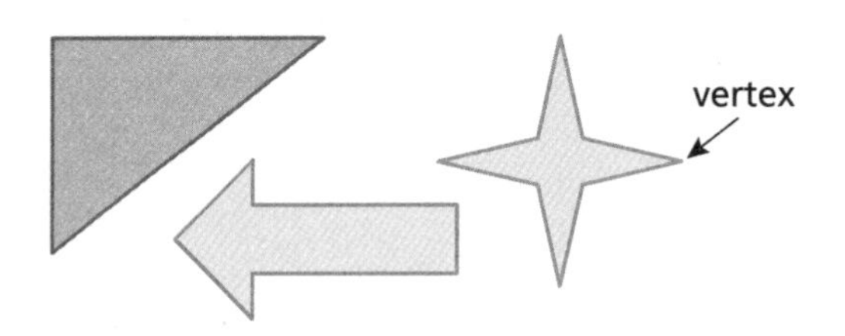

polyhedron

A solid whose faces are all polygons

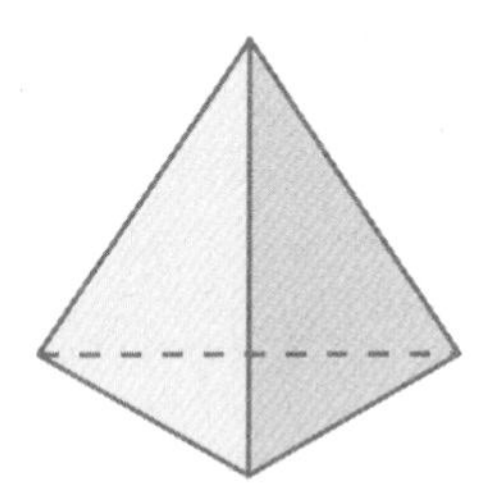

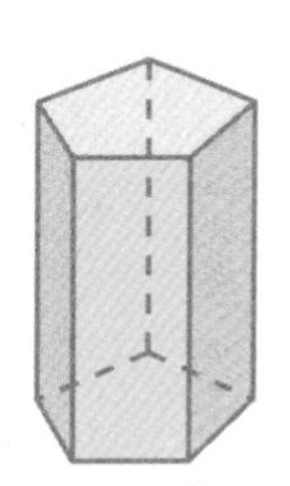

population

An entire group of people or objects

Population: All of the 14-year-old females in the United States

Sample: All of the 14-year-old females in your town

positive number

A number greater than 0

0.5, 2, 100

power

A product of repeated factors

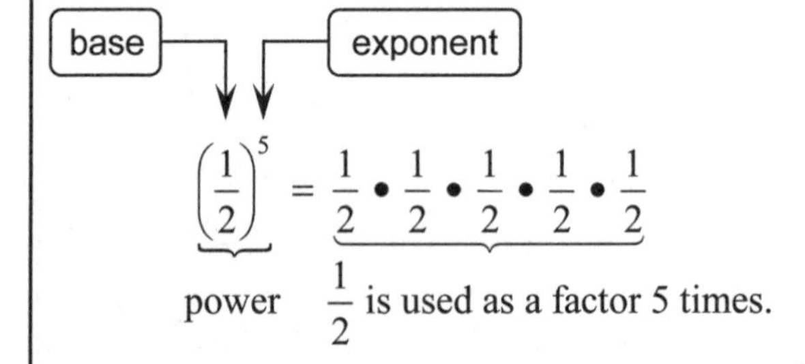

$$\left(\frac{1}{2}\right)^5 = \frac{1}{2} \bullet \frac{1}{2} \bullet \frac{1}{2} \bullet \frac{1}{2} \bullet \frac{1}{2}$$

$\frac{1}{2}$ is used as a factor 5 times.

Power of a Power Property

To find a power of a power, multiply the exponents.

$$(3^4)^2 = 3^{4 \bullet 2} = 3^8$$

$$(a^m)^n = a^{mn}$$

Power of a Product Property

To find a power of a product, find the power of each factor and multiply.

$$(5 \bullet 7)^4 = 5^4 \bullet 7^4$$

$$(ab)^m = a^m b^m$$

prism

A polyhedron that has two parallel, congruent bases; The lateral faces are parallelograms.

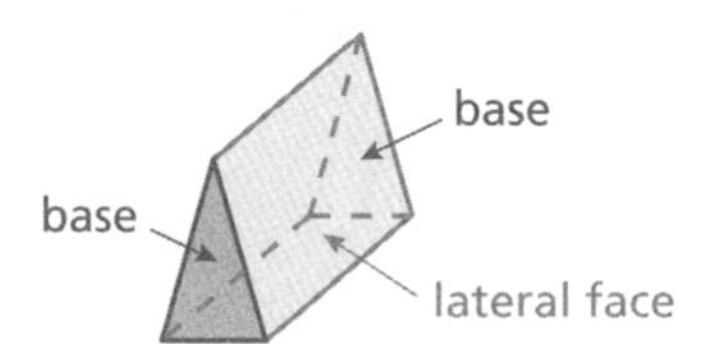

probability

A number from 0 to 1 that measures the likelihood that an event will occur

See experimental probability and theoretical probability.

Product of Powers Property

To multiply powers with the same base, add their exponents.

$$3^7 \bullet 3^{10} = 3^{7+10} = 3^{17}$$

$$a^m \bullet a^n = a^{m+n}$$

proportion

An equation stating that two ratios are equivalent

$$\frac{3}{4} = \frac{12}{16}$$

proportional

Two quantities that form a proportion are proportional.

Because $\frac{3}{4}$ and $\frac{12}{16}$ form a proportion, $\frac{3}{4}$ and $\frac{12}{16}$ are proportional.

pyramid

A polyhedron that has one base; The lateral faces are triangles.

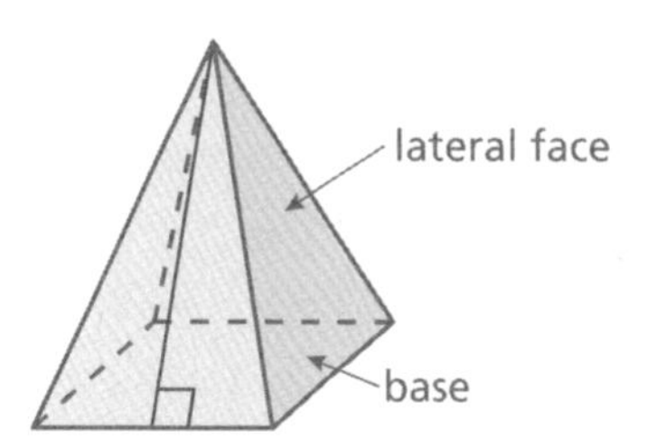

Pythagorean Theorem

In any right triangle, the sum of the squares of the lengths of the legs is equal to the square of the length of the hypotenuse.

$a^2 + b^2 = c^2$

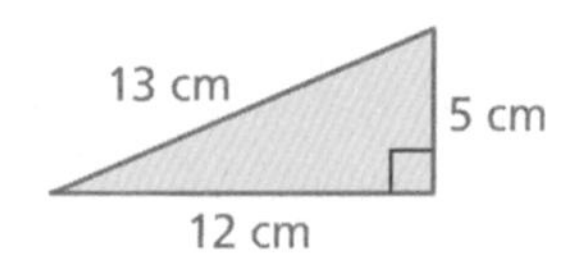

$5^2 + 12^2 = 13^2$

quadrilateral

A polygon with four sides

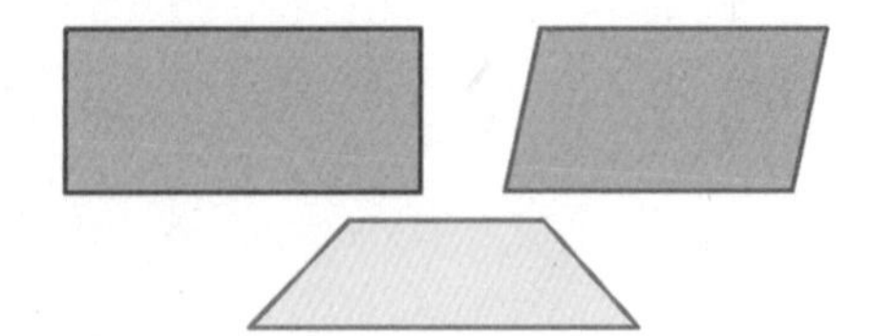

Quotient of Powers Property

To divide powers with the same base, subtract their exponents.

$$\frac{9^7}{9^3} = 9^{7-3} = 9^4$$

$$\frac{a^m}{a^n} = a^{m-n}, \text{ where } a \neq 0$$

radical sign

The symbol $\sqrt{}$ which is used to represent a square root

$$\sqrt{25} = 5$$

$$-\sqrt{49} = -7$$

$$\pm\sqrt{100} = \pm 10$$

radicand

The number under a radical sign

The radicand of $\sqrt{25}$ is 25.

radius (of a circle)

The distance from the center of a circle to any point on the circle

See circumference.

radius of a sphere

The distance from the center of a sphere to any point on the sphere

See sphere.

rate

A ratio of two quantities with different units

You read 3 books every 2 weeks.

ratio

A comparison of two quantities using division; The ratio of a to b (where $b \neq 0$) can be written as a to b, $a : b$, or $\frac{a}{b}$.

$$4 \text{ to } 1,\ 4 : 1, \text{ or } \frac{4}{1}$$

rational number

A number that can be written as $\frac{a}{b}$ where a and b are integers and $b \neq 0$

$$3 = \frac{3}{1}, \qquad -\frac{2}{5} = \frac{-2}{5}$$

$$0.25 = \frac{1}{4}, \qquad 1\frac{1}{3} = \frac{4}{3}$$

ray

A part of a line that has one endpoint and extends without end in one direction

real numbers

The set of all rational and irrational numbers

$$4, -6.5, \pi, \sqrt{14}$$

reduction

A dilation with a scale factor greater than 0 and less than 1

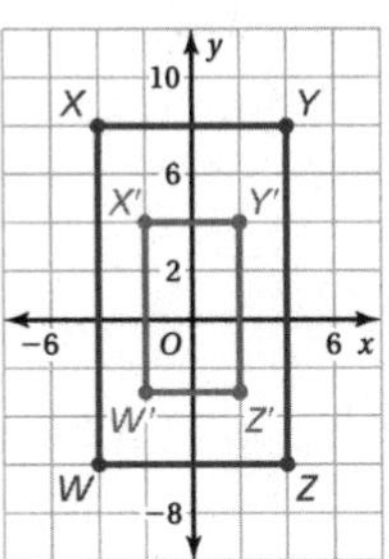

$W'X'Y'Z'$ is a reduction of $WXYZ$.

reflection

A transformation in which a figure is reflected in a line called the line of reflection; A reflection creates a mirror image of the original figure.

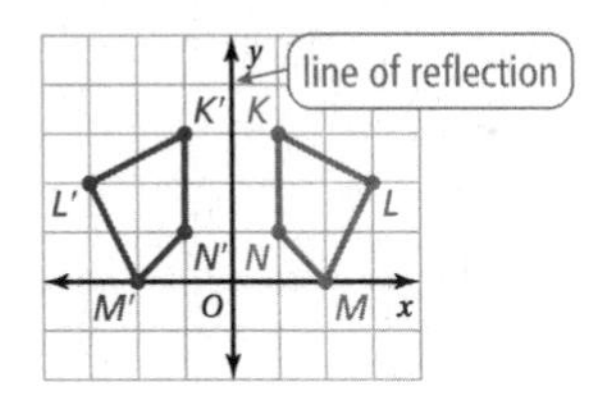

$K'L'M'N'$ is a reflection of $KLMN$ over the y-axis.

regular polygon

A polygon in which all the sides are congruent, and all the interior angles are congruent

regular pyramid

A pyramid whose base is a regular polygon

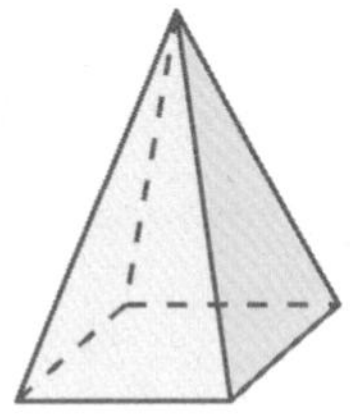

relation

A relation pairs inputs with outputs and can be represented by ordering pairs on a mapping diagram.

Ordered Pairs
(0, 1)
(1, 2)
(2, 4)

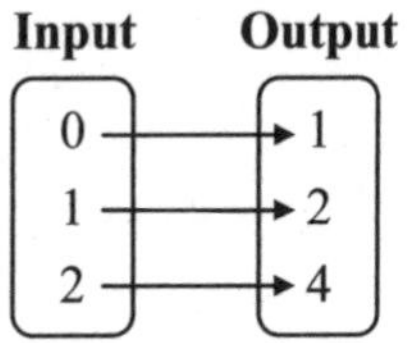

relative frequency

The fraction or percent of the time that an event occurs in an experiment

You flip a coin 20 times. If you flip heads 11 times, the relative frequency of flipping heads is $\frac{11}{20}$, or 55%.

right angle

An angle whose measure is 90°

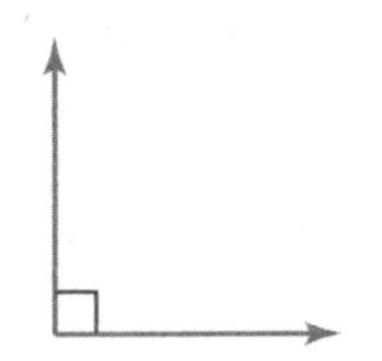

right triangle

A triangle that has one right angle

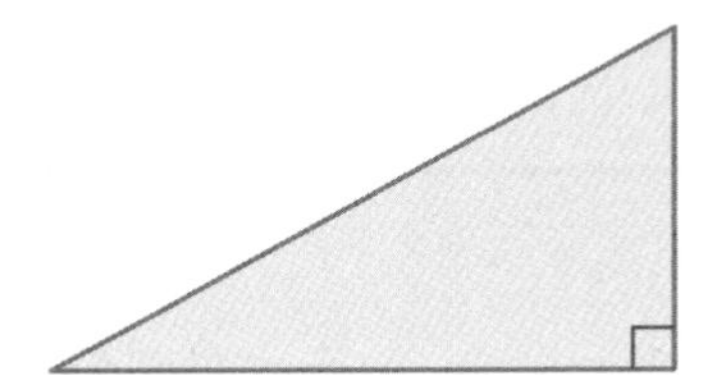

rise

The change in y between any two points on a line

See slope.

rotation

A transformation in which a figure is rotated about a point called the center of rotation; The number of degrees a figure rotates is the angle of rotation.

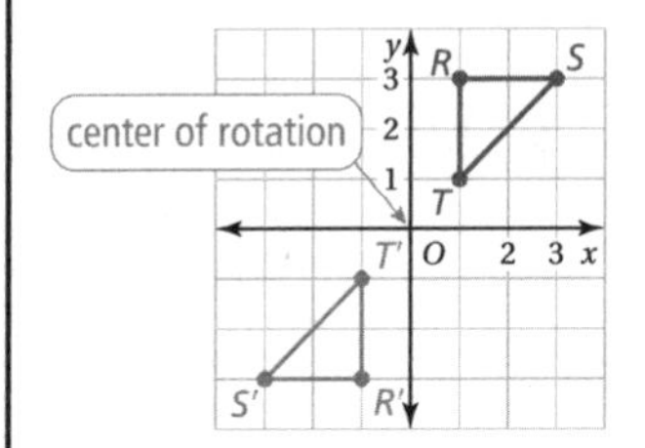

$\triangle RST$ has been rotated about the origin O to $\triangle R'S'T'$.

run

The change in x between any two points on a line

See slope.

sample

A part of a population

See population.

sample space

The set of all possible outcomes of one or more events

You flip a coin twice. The outcomes in the sample space are HH, HT, TH, and TT.

scale

A ratio that compares the measurements of a drawing or model with the actual measurements

12 cm : 1 cm

$\frac{2 \text{ in.}}{15 \text{ ft}}$

scale drawing

A proportional, two-dimensional drawing of an object

A blueprint or a map

scale factor (of a dilation)

The ratio of the side lengths of the image of a dilation to the corresponding side lengths of the original figure

See dilation.

scale factor (of a scale drawing)

A scale without units

See ratio.

scale model

A proportional, three-dimensional model of an object

scatter plot

A graph that shows the relationship between two data sets using ordered pairs in a coordinate plane

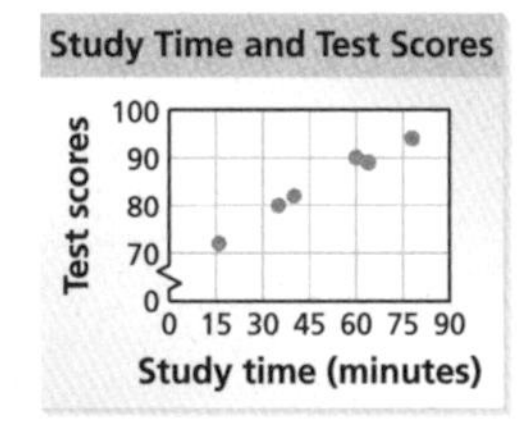

scientific notation

A number is written in scientific notation when it is represented as the product of a factor and a power of 10. The factor must be greater than or equal to 1 and less than 10.

8.3×10^4

4×10^{-3}

semicircle

One-half of a circle

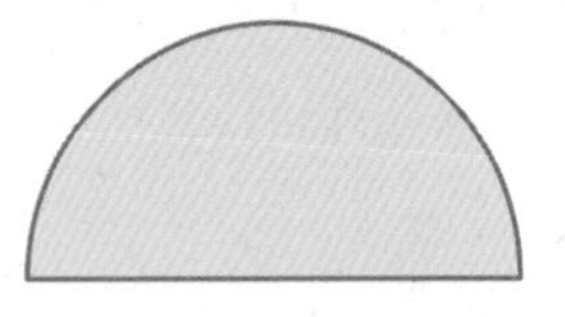

similar figures

Figures that have the same shape but not necessarily the same size; Two figures are similar when corresponding side lengths are proportional and corresponding angles are congruent.

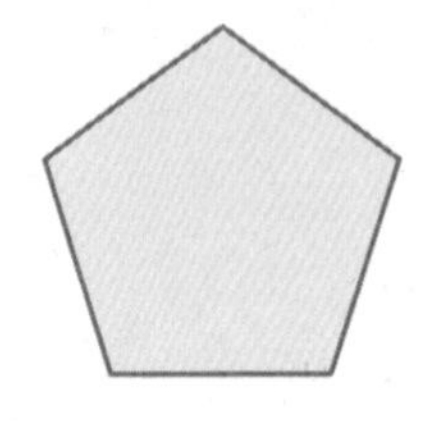

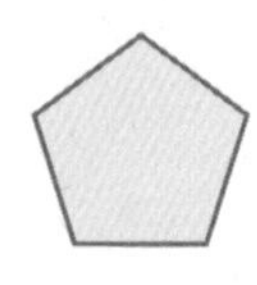

similar solids

Solids that have the same shape and proportional corresponding dimensions

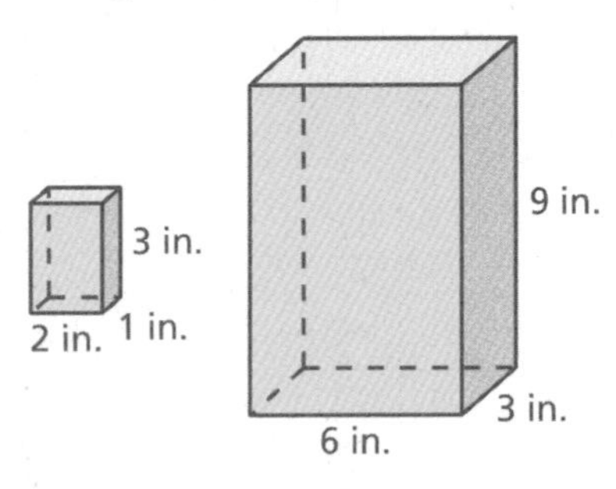

simulation

An experiment that is designed to reproduce the conditions of a situation or process

slant height (of a pyramid)

The height of each triangular face of a pyramid

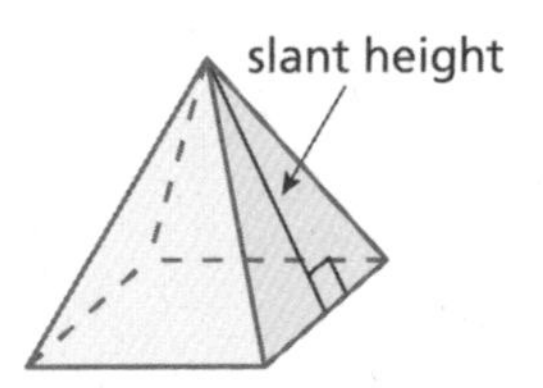

slope

The slope m of a line is a ratio of the change in y (the rise) to the change in x (the run) between any two points (x_1, y_1) and (x_2, y_2) on a line. It is a measure of the steepness of a line.

$$m = \frac{\text{rise}}{\text{run}} = \frac{\text{change in } y}{\text{change in } x}$$

$$= \frac{y_2 - y_1}{x_2 - x_1}$$

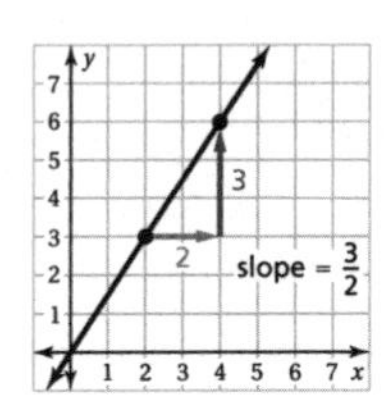

slope-intercept form

A linear equation written in the form $y = mx + b$ is in slope-intercept form. The slope of the line is m, and the y-intercept of the line is b.

The slope is 1 and the y-intercept is 2.

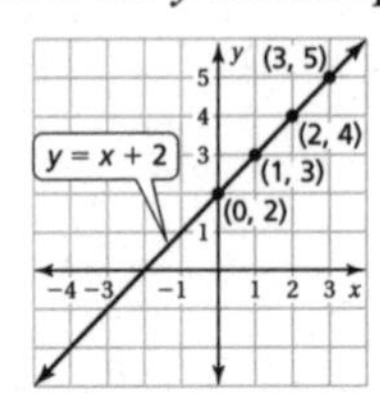

solid

A three-dimensional figure that encloses a space

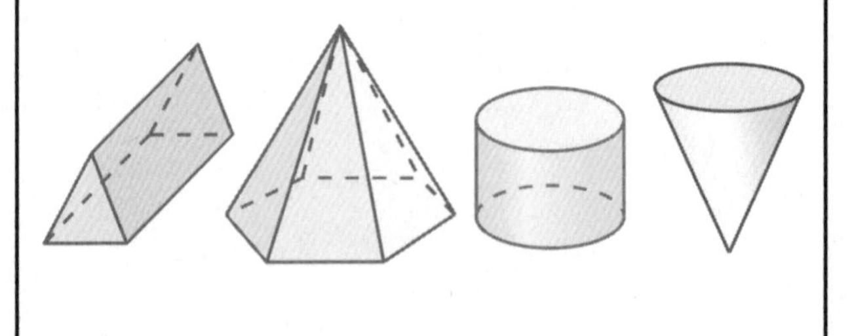

solution of an equation

A value that makes an equation true

6 is the solution of the equation $x - 4 = 2$.

solution of an inequality

A value that makes an inequality true

A solution of the inequality $x + 3 > -9$ is $x = 2$.

solution of a linear equation

All of the points on a line

solution set

The set of all solutions of an inequality

solution of a system of linear equations

An ordered pair that is a solution of each equation in a system

$(1, -3)$ is the solution of the following system of linear equations.

$$4x - y = 7$$
$$2x + 3y = -7$$

sphere

The set of all points in space that are the same distance from a point called the center

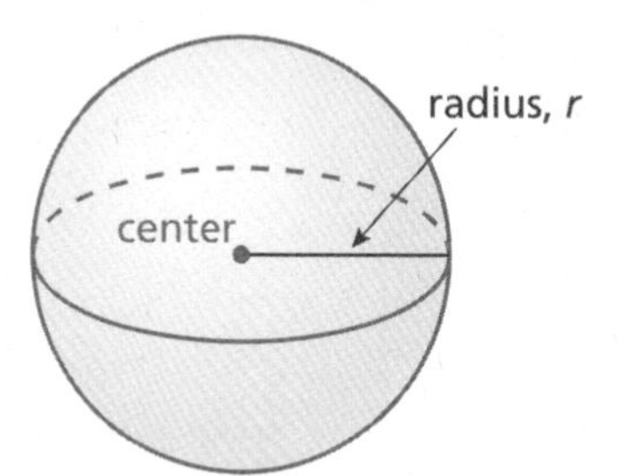

square root

A number that, when multiplied by itself, equals a given number

The two square roots of 100 are 10 and -10.

$\pm\sqrt{100} = \pm 10$

standard form

The standard form of a linear equation is $ax + by = c$, where a and b are not both zero.

$$-2x + 3y = -6$$

Subtraction Property of Equality

Subtracting the same number from each side of an equation produces an equivalent equation.

$$\begin{aligned} x + 10 &= -12 \\ \underline{-\ 10} \quad & \quad \underline{-\ 10} \\ x &= -22 \end{aligned}$$

Subtraction Property of Inequality

When you subtract the same number from each side of an inequality, the inequality remains true.

$$\begin{aligned} x + 7 &> -20 \\ \underline{-\ 7} \quad & \quad \underline{-\ 7} \\ x &> -27 \end{aligned}$$

supplementary angles

Two angles whose measures have a sum of 180°

system of linear equations

A set of two or more linear equations in the same variables, also called a linear system.

$y = x + 1$ Equation 1

$y = 2x - 7$ Equation 2

theorem

A rule in mathematics

The Pythagorean Theorem

theoretical probability

The ratio of the number of favorable outcomes to the number of possible outcomes when all possible outcomes are equally likely

$$P(\text{event}) = \frac{\text{number of favorable outcomes}}{\text{number of possible outcomes}}$$

When rolling a number cube, the theoretical probability of rolling a 4 is $\frac{1}{6}$.

three-dimensional figure

A figure that has length, width, and depth; also known as a solid

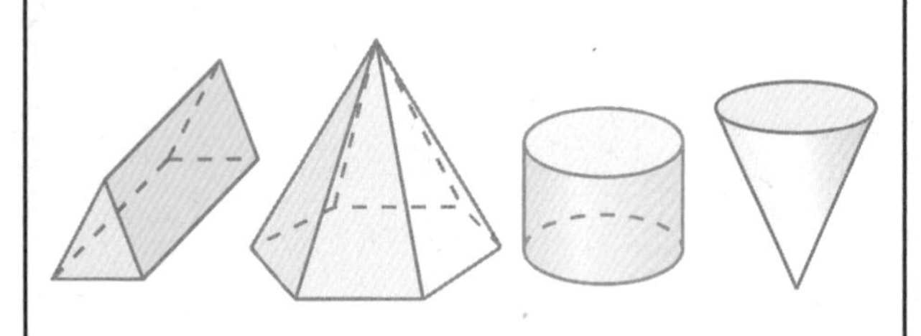

transformation

A transformation changes a figure into another figure.

See translation, reflection, rotation, and dilation.

translation

A transformation in which a figure slides but does not turn; Every point of the figure moves the same distance and in the same direction.

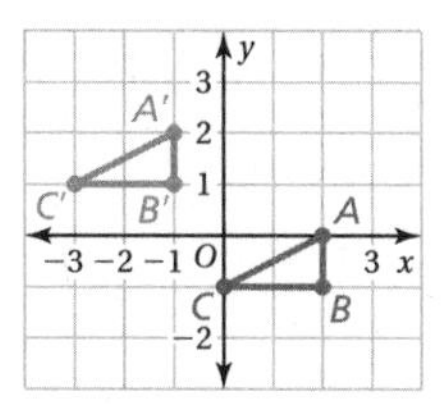

ABC has been translated 3 units left and 2 units up to $A'B'C'$.

transversal

A line that intersects two or more lines

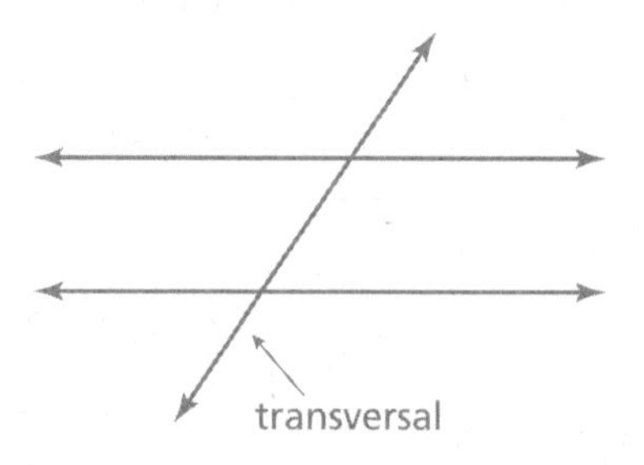

two-dimensional figure

A figure that has only length and width

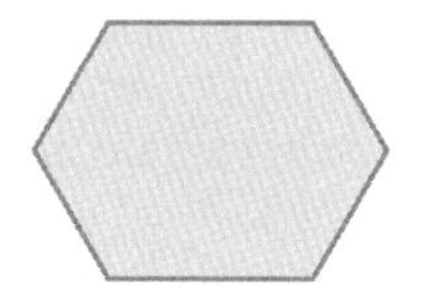
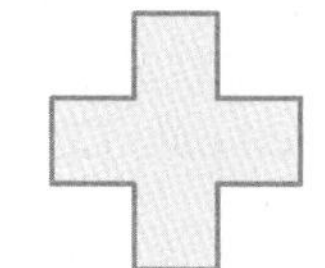

two-way table

Displays two categories of data collected from the same source

		Fundraiser	
		No	Yes
Gender	Female	22	51
	Male	30	29

unbiased sample

A sample that is representative of a population; It is selected at random and is large enough to provide accurate data.

You want to estimate the number of students in your school who like to play basketball. You survey 100 students at random during lunch.

variable

A symbol that represents one or more numbers

x is a variable in $2x + 1$.

vertex (of an angle)

The point at which the two sides of an angle meet

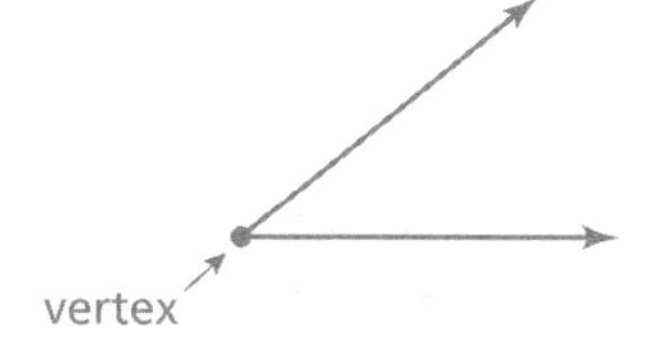

vertex (of a polygon)

A point at which two sides of a polygon meet; The plural of vertex is vertices.

See polygon.

vertical angles

The angles opposite each other when two lines intersect; Vertical angles are congruent angles.

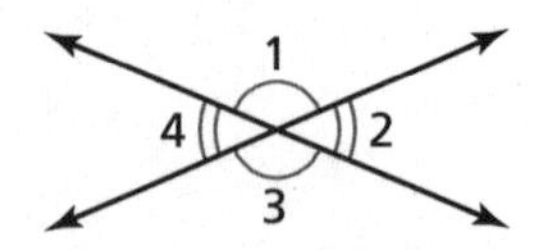

$\angle 1$ and $\angle 3$ are vertical angles.

$\angle 2$ and $\angle 4$ are vertical angles.

whole numbers

The numbers $0, 1, 2, 3, 4, \ldots$

***x*-axis**

The horizontal number line in a coordinate plane

See coordinate plane.

***x*-coordinate**

The first coordinate in an ordered pair, which indicates how many units to move to the left or right from the origin

In the ordered pair $(3, 5)$, the x-coordinate is 3.

***x*-intercept**

The x-coordinate of the point where a line crosses the x-axis

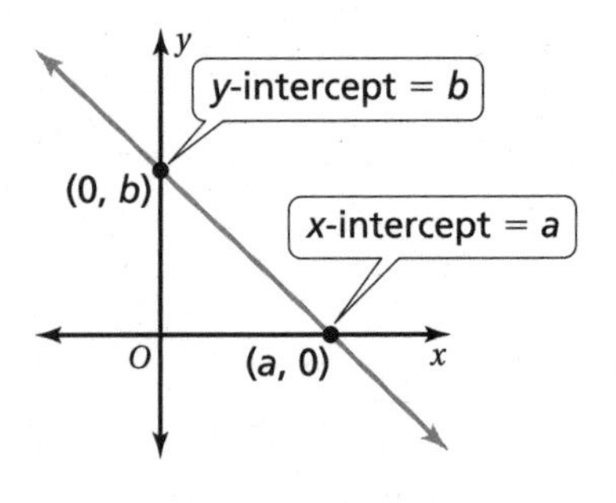

***y*-axis**

The vertical number line in a coordinate plane

See coordinate plane.

***y*-coordinate**

The second coordinate in an ordered pair, which indicates how many units to move up or down from the origin

In the ordered pair $(3, 5)$, the y-coordinate is 5.

***y*-intercept**

The y-coordinate of the point where a line crosses the y-axis

See x-intercept.

Photo Credits

37 *top* ©iStockphoto.com/Viatcheslav Dusaleev; *bottom left* ©iStockphoto.com/Jason Mooy; *bottom right* ©iStockphoto.com/Felix Möckel; **49** Elena Elisseeva/Shutterstock.com.; **52** Estate Craft Homes, Inc.; **88** ©iStockphoto.com/biffspandex; **106** Talvi/Shutterstock.com; **134** ©iStockphoto.com/PeskyMonkey; **151** claudio zaccherini/Shutterstock.com; **185** *baseball* Kittisak/Shutterstock.com; *golf ball* tezzstock/Shutterstock.com; *basketball* vasosh/Shutterstock.com; *tennis ball* UKRID/Shutterstock.com; *water polo ball* John Kasawa/Shutterstock.com; *softball* Ra Studio/Shutterstock.com; *volleyball* vberla/Shutterstock.com; **189** Gina Brockett; **198** Larry Korhnak; **199** Photo by Andy Newman; **204** ©iStockphoto.com/Franck Boston; **205** Stevyn Colgan; **219** ©iStockphoto.com/Kais Tolmats; **220** *top right* ©iStockphoto.com/Kais Tolmats; *Activity 3a and d* Tom C Amon/Shutterstock.com; *Activity 3b* Olga Gabay/Shutterstock.com; *Activity 3c* NASA/MODIS Rapid Response/Jeff Schmaltz; *Activity 3f* HuHu/Shutterstock.com; **221** *Activity 4a* PILart/Shutterstock.com; *Activity 4b* Matthew Cole/Shutterstock.com; *Activity 4c* Yanas/Shutterstock.com; *Activity 4e* unkreativ/Shutterstock.com; **224** NASA **257** Scott J. Carson/Shutterstock.com; **297** *top left* ©iStockphoto.com/Luke Daniek; *top right* ©iStockphoto.com/Jeff Whyte; *bottom left* ©Michael Mattox. Image from BigStockPhoto.com; *bottom right* ©iStockphoto.com/Hedda Gjerpen; **318** ryasick photography/Shutterstock.com; **326** Warren Goldswain/Shutterstock.com; **329** *top right* John McLaird/Shutterstock.com; *center right* Robert Asento/Shutterstock.com; *bottom right* Mark Aplet/Shutterstock.com; **339** *Activity 1a left* ©iStockphoto.com/Shannon Keegan; *Activity 1a right* ©iStockphoto.com/Lorelyn Medina; *Activity 1b left* Joel Sartore/joelsartore.com; *Activity 1b right* Feng Yu/Shutterstock.com; *Activity 1c left* ©iStockphoto.com/kledge; *Activity 1c right* ©iStockphoto.com/spxChrome; *Activity 1d* ©iStockphoto.com/Alex Slobadkin;

Cartoon Illustrations: Tyler Stout

Cover Image: foxie/Shutterstock.com

For use with Chapter 2 Section 3*

a.

d.

*Available at *BigIdeasMath.com.*

Pattern Blocks – Chapter 2 Section 6*

*Available at *BigIdeasMath.com.*

Pattern Blocks – Chapter 2 Section 6*

*Available at *BigIdeasMath.com.*

Algebra Tiles*

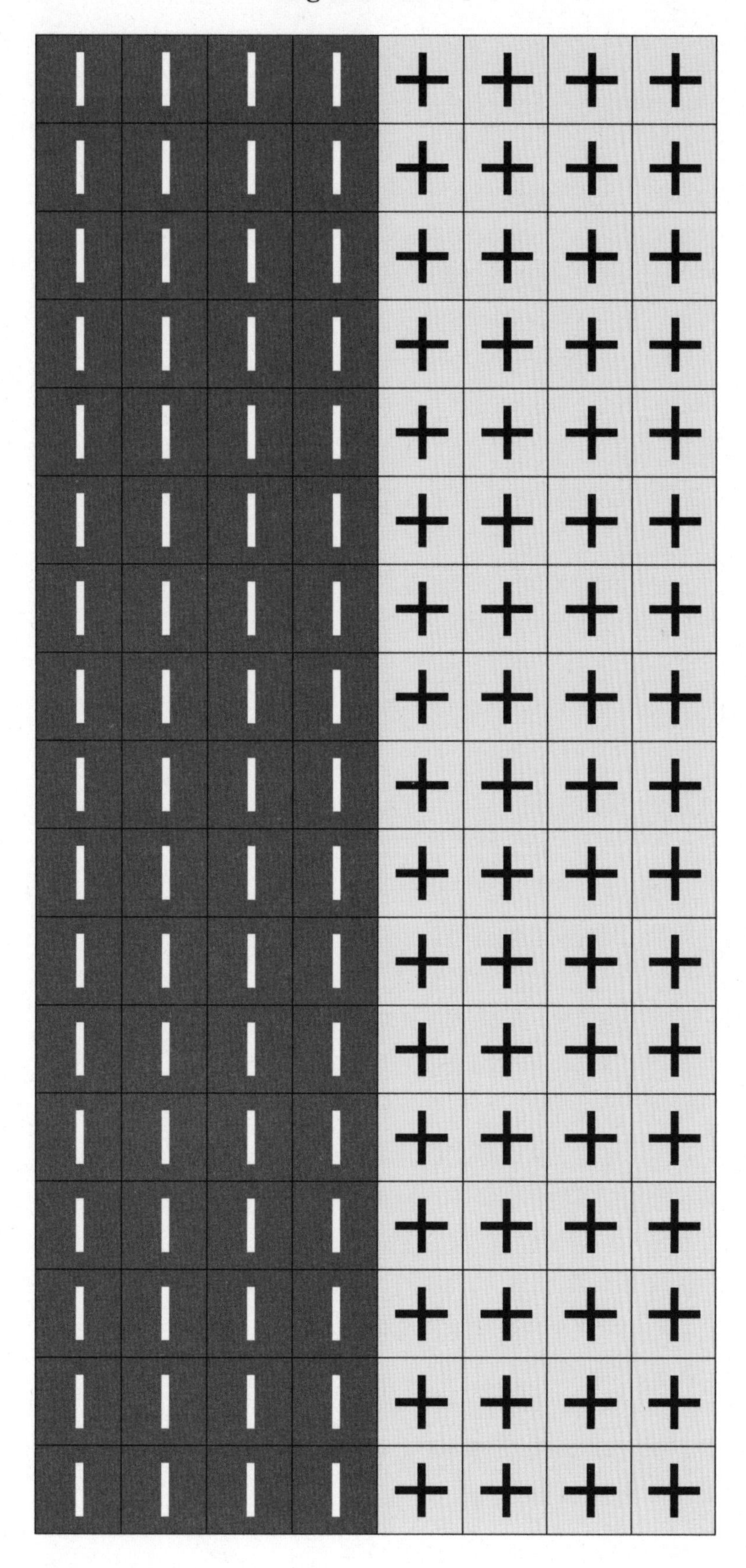

*Available at *BigIdeasMath.com.*

Algebra Tiles*

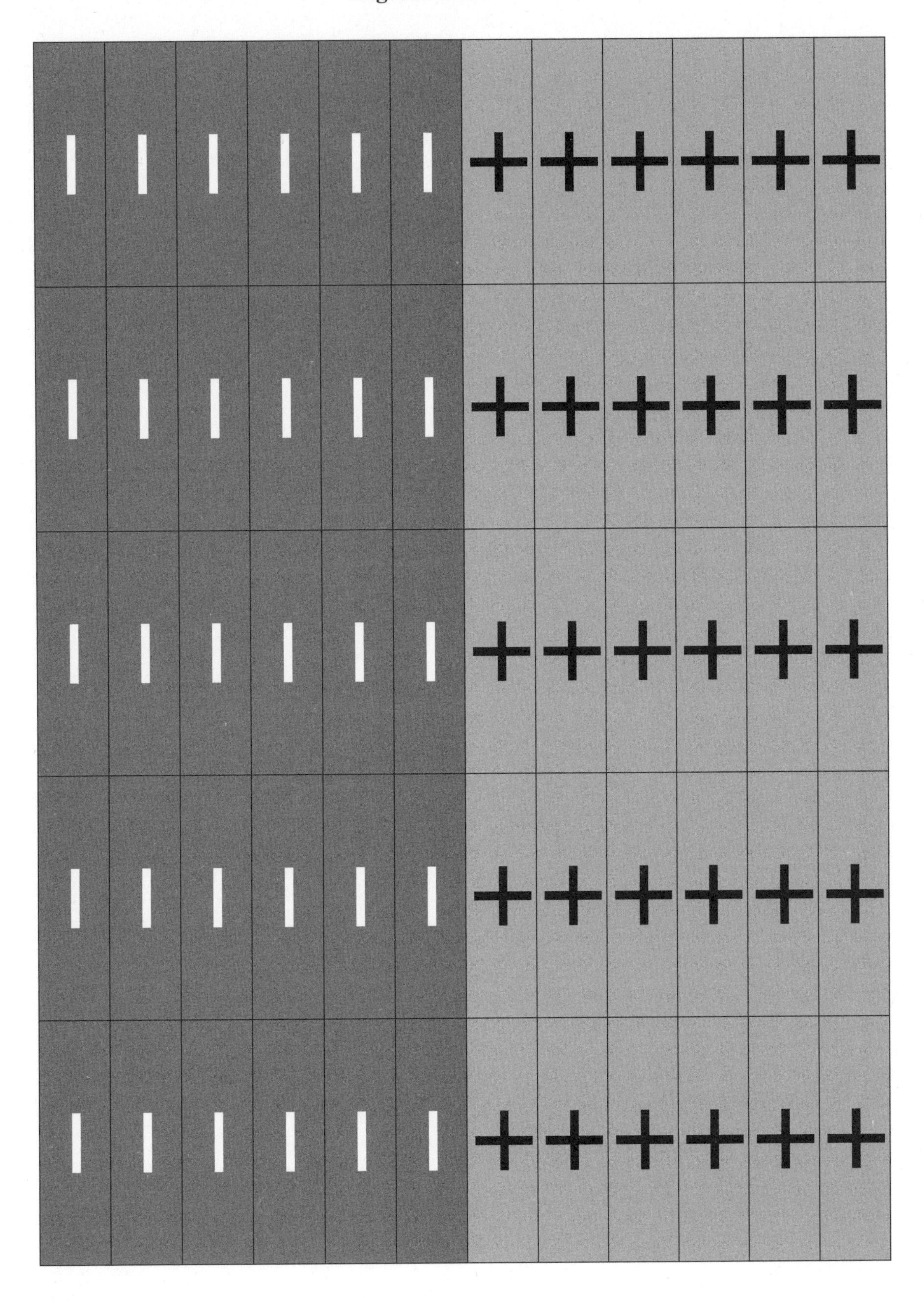

*Available at *BigIdeasMath.com.*

For use with Chapter 13 Section 3*

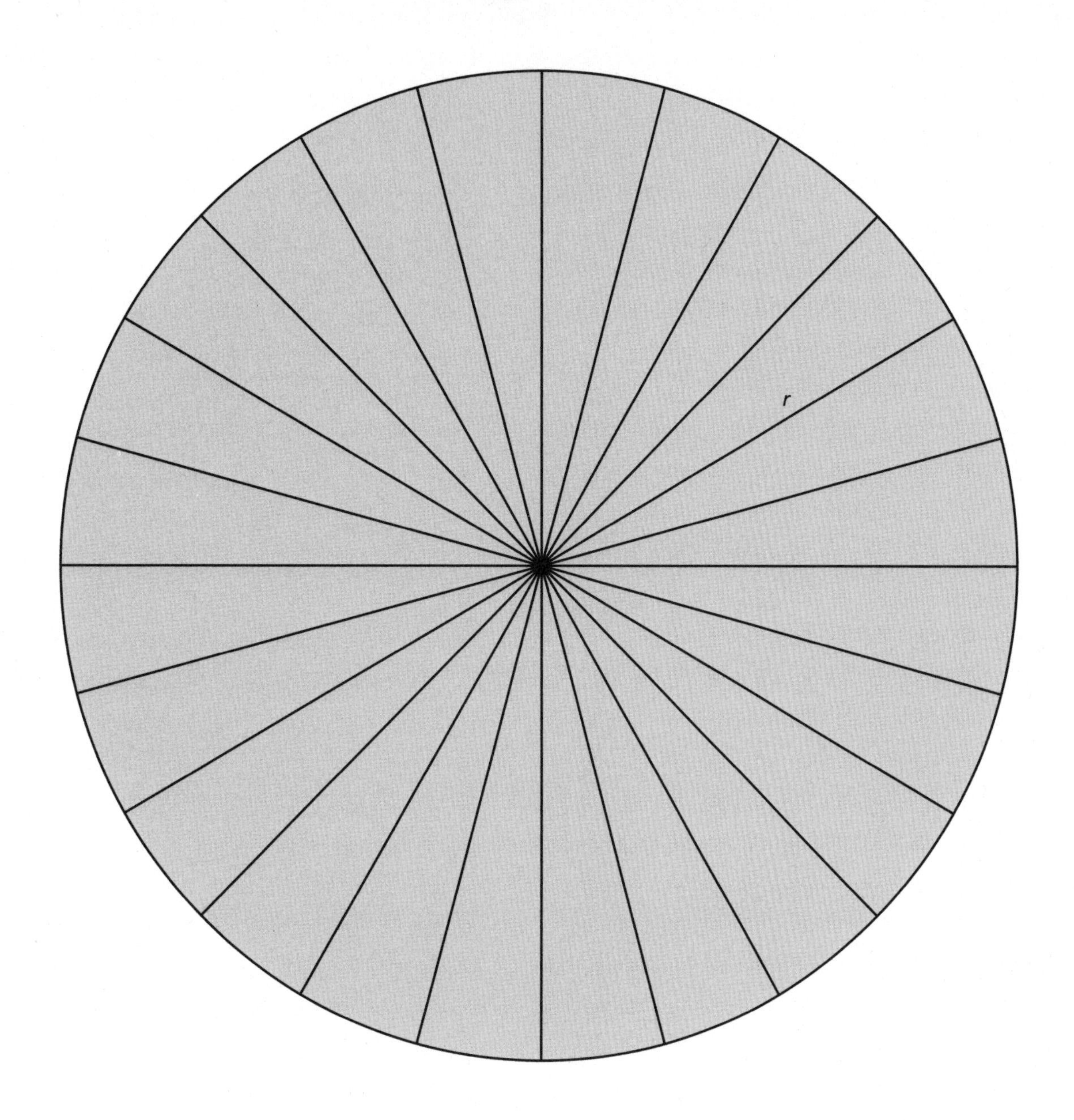

*Available at *BigIdeasMath.com.*

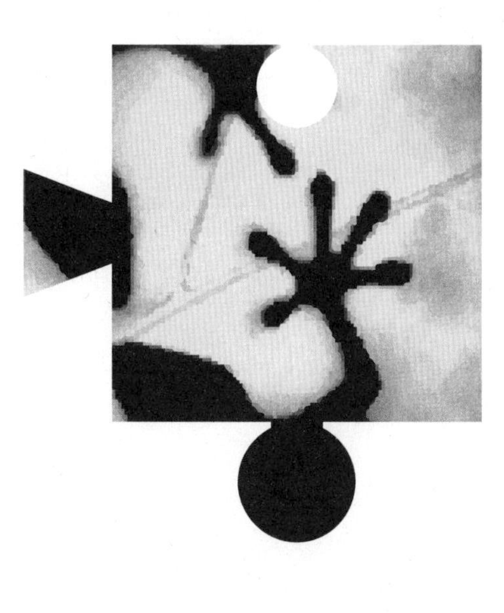

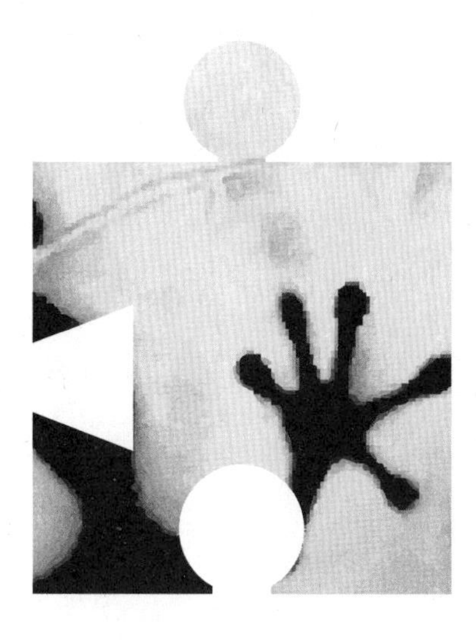

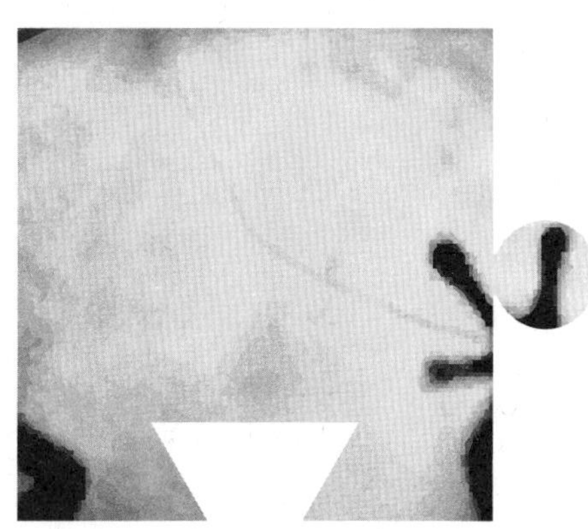

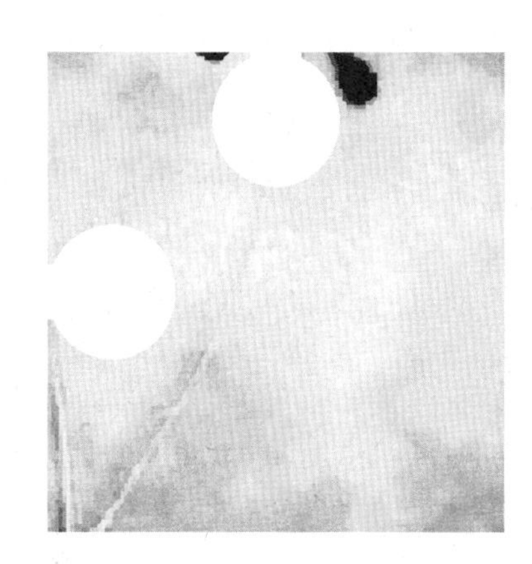

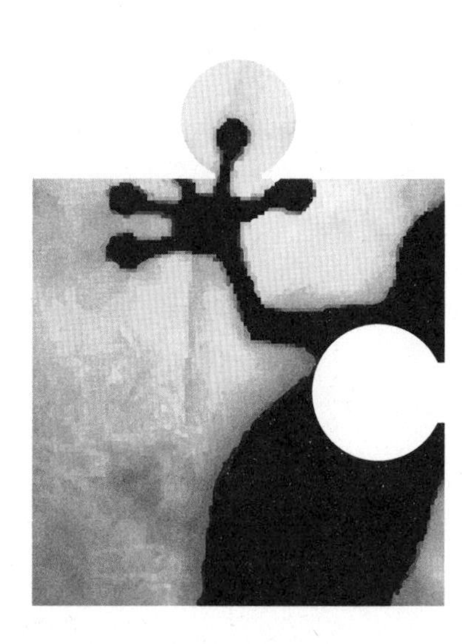

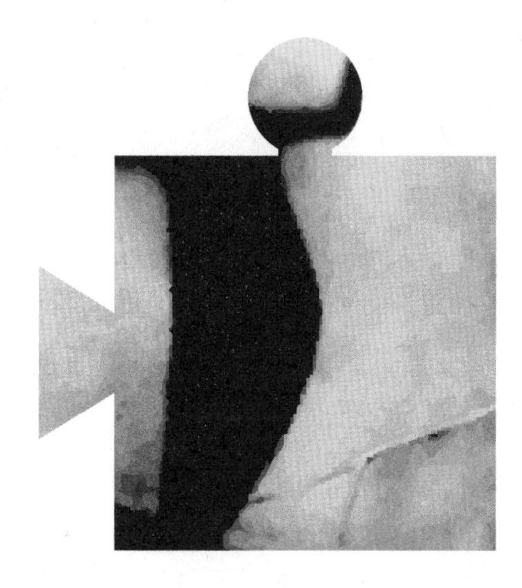

*Available at *BigIdeasMath.com.*

For use with Chapter 14 Section 1*

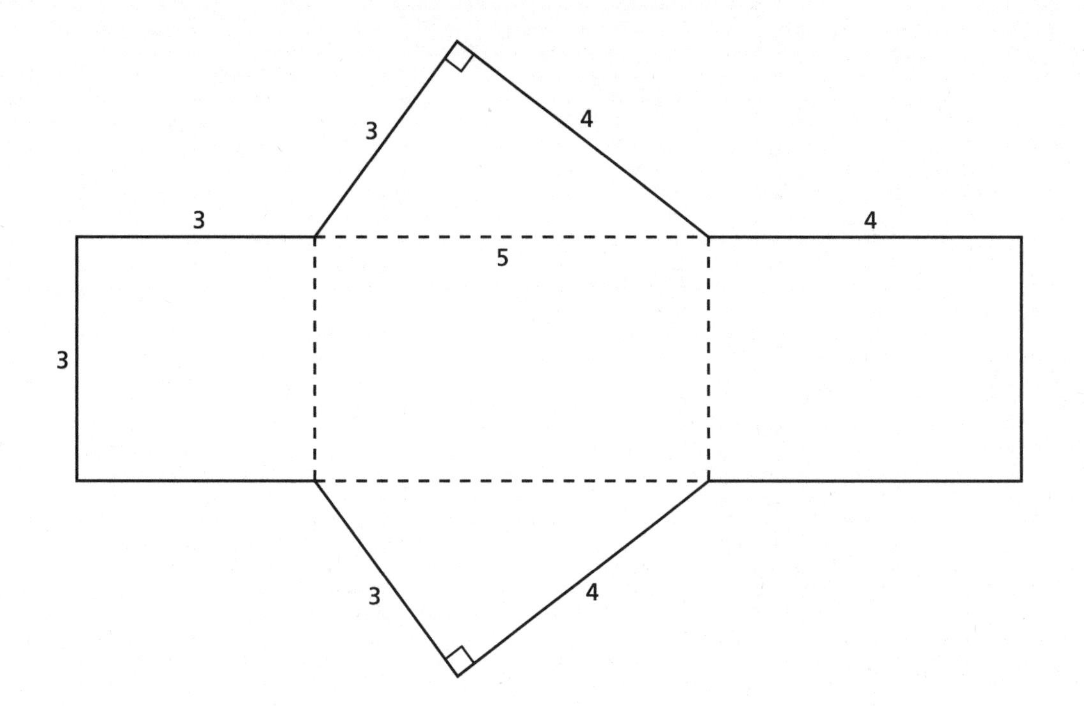

*Available at *BigIdeasMath.com.*

For use with Chapter 14 Section 5*

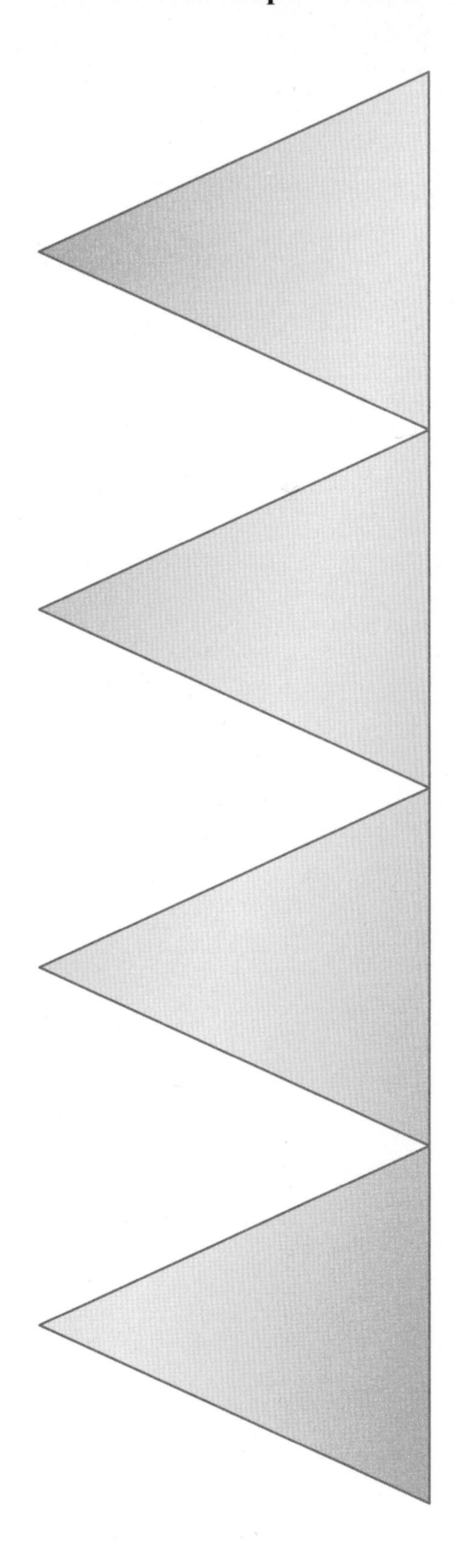

*Available at *BigIdeasMath.com.*

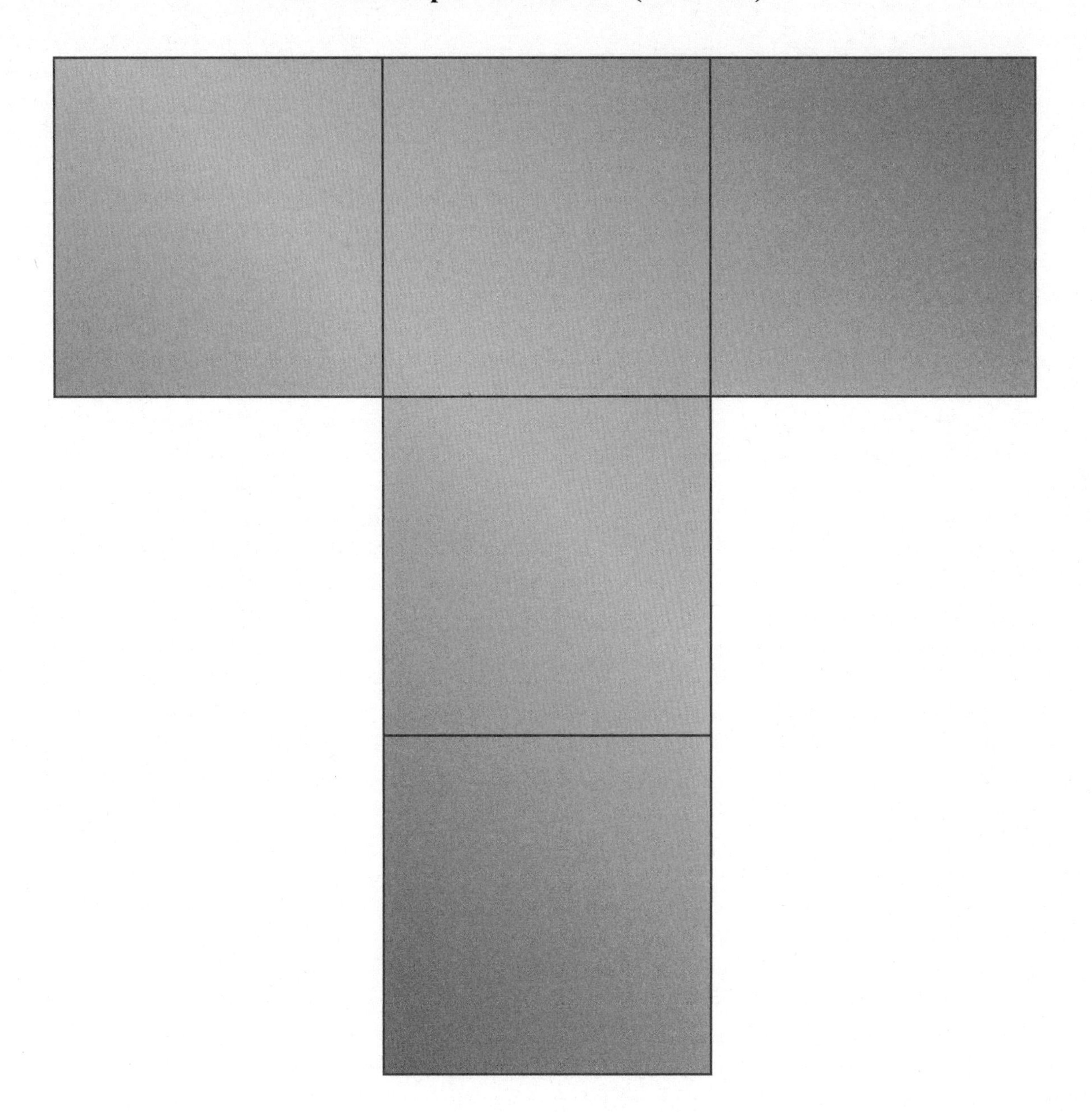

*Available at *BigIdeasMath.com.*